BRASS DIAL
CLOCKS

BRASS DIAL CLOCKS

BRIAN LOOMES

Antique Collectors' Club

ISBN 1 85149 221 6

British Library Cataloguing-in-Publication Data
A catalogue record for this book is available from the British Library

*Frontispiece: Sophisticated basket top timepiece bracket clock c.1710 signed 'Gabl Smith Barthomley', five
pillar single fusee movement with verge escapement pull quarter repeating on two bells. See page 365.
(Photograph by courtesy of Christie's.)*

Title-page: Twelve-inch longcase dial of about 1700-1705 by Thomas Meekings of Dublin. See page 317.

Printed in England
by the Antique Collectors' Club Ltd., Woodbridge, Suffolk
on Consort Royal Satin paper
supplied by the Donside Paper Company, Aberdeen, Scotland

ANTIQUE COLLECTORS' CLUB

The Antique Collectors' Club was formed in 1966 and quickly grew to a five figure membership spread throughout the world. It publishes the only independently run monthly antiques magazine, *Antique Collecting*, which caters for those collectors who are interested in widening their knowledge of antiques, both by greater awareness of quality and by discussion of the factors which influence the price that is likely to be asked. The Antique Collectors' Club pioneered the provision of information on prices for collectors and the magazine still leads in the provision of detailed articles on a variety of subjects.

It was in response to the enormous demand for information on 'what to pay' that the price guide series was introduced in 1968 with the first edition of *The Price Guide to Antique Furniture* (completely revised 1978 and 1989), a book which broke new ground by illustrating the more common types of antique furniture, the sort that collectors could buy in shops and at auctions rather than the rare museum pieces which had previously been used (and still to a large extent are used) to make up the limited amount of illustrations in books published by commercial publishers. Many other price guides have followed, all copiously illustrated, and greatly appreciated by collectors for the valuable information they contain, quite apart from prices. The Price Guide Series heralded the publication of many standard works of reference on art and antiques. *The Dictionary of British Art* (now in six volumes), *The Pictorial Dictionary of British 19th Century Furniture Design, Oak Furniture* and *Early English Clocks* were followed by many deeply researched reference works such as *The Directory of Gold and Silversmiths,* providing new information. Many of these books are now accepted as the standard work of reference on their subject.

The Antique Collectors' Club has widened its list to include books on gardens and architecture. All the Club's publications are available through bookshops world wide and a full catalogue of all these titles is available free of charge from the addresses below.

Club membership, open to all collectors, costs little. Members receive free of charge *Antique Collecting*, the Club's magazine (published ten times a year), which contains well-illustrated articles dealing with the practical aspects of collecting not normally dealt with by magazines. Prices, features of value, investment potential, fakes and forgeries are all given prominence in the magazine.

Among other facilities available to members are private buying and selling facilities and the opportunity to meet other collectors at their local antique collectors' clubs. There are over eighty in Britain and more than a dozen overseas. Members may also buy the Club's publications at special pre-publication prices.

As its motto implies, the Club is an organisation designed to help collectors get the most out of their hobby: it is informal and friendly and gives enormous enjoyment to all concerned.

For Collectors — By Collectors — About Collecting

ANTIQUE COLLECTORS' CLUB
5 Church Street, Woodbridge Suffolk IP12 1DS, UK
Tel: 01394 385501 Fax: 01394 384434
—————— or ——————
Market Street Industrial Park, Wappingers' Falls, NY 12590, USA
Tel: 914 297 0003 Fax: 914 297 0068

CONTENTS

SELECT BIBLIOGRAHY

Allan, Charles. *Old Stirling Clockmakers* (Stirling 1990)

Bailey, Chris. *Two Hundred Years of American Clocks & Watches* (Prentice-Hall 1978)

Baillie, G.H. *Watchmakers & Clockmakers of the World, Volume One* (NAG Press 1929)

Barder, R.C.R. *English Country Grandfather Clocks* (David & Charles 1983)

Barder, R.C.R. *The Georgian Bracket Clock* (Antique Collectors' Club 1993)

Barker, David. *The Arthur Negus Guide to Clocks* (Hamlyn 1980)

Bates, Keith. *Clockmakers of Northumberland & Durham* (Northumberland 1980)

Beeson, C.F.C. *Clockmaking in Oxfordshire* (Museum of the History of Science 1989)

Bellchambers, J.K. *Somerset Clockmakers* (Torquay 1969)

Beney, David R. *Beaminster Clocks* (Beaminster 1996)

Bird, Clifford & Yvonne. *Norfolk & Norwich Clocks & Clockmakers* (Phillimore 1996)

Britten, F.J. *Old Clocks & Watches* (Methuen 1982)

Brown, Colin & Mary. *The Clockmakers of Llanrwst* (Bridge Books 1993)

Bruton, Eric. *The Longcase Clock* (Hart-Davis 1970)

Craven, Maxwell. *John Whitehurst of Derby* (Mayfield Books 1996)

Daniel, John. *Leicestershire Clockmakers* (Leicestershire Museums 1975)

Darken, Jeff & Hooper, John. *English 30 Hour Clocks* (Woking 1997)

Dawson, P., Drover, C.B. & Parkes, D.W. *Early English Clocks* (Antique Collectors' Club 1982)

Dowler, Graham. *Gloucestershire Clock & Watch Makers* (Phillimore 1984)

Edwardes, Ernest L. *The Grandfather Clock* (Sherrat 1980)

Edwardes, Ernest L. *The Story of the Pendulum Clock* (Sherrat 1977)

Elliot, Douglas. *Shropshire Clock & Watch Makers* (Phillimore 1979)

Greenlaw, Joanna. *Swansea Clocks* (Llandybie 1997)

Haggar, A.L. & Miller, L.F. *Suffolk Clocks & Clockmakers* (Ticehurst 1974)

Hughes, R.G. *Derbyshire Clock & Watch Makers* (Derby Museum 1976)

Legg, Edward. *Clock & Watch Makers of Buckinghamshire* (Bradwell Abbey Field Centre 1976)

Loomes, Brian. *White Dial Clocks, the complete guide* (David & Charles 1981)

Loomes, Brian. *Antique British Clocks – A Buyer's Guide* (Robert Hale 1991)

Loomes, Brian. *British Clocks Illustrated* (Robert Hale 1992)

Loomes, Brian. *Clockmakers of Northern England* (Mayfield Books 1997)

Loomes, Brian. *Complete British Clocks* (David & Charles 1978)

Loomes, Brian. *Country Clocks & their London Origins* (David & Charles 1976)

Loomes, Brian. *Grandfather Clocks & their Cases* (David & Charles 1985)

Loomes, Brian. *Lancashire Clocks & Clockmakers* (David & Charles 1975)

Loomes, Brian. *Painted Dial Clocks* (Antique Collectors' Club 1995)

Loomes, Brian. *The Concise Guide to British Clocks* (Barrie & Jenkins 1992)

Loomes, Brian. *The Early Clockmakers of Great Britain* (NAG Press 1982)

Loomes, Brian. *Watchmakers & Clockmakers of the World. Volume Two* (NAG Press 1976)

Loomes, Brian. *Westmorland Clocks & Clockmakers* (David & Charles 1974)

Loomes, Brian. *Yorkshire Clockmakers* (2nd edition Littleborough 1985)

Ly, Tran Duy. *American Clocks* (Arlington Book Co 1989)

Ly, Tran Duy. *Longcase Clocks & Standing Regulators* (Arlington Book Co 1994)

Mason, Bernard. *Colchester Clockmakers* (Country Life 1969)

Mather, H. *Clock & Watch Makers of Nottinghamshire* (Nottingham 1979)

McKenna, Joseph. *Watch & Clockmakers of Birmingham* (Birmingham 1986)

McKenna, Joseph. *Watch & Clockmakers of Warwickshire* (Birmingham 1985)

Miles-Brown, H. *Cornish Clocks & Clockmakers* (David & Charles 1970)

Moore, A.L., Rice, R.W. & Hucker, E. *Bilbie & the Chew Valley Clockmakers* (Weston 1995)

Moore, Nicholas. *Chester Clocks & Clockmakers* (Chester Museum 1976)

Norgate, J. & M. & Hudson, F. *Dunfermline Clockmakers* (Dunfermline 1982)

Ord-Hume, Arthur W.J.G. *The Musical Clock* (Mayfield Books 1995)

Pearson, Michael. *Kent Clocks & Clockmakers* (Mayfield Books 1997)

Peate, Ioweth C. *Clock & Watch Makers in Wales* (Welsh Folk Museum 1960)

Penfold, John. *The Clockmakers of Cumberland* (Ashford 1977)

Pickford, Chris. *Bedfordshire Clocks & Watchmakers* (Beds. Historical Record Soc. 1991)

Ponsford, Clive. *Devon Clocks & Clockmakers* (David & Charles 1985)

Ponsford, Clive. *Time in Exeter* (Exeter 1978)

Ponsford, Scott & Authers. *Clocks & Clockmakers of Tiverton* (Tiverton 1977)

Pryce, W.T.R. & Davies, T. Alun. *Samuel Roberts. Clockmaker* (St. Fagans 1985)

Roberts, Derek. *British Longcase Clocks* (Schiffer 1990)

Robinson, Tom. *The Longcase Clock* (Antique Collectors' Club 1981)

Seaby, W.A. *Clockmakers of Warwick & Leamington* (Warwick Museum 1981)

Smith, John. *Old Scottish Clockmakers* (E.P. Publishing 1975)

Snell, Michael. *Clocks & Clockmakers of Salisbury* (Salisbury 1986)

Stuart, Susan E. *Clockmakers of North Lancs & South Westmorland* (Lancaster University 1996)

Tebbutt, Laurence. *Stamford Clocks & Watches* (Stamford 1975)

Treherne, Alan. *Nantwich Clockmakers* (Nantwich Museum 1986)

Tribe, T. & Whatmoor, P. *Dorset Clocks & Clockmakers* (Oswestry 1981)

Tyler, E.J. *The Clockmakers of Sussex* (Ashford 1986)

Walker, J.E.S. *Hull & East Riding Clocks* (Hornsea Museum 1982)

Wallace, William. *Marking Time in Hamilton* (Hamilton 1981)

White, George. *English Lantern Clocks* (Antique Collectors' Club 1989)

White, Ian. *Watch & Clock Makers in the City of Bath* (Ticehurst 1990)

Wood, Stacy B.C. Jr. *Clockmakers & Watchmakers of Lancaster County, Penn.* (Masthof Press 1995)

Introduction and Acknowledgements

This book is a detailed study of domestic clocks in Britain and America in the brass dial period, that is from about 1660 to about 1800. Many books concentrate on clocks made in London, which was after all *the* major centre of excellence in clockmaking in the world during this period. The emphasis in this book is on clocks made in the provinces and London clocks are discussed only when their styles have a bearing on those of the provincial makers.

Thirty-hour clocks are given considerable coverage, especially in the earlier periods up to about 1750. Here the earliest source of influence is from the lantern clock, a type of clock not discussed in detail in this book but covered in so far as its stylistic features can be followed through into other kinds of later clock, principally thirty-hour longcase clocks, hook-and-spike clocks and hooded clocks.

Bracket clocks were relatively uncommon in provincial work until the nineteenth century and even then some provincial makers bought their spring clocks from London specialists and had them lettered with their own names, more as retailers. Examples are known in the first half of the eighteenth century by very few provincial makers though these men do seem to have carried out their own work. The studies of individual clockmakers in Chapter Sixteen give an indication of the proportion of such clocks made by some of the clockmakers so covered. The higher price and lower accuracy and lower reliability of early spring clocks when compared with longcase examples is probably the reason for their scarcity.

London clocks had their own very distinctive styles, which can be seen as a focal point from which provincial clocks varied to a large or small degree according to certain trends we shall examine in the book. London clock-makers worked under a very strict discipline of quality, the standards of which were controlled by the Worshipful Company of Clockmakers whose authority was limited to the capital itself. This strong discipline of quality may have been the reason a similarly strong tradition of style evolved there, so strong that it is predictable. With a little experience we can easily describe the typical London clock of a certain period without having seen it. Of course there are exceptions, most notably with clocks of more complicated or special nature or by exceptional makers, but it is generally true that if a given period is specified we can anticipate the style of clock we are likely to find in London. Such is certainly not the case with provincial clocks.

Provincial clocks offer a much more varied field of study, that variety increasing with distance from London as the London influence waned and increasing too as time progressed. For this reason the greatest variety of styling appears in clocks from north-west England, the two poles of strongest influence being London and Lancashire. It is for this reason that my illustrations of clocks from north-western counties outnumber those from south-eastern ones. Those counties closest to London were often counties with fewer clockmakers, partly because of the smaller size of some of these counties, partly because buyers living at an accessible distance may have preferred to buy their clocks from London itself. Clocks from the south-eastern counties, and to some degree those from central southern England too, were often modelled along London principles and so lack that diversity found further north.

The influence of Lancashire spread into other nearby counties, most obvious being Cumberland, Westmorland and Cheshire. The counties of the North-east were not nearly so strongly influenced. This influence is predominantly seen in the style of the clocks themselves but is also apparent in casework from those areas. This casework influence can be seen too in certain Irish clocks, principally later eighteenth century Dublin clocks, and that influence spread even to America where some cases clearly show an Irish 'Chippendale' styling.

As an Appendix I have included a brief survey of the earliest clockmakers of each county and their work, a group I refer to as the 'A Team', being the first pioneer group of provincial clockmakers to set up in business where none had gone before. This is the first time such a survey has ever been attempted and must contain errors and omissions. However, it indicates several factors such as the dates of the spread of the trade, the relative numbers of varying types of clock made within the county and their distribution throughout the land. It becomes very obvious from even a passing glance at such a survey how lantern clocks were heavily concentrated in the South and how few early makers even attempted bracket clock making.

London clockmakers were vastly more numerous at this early period. Details of over four thousand London clockmakers working before the year 1700 (plus their numerous apprentices) are set out in my book *The Early Clockmakers of Great Britain*.

The photographs in this book were mostly taken by myself and my son, Robert, and were processed by my wife, Joy, who made respectable prints from our poor negatives. The clocks are mostly examples we have handled as dealers during thirty years in the trade. A few photographs were loaned, especially those of American clocks.

Clock enthusiasts, collectors, dealers and restorers are usually very willing to share their knowledge and experience and the following people have helped in varying ways:

from the United Kingdom: Granville Barrett, Lee Borrett, Bill Eyre, Edward Legg, Stuart Walker, John Thornton, Clifford Bird, Brian Stedeford, Tom Robinson, Susan Stuart, Gordon Morris, Robert Woodhouse, Michael Bulleid, Alan Treherne, Mike Williamson, Brian Mitton, Paul Sykes, Ian Haigh, Anthony Barthorpe, David Firth, Brian Wilson, Barnaby Smith

from the United States of America: Tom Spittler, Stacy Wood, Chris. Bailey, Ed. Lafond, Richard Mones, Graham H. Jeffries, Patricia A. Tomes, Curator of the Museum of the National Association of Watch and Clock Collectors Inc.

from Australia: Neil Keene, Richard Heher

from New Zealand: Simon Johnson

To anyone I have forgotten, I apologise in advance.

Brian Loomes
Pateley Bridge
North Yorkshire

PLATE 1. *Typical late 17th century lantern clock made in 1692 by Richard Savage of Shrewsbury, a 'marriage clock' to mark the marriage of Edward and Millicent Pardo named on the front fret. Height about 15in.*

CHAPTER ONE

THE FIRST CLOCKS,
WHAT THEY DID AND HOW THEY DID IT

The pendulum was introduced into Britain in 1658 by Ahasuerus Fromanteel, who made the first pendulum clocks in the land and who set the pattern for all future British clockmaking. Domestic clocks were made in Britain before 1658, however, and these were of a kind we now call lantern clocks. The name lantern clock is relatively recent (probably not much more than a century old) and its origin is unknown but probably stems from the fact that they vaguely resemble in shape an old hand lantern. At the time they were first made lantern clocks were the only kind of domestic clock and were referred to simply as 'clocks', occasionally as 'house clocks' or 'chamber clocks'. This book will not deal in depth with lantern clocks, for a detailed examination of which the reader should consult George White's *English Lantern Clocks*. It is, however, necessary to mention them here briefly in so far as they played some part in the development of the longcase clock and to

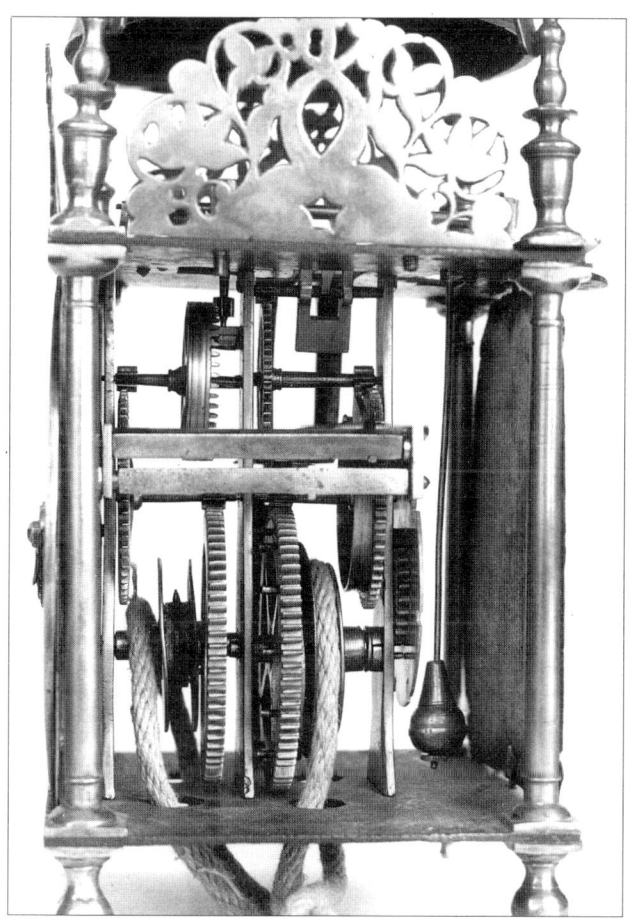

PLATE 2. *Movement of the Savage lantern clock showing verge escapement with bob pendulum, here positioned inside the backplate. Typical posted construction with brass pillars, feet and finials, but unusual in having iron top and bottom plates.*

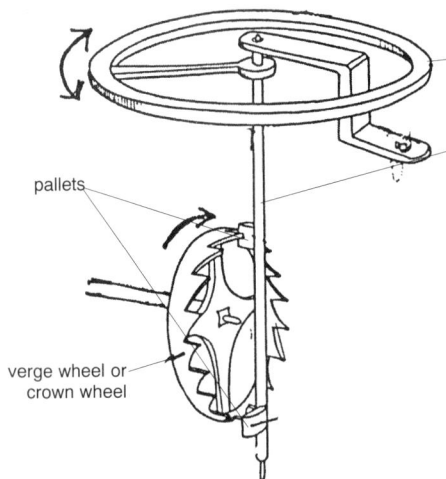

FIGURE 1. *The balance wheel principle. The verge wheel is pulled round by the weight and pushes alternately against each flag or 'pallet', thus forcing the balance wheel itself to swing backwards and forwards, thereby regulating the speed of the clock.*

understand that part we need at least to know what a lantern clock was, what it did, and how it did it.

Lantern clocks were relatively simple clocks, though individual examples may sometimes be anything but simple for us to understand today because of the changes they may have undergone over what might be more than three centuries of use and abuse. They were simple not because the men who made them were of limited intellect or incapable of making more complicated clocks; on the contrary, many clockmakers even at this very early period could produce work of an amazingly complex nature when the occasion arose. The clocks they made were simple partly because the general lack of clocks in the home meant that the customers were by and large unfamiliar with and perhaps even incapable of understanding a more complex time recorder, and partly because, even with those customers more educated in time telling, the majority would be incapable of handling, winding and regulating a clock of any degree of complexity.

The mechanical principles employed in a lantern clock are much the same as those in other later clocks and an understanding of the basic mechanics will help us identify and appreciate other types and the ways in which they developed.

TRAINS OF WHEELS

A typical lantern clock consists of two sections, two sets of wheels, each known as a 'train' of wheels. One train was to drive what we now call the clock part (in ancient times called the 'watch' part) and this is known as the 'going train'; the other train was to drive the strikework (counting the number of each hour) and is called the 'strike train'. Occasionally a lantern clock might chime the quarter hours, 'chiming' indicating the striking of any number of blows at each of the three (sometimes four) quarters. A chiming clock had a third train called the 'chiming train'. Rather than chiming quarters some clocks (though seldom lantern clocks) played a tune at intervals, often three hourly or four hourly, and such a musical clock would have a 'musical train'. Some examples also had alarmwork, and in those examples a further train of wheels drove the alarm – known as the 'alarm train'. As alarm lantern clocks might well have been intended for use close to the sleeping quarters, some, but not all, alarm examples were built without strikework, that is they consist of going train and alarm train only.

SINGLE HAND TIME

The seventeenth century lantern clock had only one hand, which indicated hours. The dial was calibrated on its numbers ring, known as the chapter ring, in such a way that the time could be read in units of hours, half-hours and quarter-hours. Minutes do not appear on early lantern clock dials, nor is there any minute hand, except in examples which have been altered later to show minutes. (For our purposes we can ignore those rare exceptions which chime the quarter-hours and therefore have a third 'chiming train' of wheels, some of which do register minutes by a minute hand. Two hands also appear on some arched dial lantern clocks about the second quarter of the eighteenth century and on that type made with pseudo-Turkish numerals and known as 'Turkish Market' clocks.)

EARLY ESCAPEMENTS

The first lantern clocks in Britain were regulated by what is called the balance wheel escapement.

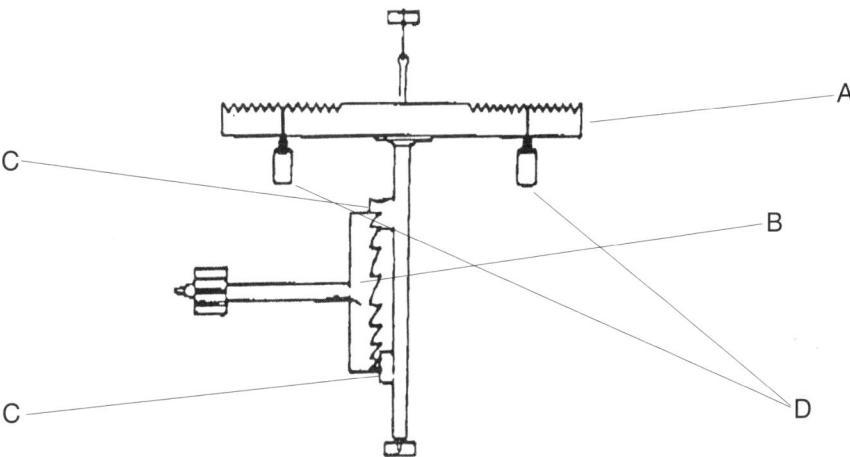

A. The foliot bar; B. Escape wheel (verge wheel) on its arbor; C. Pallets; D. Riding or cursor weights

FIGURE 2. *The foliot escapement. The verge wheel is pulled round by the weight, 'escaping' one tooth at a time by pushing alternately against each of the pallets which are attached to the foliot arbor, thus causing the foliot bar to swing to and fro. The clock is slowed down to the pace determined by the swing of the foliot bar. Adjustable riding weights (cursors) allow some regulation in the pace of the foliot swing. Further timekeeping adjustments were made by adding or removing lead shot to the weight canister.*

The escapement is that inbuilt device which slows down the clock's running pace to one which can be controlled so that it can be made to register twenty four hours a day on its dial. Without any escapement the pull of the weight(s) would spin the clock's wheels round rapidly so that it would run down instantly. It is the improvements in the type of escapement which increasingly led to greater accuracy of timekeeping over the years. The balance wheel served its primitive purpose as a speed regulator but had the major disadvantage that it was not adjustable. The only means of altering the rate at which a balance wheel clock ran was by adding or removing lead shot from the weight canister; the heavier the weight, the faster it ran, and vice versa. This may not seem to have been a problem once the clock was correctly set for timekeeping, but running conditions such as changes in the weather or lack of oil would mean that a balance wheel clock might gain or lose as much as a quarter of an hour a day.

Before the introduction to clockwork of the pendulum in 1658 the balance wheel was the only method used in domestic clocks in Britain to regulate a clock's speed (with the exception of two or three near unique British foliot examples).

In mainland Europe the foliot was used instead, a system very similar to the balance wheel though the foliot was capable of some adjustment. The main point for our purposes here is that balance wheel lantern clocks were inaccurate and unreliable timekeepers and because of that they registered time on their single-handed dials in hours, half-hours and quarter-hours, not in minutes. Minutes were known, and were shown on occasional clocks in mainland Europe as early as the mid-sixteenth century, normally by means of a separate minute hand. But in Britain minutes were not registered on the dials of lantern clocks and the reason was probably on account of the variability of timekeeping. What was the point in showing units of less than a quarter of an hour when the clocks might vary by a quarter of an hour a day?

SOLAR AND MEAN TIME

In considering this point further we might add the factor that a clock was set to time in the first place by reference to a sundial. The gnomon of a sundial casts a single shadow and what that signified was understood by everyone. We must remember that during the seventeenth century and even as late as the mid-eighteenth century the clock a family might purchase was the first clock that family had *ever* owned and as such was a new and puzzling instrument. But even a public largely new to clocks could understand that the single hand on a clock did the same thing exactly as the single shadow on a sundial, namely it pointed to hours and quarters.

Sundials are known which register time units as small as minutes, but the great majority, especially at this pre-pendulum period, show markings in units of quarter-hours. Sundials register solar time, whereby the day length varies at different seasons. To correct solar time to mean time, that is 'clock time' of twenty-four hours each day, involved adding or subtracting a number of *minutes* each day according to the month and day of the year, the exact amount being determined by reference to a printed table known as an Equation of Time table. The chances are that with pre-pendulum clocks most owners would never go to such lengths, even assuming each had such a table available, but would set their balance wheel clock by solar time direct from a sundial which itself indicated only the nearest quarter of an hour. In other words they would set their clocks wrongly by plus or minus several minutes after which the clock itself would gain or lose by as much as a quarter of an hour a day!

HOURS, QUARTERS AND MINUTES

It follows that for most people who then owned a balance wheel lantern clock our modern concept of accuracy would have been totally foreign. Most of those who owned a clock, and probably all those who did not, thought of time in units of quarter hours. Even today, in our very time-conscious society, we think in terms of those same units throughout our everyday lives, planning our appointments by the hour, or the half or the quarter. Only for a specific purpose would we ever think in terms of minutes – such as when catching a train or a TV programme.

To a seventeenth century population time was measured in quarter-hour units. The word 'minute' may have been used by scientists and astronomers, but was simply not in use in everyday life. The smallest time unit in popular usage was the 'half-quarter', which we would now describe as seven and a half minutes, though today this unit has fallen completely from use.

An indication of this fact is seen ironically within the advertisement by Ahasuerus Fromanteel in 1658 for his pendulum clocks, the first to be sold in Britain. In that same advertisement he also offered fire engines of his own unique design, whose advantages included the fact that they could be easily and speedily cleaned, so speedily that he claimed 'they may be presently cleansed without charge in half a quarter of an hour's time'. Even Fromanteel, who was advertising his new two-handed clocks with minute-marked dials, did not refer to minutes, as he knew this was a term not understood by the popular imagination, which thought in quarters and half-quarters. It is very important that we appreciate this fact as it has considerable bearing on our understanding of the styles of clock dials for over a century after the introduction of the pendulum (1658), which was the date when minutes were first introduced on British clock dials.

The normal method of time indication on a lantern clock was to show a

PLATE 3. *Sundial dated 1775 dividing the hours into halves, quarters and (unusually for this late date) half-quarters*

PLATE 4. *Dial from a lantern clock dated 1627 and with the unidentified (maker's?) monogram R K. Only half-hour divisions are marked by asterisks.*

double track marked off in quarter-hour units, the half-hour point being almost always indicated by a pronounced marker such as meeting arrow-heads or a fleur-de-lis or a trident symbol. With some very early examples, however, the marking was limited only to the half-hours and not the quarters, and this may perhaps be an indication of the less accurate expectation of pre-pendulum timekeeping to the nearest half-hour rather than the nearest quarter.

DURATION BETWEEN WINDING

Balance wheel lantern clocks were wound twice a day, the duration of run varying anywhere between twelve and fifteen hours at each winding, even though on occasion such clocks were hung very close to the ceiling in order to gain a little extra duration. It is probably a measure of the inaccuracy of these clocks that longer duration was not aimed at, as the reverse certainly applied in that when the new and more accurate pendulum clocks appeared longer duration between winding became immediately available. According to his own advertisement of 1658, Fromanteel's new pendulum clocks were available to run a week, a month, or even a year between windings. The greater accuracy of the verge escapement and bob pendulum now made it worth while to offer both minute registration for finer indication of time together with long duration for a clock which no longer needed timekeeping adjustment daily. The lantern type of clock with a single hand was still available to those who wanted it (now controlled by the verge pendulum instead of the older balance wheel), but even on those the duration was now increased to once a day winding rather than the twice a day which had previously been the norm.

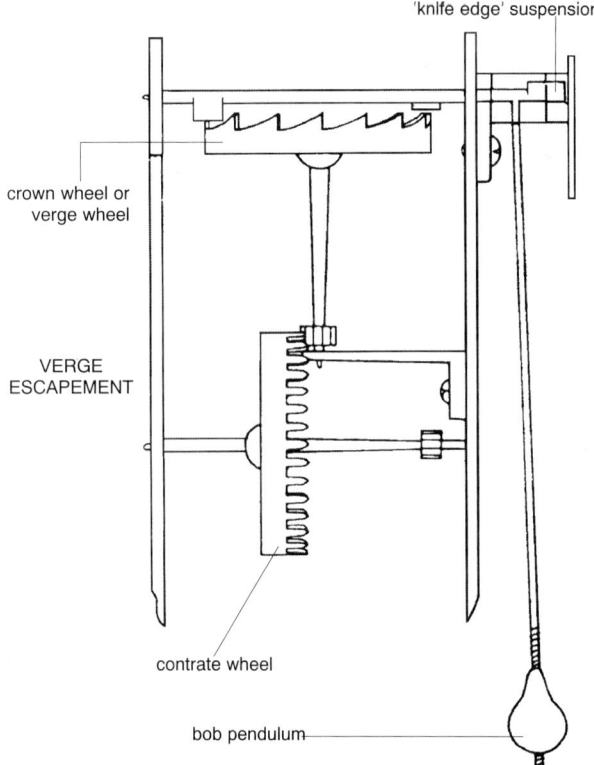

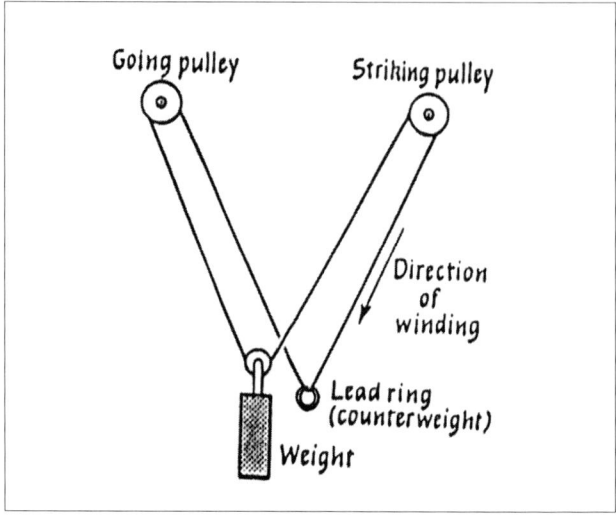

FIGURE 4. *Endless rope winding. The endless rope winding system is believed to have been devised by Huygens about 1658. Winding was through the striking ratchet and therefore the going ratchet was never disengaged from drive during winding, which meant that this system had a kind of inbuilt maintaining power. On many clocks chain was used as an alternative to rope.*

FIGURE 3. *The verge pendulum principle. The verge escapement was controlled by a short pendulum, sometimes known as a 'bob' pendulum, which was capable of fast/slow regulation by raising or lowering the bob on its screw thread. The pendulum is made to swing by pressure from the teeth of the rotating crown wheel (or verge wheel) against two projecting pallets (or 'flags') on the staff of the pendulum.*

This was done by employing the endless rope system devised by Christiaan Huygens (thought to have been introduced into England by Fromanteel) and used ever after for what we now call thirty-hour clocks, whether of lantern or longcase type.

AVAILABILITY OF ONE HAND OR TWO
The arrival of the pendulum in 1658 instantly made available clocks which had two hands, one for hours and one for minutes. In fact an additional feature available very soon after 1658 was a third hand to show seconds, although this was seldom used on clock dials till after 1670. (The term 'second' derived from 'second minute', i.e. the second subdivision of the dial into units of sixty.) But, to ignore the seconds factor for the present, the public were now presented with the availability of two hands and it might be anticipated that all clocks from that point forward would become two-handers, but this was not the case. The concept of minutes was still quite foreign to the public at large and there must have been considerable reluctance at first to accept the confusing, new-fangled, two-handed, minute-reading concept. Probably on account of this conservatism, single-handed clocks were still produced for many years (over a century) after two-handers were introduced.

LONGCASE CLOCKS
The arrival of the new pendulum clock brought with it the long case itself, a convenient standing cabinet in which to house and conceal weights and lines and which held the clock dial and movement at such a height from the floor as was convenient for viewing the dial and adequate for the required weight-drop to give the due length of duration.

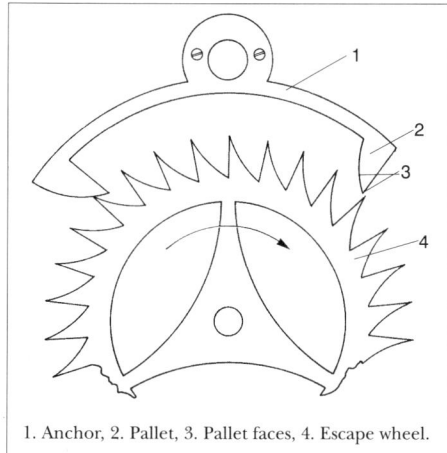

1. Anchor, 2. Pallet, 3. Pallet faces, 4. Escape wheel.

FIGURE 5. *The anchor escapement.*

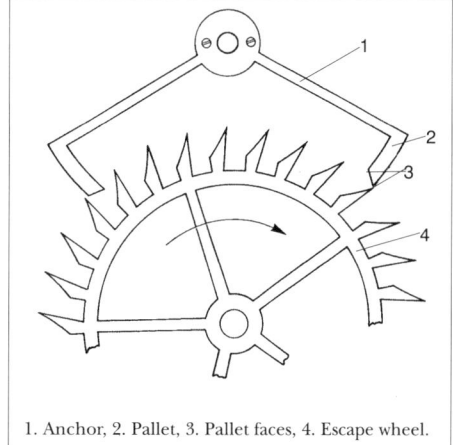

1. Anchor, 2. Pallet, 3. Pallet faces, 4. Escape wheel.

FIGURE 6. *The deadbeat escapement.*

In 1658 it was possible to have a single-handed longcase clock, but in fact these are very unusual in the first few years of the (verge pendulum) longcase clock, probably because the lantern clock was the customary method of showing single-handed time. The earliest verge escapement longcase clocks therefore were almost exclusively two-handers of eight-day duration (occasionally longer). Nevertheless, it is important to note that single-handed or two-handed longcase clocks were available from this time according to the choice of the customer. The production of these very first (verge) longcase clocks was limited to London and almost exclusively to the work of Fromanteel himself.

THE ANCHOR ESCAPEMENT

About 1670 a further modification occurred in the pendulum system, the anchor escapement with its long pendulum. The standard long pendulum beating once a second has a length of 39.14 inches, but thirty-hour clocks *not* showing seconds may have very varying pendulum lengths. The exact date of introduction and its originator remain an area of dispute amongst scholars, but to most of us this is merely an academic matter. Longcase clocks of this extreme age are very scarce and the average clock enthusiast is unlikely ever to see one. The anchor escapement was capable of greatly improved accuracy of timekeeping, its long pendulum being adjustable for fine tuning. Accuracy could now be measured in terms of a minute a week or less. By the 1680s the longcase was beginning to establish itself to the degree that some were being made in the provinces, though until about 1700 London was *the* centre of production.

The anchor escapement (sometimes called a recoil escapement) was the secret of the success of the longcase clock. It established itself as the standard escapement on almost all longcase clocks hereafter. Other forms of escapement which followed and are occasionally met with in longcase clocks are really only variants of the anchor principle (i.e. the deadbeat, tic-tac and pinwheel) and were normally only used for special reasons in clocks of a specialised nature. The great majority of longcase clocks made after 1680 will prove to have the standard anchor escapement.

THE DEADBEAT ESCAPEMENT

The deadbeat escapement is said to have been invented about 1720 by George Graham, the renowned London clockmaker. The deadbeat was an improved form of anchor escapement, having wheel teeth and pallet faces of different

PLATE 5. *Eight-and-a-half inch longcase clock dial of about 1670 by Ahasuerus Fromanteel of London. This clock has bolt-and-shutter maintaining power, as can be seen by the shutters covering the winding holes. Note the very plain and formal style.*

profile. Such an escapement ran without recoil, the escape wheel stopping 'dead' at each tick. The deadbeat was more accurate than the anchor, particularly when used with other timekeeping aids such as a compensated pendulum and maintaining power (to keep the clock running during winding). However, the deadbeat escapement was more delicate and damage could result to the escape wheel teeth from abuse or if the clock were allowed to tick backwards when the weight drive was temporarily removed during winding – many anchor escapement longcase clocks do that without harm. It was probably more for this reason than for the timekeeping aspect that maintaining power was fitted to most deadbeat escapement clocks.

The deadbeat escapement, usually with maintaining power, was used principally on one particular type of longcase clock made for precision timekeeping known as a regulator – made mainly for scientists and astronomers or for a clockmaker's own use as a master clock. Such precision clocks of the nineteenth century sometimes had jewelled pallets, which were more hardwearing than steel. A deadbeat escapement was fine in the hands of a capable owner, but was rarely used on normal domestic longcase clocks involving the less than careful use of the everyday household, except for those which had a centre seconds hand (sometimes called a sweep seconds clock).

Centre seconds clocks showed seconds by a long third hand (usually counterbalanced to avoid drag) fitted concentrically with the hour and minute hands. The main fashion for centre seconds hands was limited to a small proportion of the work of some clockmakers (many clockmakers never bothered to use it) during the latter period of brass dial clockmaking (about 1760 to 1790) and the early period of white dial clockmaking (about 1770 to

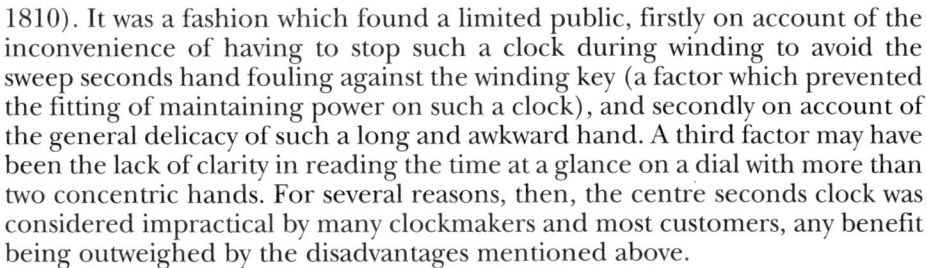

PLATE 6. *The Fromanteel (plated) movement has latched dial feet and latched movement pillars and outside countwheel striking, the strike train here being unusually positioned on the right. One of the shutters can be seen behind the dial.*

PLATE 7. *Ebony case of the Fromanteel clock with architectural hood pediment, long panelled door and panelled sides, glass windows full height of the hood, early style of convex mouldings.*

1810). It was a fashion which found a limited public, firstly on account of the inconvenience of having to stop such a clock during winding to avoid the sweep seconds hand fouling against the winding key (a factor which prevented the fitting of maintaining power on such a clock), and secondly on account of the general delicacy of such a long and awkward hand. A third factor may have been the lack of clarity in reading the time at a glance on a dial with more than two concentric hands. For several reasons, then, the centre seconds clock was considered impractical by many clockmakers and most customers, any benefit being outweighed by the disadvantages mentioned above.

THE PINWHEEL ESCAPEMENT

The pinwheel escapement was rarely, if ever, used on longcase clocks (except occasionally on regulators) and was only occasionally used on bracket clocks, mostly of the late eighteenth century or early nineteenth. It was a form of

PLATE 8. *View into the movement of a late 18th century bracket clock by Thomas Lister of Halifax showing the pinwheel escapement, a variation of the anchor escapement but using a wheel with pins instead of the normal teeth.*

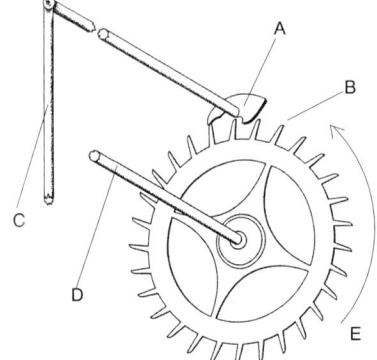

A. Pallets; B. Escape wheel; C. Pendulum rod; D. Wheel arbor; E. Direction of wheel in drive

FIGURE 7. *The tic-tac escapement. The action of the tic-tac escapement is similar to that of the anchor, but the pallets span only two or three teeth of the escape wheel. The escape wheel teeth may vary considerably in profile.*

deadbeat escapement whereby the escape wheel had horizontal pins around its rim instead of teeth. The action is similar to that of a deadbeat escapement. For our purposes in this book it is of little more than academic interest.

THE TIC-TAC ESCAPEMENT

The tic-tac escapement resembles the anchor escapement in principle but the pallets embrace only two or three teeth. Its advantage appears to be that its action resembles that of the verge in being less sensitive to levels than the anchor whilst being more accurate than the verge – a sort of half-way house between verge and anchor. It produces a wide arc of swing, much wider than that of the anchor, and it was probably for that reason it seems not to have been used for longcase clocks (except in one or two examples of an experimental nature made in the 1670s by Joseph Knibb of London and in occasional spring clocks of the same period by Thomas Tompion of London). The pendulum is longer than that of the verge but usually less than half that of the standard longcase one-second pendulum and this can mean a tic-tac pendulum length of anywhere between about twelve and twenty inches. The one-second 'Royal' pendulum with anchor escapement proved far more suitable for longcase use. The tic-tac was sometimes used for wall clocks of the hook-and-spike type and just occasionally on late seventeenth century lantern clocks.

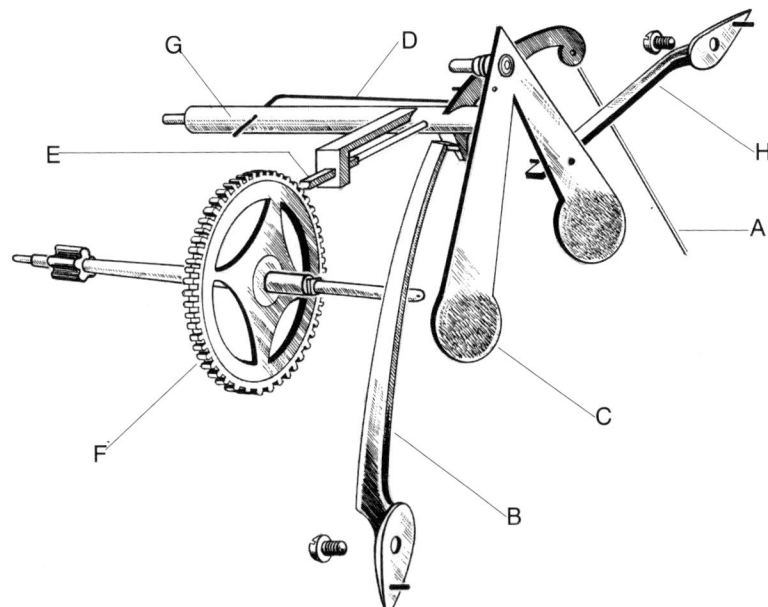

FIGURE 8. *Maintaining power – bolt-and-shutter type. The winding key cannot be inserted until the shutters (C) are drawn aside by pulling on a cord or lever (A). This action tensions a bladespring (B). A spring (D) then forces a bolt (E) to press down on the teeth of the wheel (F), keeping the clock ticking during winding. The shutter spring (H) forces the shutters back into position after disengagement. Based on a drawing by David Barker.*

MAINTAINING POWER

In principle maintaining power was a separate spring-drive mechanism which came into play to keep a clock running during the time it was being wound. Its purpose initially may have been to avoid a loss of time, but it is more likely it was to avoid potential damage to the wheel teeth if the escape wheel should jam against the pallets or skip backwards during winding, either of which can sometimes happen.

The earliest clocks of eight-day duration or longer often had maintaining power of what is called the bolt-and-shutter type. Shutters covered the winding holes to force the owner to apply the maintaining power before he was able to insert the winding key. The pulling of a lever shunted the shutters aside to give access to the winding squares. At the same time it brought into action a spring-loaded bolt which pushed against one of the wheels to keep the train running for a minute or two. As the wheels turned further the action of the spring gradually lessened, removing the bolt and returning the shutters to the closed position. Of course there was no real need for a shutter on the striking side, but this was usually present just as a match for the going side. Occasionally this type of maintaining power had no shutters at all (see Plate 43).

It soon appears to have been realised that maintaining power was not necessary and by about 1700 it had fallen from use on normal domestic clocks. Often it was removed at a later period as a needless complication from those clocks originally built with it, and those examples we see today are often modern replacements.

Bolt (with or without shutters) maintaining power was used by the early London makers but seldom by provincial clockmakers, with the possible exception of those who were London trained. Hence it is barely relevant to provincial clockwork at all. As it happens, the Huygens endless rope drive of a thirty-hour longcase clock kept the wheels turning during winding anyway, so that thirty-hour clocks have this inbuilt maintaining power already. It is for this reason that some provincial clockmakers made their regulators in daily winding form (see Plate 283).

A different type of maintaining power came into use later for precision clocks, particularly regulators. This is known as Harrison's maintaining power, invented in the 1730s by John Harrison of London, the famous maker of marine timekeepers. A spring within the winding ratchet keeps the going train wheels turning during winding.

STRIKEWORK, CHIMING, MUSIC

Striking is the term used to describe the action of a clock when it counts out the number of the hour on a bell by means of a separate wheel train additional to that for the going of the clock (see page 14). Chiming refers to the use of any number of bells to indicate the quarter-hours, usually by means of a peal of several notes. A clock which plays tunes, however often, is referred to as a musical clock. A chiming or musical clock would normally require a separate, third train of wheels. Striking of hours was usual with lantern, longcase and bracket clocks, but only a small minority of any of these chimed quarters or played tunes.

COUNTWHEEL STRIKING

Early strikework (or chiming or music) was controlled by a system known as a countwheel, alternatively called locking plate striking. The locking plate or countwheel itself was a disc with notches set around its rim at ever wider intervals. The clock continued to strike as long as a lever (called a detent) was held above the rim of the turning countwheel until the next available slot arrived. An early chiming or musical clock would have a separate countwheel to control the length of chime or music play – later ones often had a different system described on page 26.

On thirty-hour clocks (lantern, hook-and-spike and longcase) the strike countwheel was always positioned at the back of the movement, outside the posts/plates. It remained there on thirty-hour clocks throughout their entire period, so that these clocks always had an 'outside' countwheel and the position of the countwheel is not a pointer to age. With longcase clocks of month duration and longer there was also a tendency for countwheel striking to be retained, usually positioned outside the backplate (even after the invention of rack strikework, described on pages 26-28).

The position of the countwheel on eight-day longcase clocks did change with time and its position can therefore be an indicator towards period, though this is not always entirely reliable. The earliest eight-days usually had the countwheel positioned high up outside the backplate, the detent initially coming through a slot in the backplate but later positioned separately outside the backplate. By the 1680s (later in some provincial work) the countwheel was lowered in position on an extension of the greatwheel arbor and had a long detent positioned high up above it outside the backplate. Both of these early systems are referred to as having *outside* countwheel striking. There is no point in using this term when referring to countwheel-striking thirty-hour

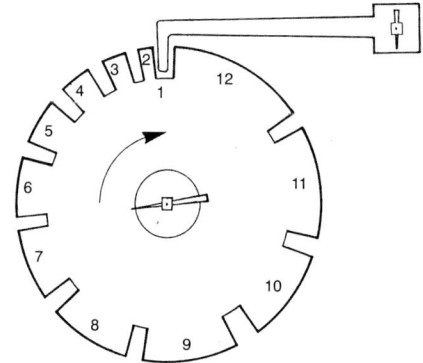

FIGURE 9. *Countwheel striking system, also known as locking plate striking. During striking the arm (detent) is held out of the notch in the countwheel an ever increasing space for each hour, returning to 1 again after 12.*

PLATE 9. *Rear view of a thirty-hour longcase clock of about 1765 by Thomas Stripling of Lichfield, the countwheel being of the form which uses a pinwheel instead of the normal notched countwheel. The pins are on the side closest to the backplate.*

PLATE 10. *Rear view of a late 18th century thirty-hour longcase clock by Gilbert Kidd of Malton showing a different form of pinwheel striking but using the same principle as that in Plate 9.*

clocks as these always had their countwheels outside.

By about 1690 the countwheel on eight-day longcase clocks began to be positioned *inside* the plates, attached to the mainwheel. This date was a very variable one depending on the whim of the individual makers, but generally *outside* countwheel striking pre-dates *inside* countwheel on eight-day clocks, and this can be a useful dating guide.

The countwheel system was used on longcase and bracket clocks too until the invention of a new system known as rack striking or rack repeating work (see below). The countwheel system had the disadvantage that its sequence of notes was permanently set and could not be varied. After seven it must necessarily strike eight. The rack strike system allowed the option of repeating the past hour (or quarter-chimes, or music) and gave a variability not possible with the countwheel.

The new rack striking system (supposedly dating from the 1670s) eventually replaced the countwheel as a preferred method on eight-day work by most, though not all, clockmakers. This replacement, however, was not as rapid as might be imagined. Some clockmakers took to it at once on eight-day longcase clocks and some provincial examples have the new system by about 1700, sometimes earlier. London longcase clock makers seem to have taken to the new system more slowly and their clocks were often still controlled by inside countwheel striking as late as 1720. Gradually all London eight-day longcase clocks turned to rack striking. London-made bracket clocks retained countwheel striking for a while after the introduction of the rack principle, which at first was used on bracket clocks for pull quarter repeating rather than hourly striking.

In the provinces there were some clockmakers who never liked the rack system. This applied principally in the North-west centred on Lancashire (and to other makers elsewhere of an independent mind), where inside countwheel striking was by no means unusual on eight-day longcases until the 1760s. Examples of inside countwheel eight-day provincial longcases are found as late as the 1820s even, though they are unusual this late.

Why some makers retained the countwheel for so long is unknown, except that rack striking involved a spring (which would sooner or later fail) and some makers may have felt the countwheel was more robust or more reliable. The countwheel system also permitted those who wished to do so to leave the strike train unwound, thereby running the clock as a non-striker – something very risky with a rack-striking clock unless fitted with a special silencing device, which most were not.

Occasional makers of eight-day clocks with inside countwheel strikework, principally in North-west England, would set pins into the greatwheel rim as an alternative to the normal notched countwheel. The principle is exactly the same but may puzzle those unaccustomed to seeing this.

On thirty-hour clocks a few clockmakers used a pinwheel as an alternative, and occasional clockmakers used a countwheel with raised bumps on its rim instead of notches. These are just variations on the theme and such a clock is still controlled by countwheel striking. These variations were used in the eighteenth century on thirty-hour longcase clocks as a preferred option by some clockmakers who must have thought their system better. Thirty-hour clocks retained the countwheel system throughout their entire period of production and did not turn over to the new rack strike system, except in those unusual instances where a thirty-hour clock was called for with repeating facility when rack striking was fitted as an optional extra at extra cost.

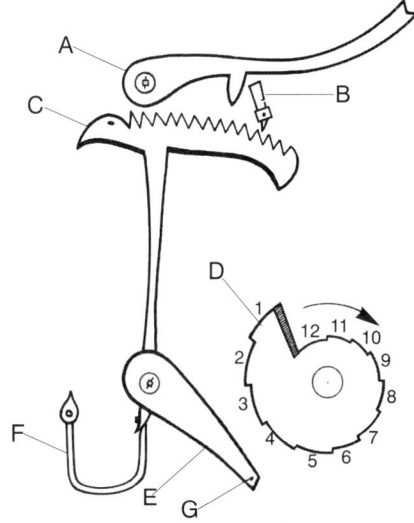

FIGURE 10. *Rack striking. On release the rack (C) is pushed to the left by the rack spring (F) making the appropriate number of teeth available to the gathering pallet (B). The number of the strike is determined by the number of rack teeth available, this being established by which step of the snail (D) the pin (G) of the rack tail (E) falls on to. A is the rack hook, which holds the rack in position until released on the hour.*

PLATE 11. *The earliest form of rack striking, using a rack with teeth on its upper and lower surfaces, used on a lantern clock of the 1680s by Henry Webster of Aughton, Lancashire. This view is seen from the back of the clock with the backplate removed.*

RACK STRIKING

Rack striking is thought to have been invented about 1676 by Edward Barlow, a Roman Catholic priest and inventor, who was born Booth but later changed his name to Barlow after his godfather. There is some uncertainty about the exact device Barlow invented but it is believed to have been a repeating mechanism governed by a toothed segment called a 'rack' and a stepped cam called a 'snail'. This was first used in London-made bracket clocks for pull-repeating work only, then very soon (before 1680) adapted in bracket clocks to govern the strike itself. As the snail was attached to the hour wheel it could not come out of sequence with the hands and the hour could be repeated as often as desired (by pulling a cord or pressing a lever).

Why anyone should want to repeat the last hour (or quarter or hour and quarter) of a clock might at first appear a mystery to us today. However, it is known that a repeater was used in several ways, apart from the sheer fun of making a clock repeat a chime to amuse guests. It was used during the night or on a dark winter morning as, for instance, with a clock in a bedroom where a pull of the repeater cord would ascertain the nearest hour or quarter without striking a light. Some longcase clocks with repeating facility had a cord which reached *upwards* through the ceiling so that they could be repeated by an owner still in bed.

Another use of repeating work was in the nature of a dinner gong, so that where a family failed to assemble at table at the prescribed time the repeater could be sounded to summon all within earshot on the loud bell, which on most longcase clocks could be heard throughout the house. The rack-striking clock had another advantage over the countwheel version in that the clock hands could be re-set without throwing out the sequence of striking, so that some clocks would have been built with rack-striking 'repeating' work even though never intended to actually repeat.

'INSIDE' RACK STRIKING

The first known form of rack striking has a rack with teeth on its upper and lower faces. These teeth were usually ratchet form but were sometimes ordinary wheel teeth. When the rack is released an extension (called the rack tail) falls on to the stepped snail, the distance of fall to the individual step on the snail determining how many teeth the rack is to reach and therefore how many blows it will strike. It is then gathered away from the snail during striking by means of a 'gathering pallet' picking up the teeth on its underside. This appears to be an experimental form, used on a few early bracket clocks and one unique lantern clock by Henry Webster of Aughton, Lancashire (Plate 11).

The system was very soon modified to provide the rack with only one set of teeth (usually ratchet teeth but sometimes wheel teeth) on its upper surface, gathering being achieved by means of a separate gathering pallet picking up along the same toothed surface as that which locks off the desired striking number. This is the normal form of the earliest type of rack striking, usually called 'inside' rack striking because it has most of the mechanism, including the rack itself and the rack hook, positioned between the plates. Inside rack strikework usually has the rack itself positioned horizontally, its impulse being activated by gravity as the rack tail 'falls' into position on the snail.

Although rack striking was known amongst London clockmakers by 1680, longcase clocks there seem to have kept the countwheel form until about 1710, sometimes as late as the 1720s or 1730s. It is not often, therefore, that inside rack striking is met on London longcase clocks. Some provincial clockmakers, however, did use the inside form of rack striking in the 1700-1740 period.

'OUTSIDE' RACK STRIKING

By the time London clockmakers adopted rack strikework on longcase clocks it had developed into the type we call 'outside' rack striking, whereby most of the mechanism, including rack and rack hook, had been brought forward on to the front of the frontplate ('outside' the plates). We do know examples of provincial eight-day clocks using external rack striking in the 1690s, so that, oddly enough, some provincial makers seem to have adopted outside rack striking sooner than was the general trend in London.

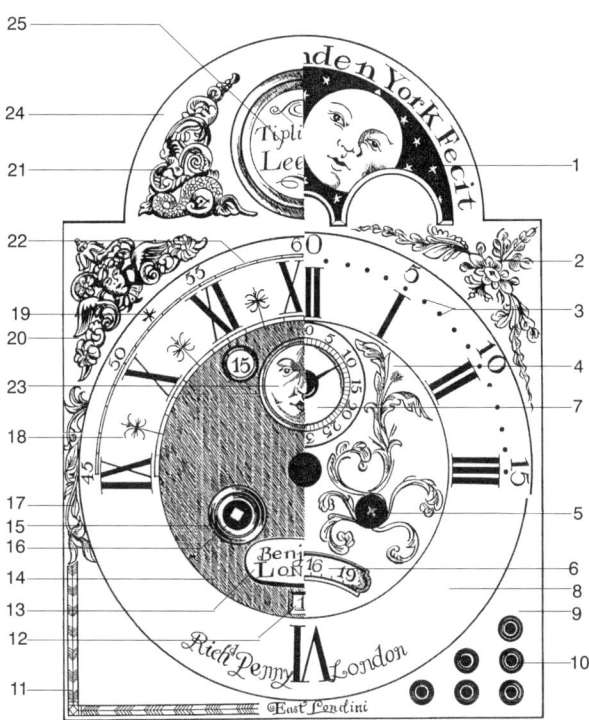

FIGURE 11. *Terms used in describing dials. A composite dial explaining features which may be met with individually at differing periods.*

1. Moon dial. 2. Engraved corners. 3. Dotted minutes. 4. Engraved dial centre on to polished ground. 5. Leaf-shaped winding arbor. 6. Curved date aperture sometimes known as a 'mouth' calendar. 7. Seconds dial. 8. Chapter ring. 9. Dial plate or dial sheet. 10. Cup-and-ring turnings. 11. Herringbone engraved border, sometimes called wheat-ear engraving. 12. Square date box. 13. Name-plate signature. 14. Matted dial centre, here with distinct vertical grain. 15. Ringed winding hole. 16. Normal plain winding arbor. 17. Engraving between spandrels. 18. Half-hour marker. 19. Half-quarter marker. 20. Corner spandrel. 21. Arch spandrel. 22. Normal full minute band, sometimes known as track minutes. 23. Penny moon with lunar date in separate box alongside. 24. Arch. 25. Name boss.

FIGURE 12. *Terms used in describing casework.*

1. Hood. 2. Trunk or body. 3. Base. 4. Swan-neck pediment. 5. Architectural pediment. 6. Blind fretting. 7. Pillar caps and bases. 8. Fluted pillar with double-reeded base. 9. Plain pillar. 10. Hood door. 11. Eagle finial. 12. Spire finial. 13. Patera (plural paterae). 14. Top-of-trunk moulding. 15. Dentil moulding (simple left, key pattern centre). 16. Trunk door. 17. Crossbanding. 18. Lenticle glass. 19. Fluted quarter-columns. 20. Pedestal for quarter-column. 21. Base panel. 22. Canted corner. 23. Ogee bracket foot. 24. Plain bracket foot. 25. Stringing line. 26. Shell inlay. 27. Fan inlay. 28. Escutcheon plate. 29. Seatboard.

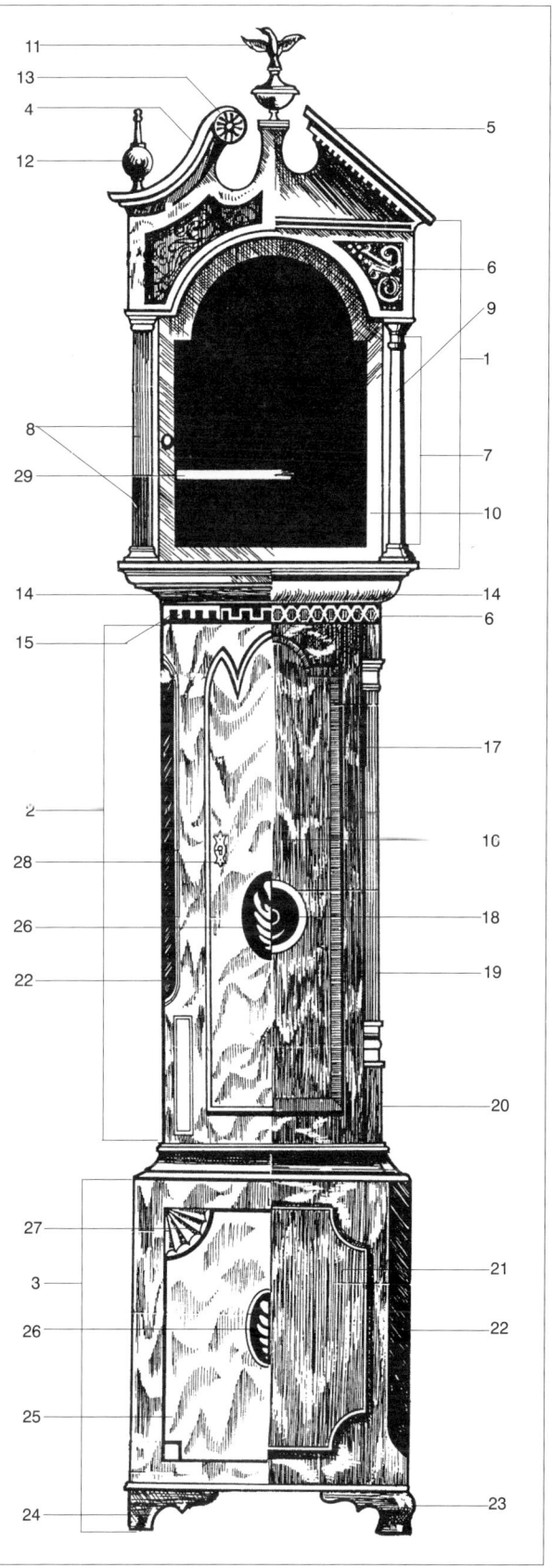

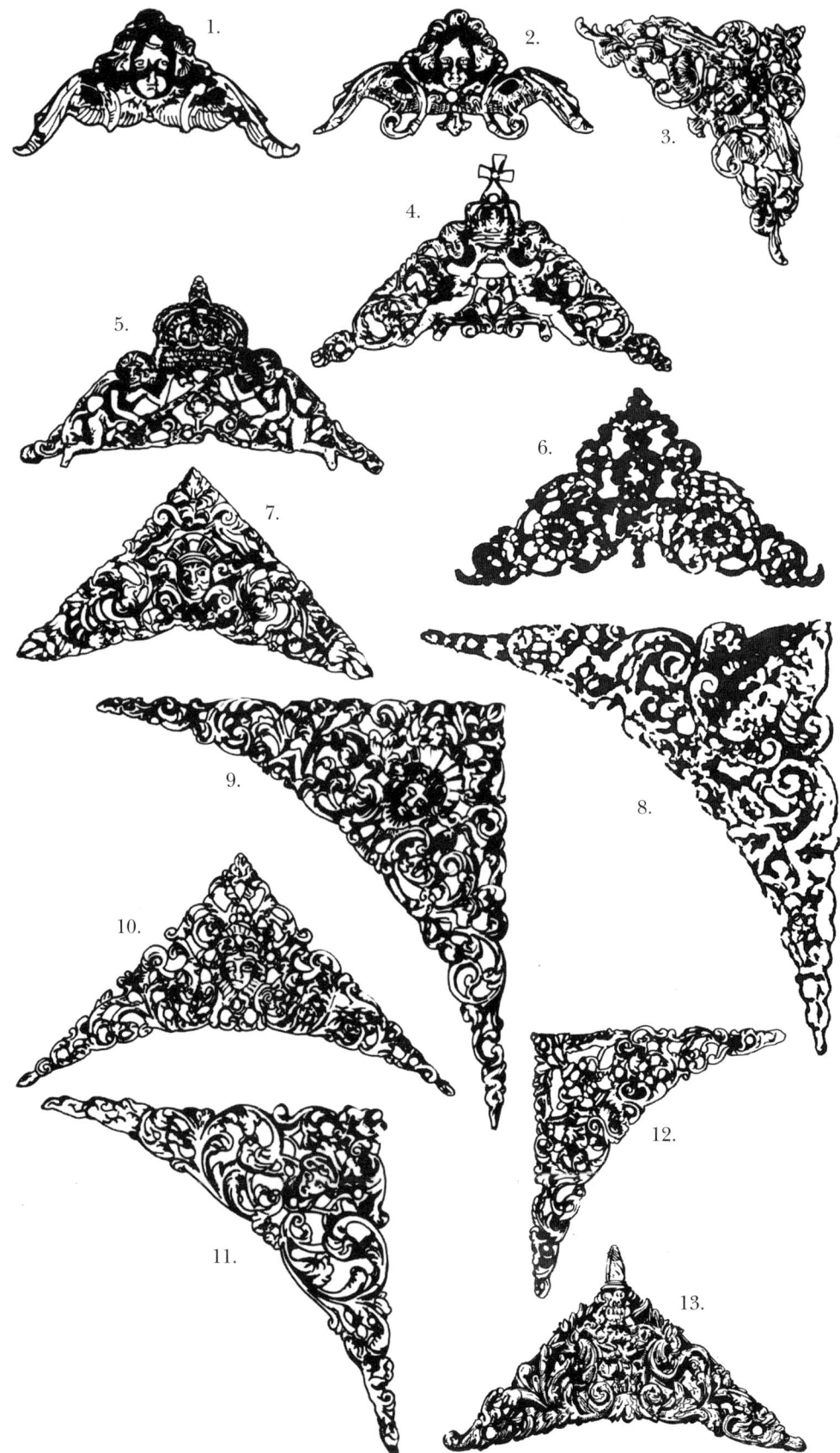

FIGURE 13. *Spandrel patterns. The majority of known spandrel patterns are set out here with the dates of their normal use. The earlier patterns were often for smaller dials than later ones. Sometimes, therefore, an early pattern may occur later than normal where an unusually small dial size has been used. There can be other exceptional reasons why a spandrel may have been used outside its normal period.*

1. *Small London cherub head. 1660-70*
2. *Large London cherub head. 1670-90*
3. *Later London cherub head. 1690-1720*
4. *Twin cherubs with crown. 1690-1720*
5. *Twin cherubs with crossed maces and large crown. 1695-1735*
6. *Tompion cherub head. 1700-30*
7. *Indian head (or male head or mask head). 1700-40*
8. *Elf head with cap. 1700-30*
9. *Female head with headdress. 1710-45*
10. *Another form of female head. 1700-45*
11. *Another form of female head. 1715-45*
12. *Small urn. 1725-40*
13. *Twist knop. 1725-40*

30

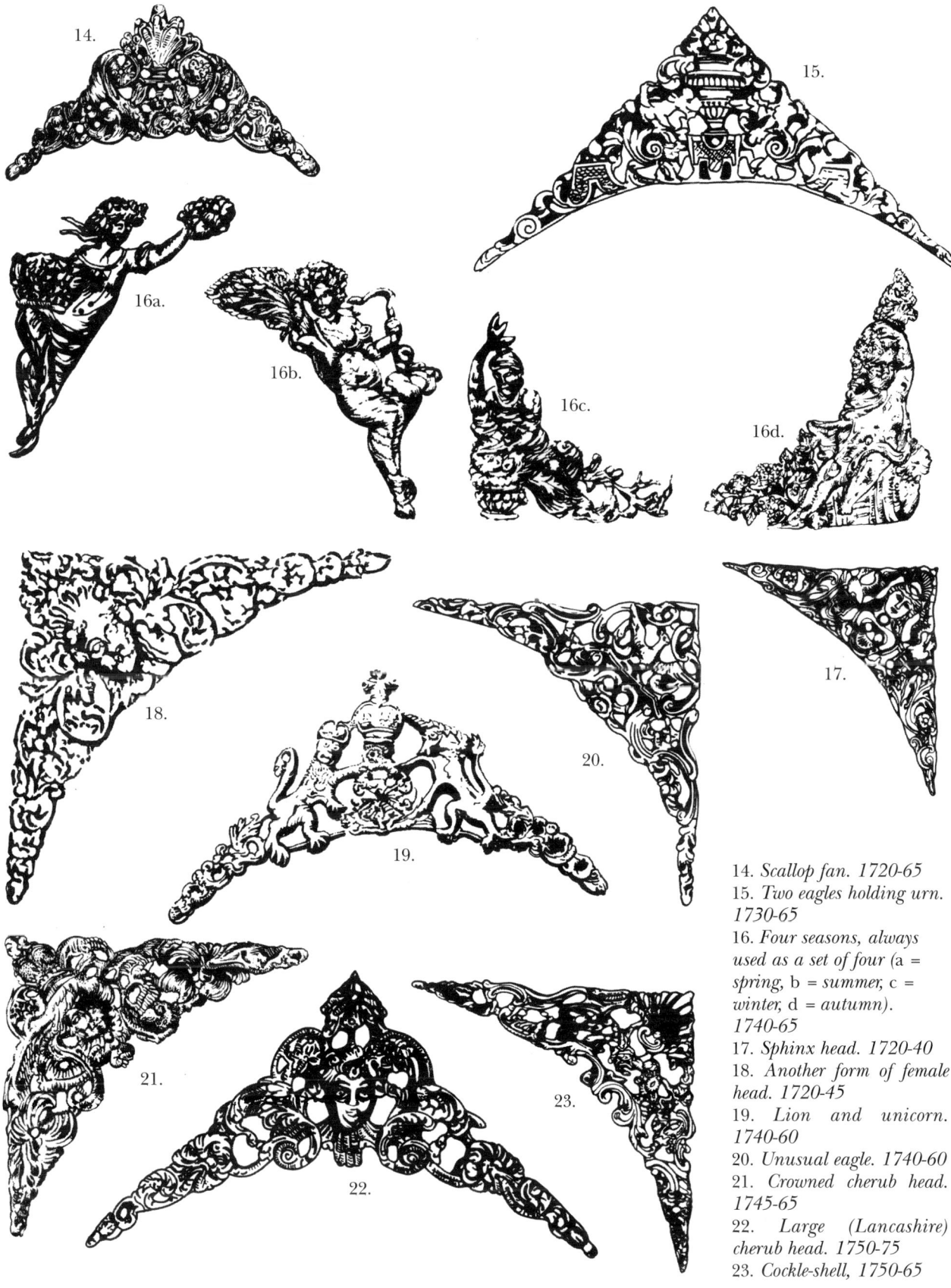

14. Scallop fan. 1720-65

15. Two eagles holding urn. 1730-65

16. Four seasons, always used as a set of four (a = spring, b = summer, c = winter, d = autumn). 1740-65

17. Sphinx head. 1720-40

18. Another form of female head. 1720-45

19. Lion and unicorn. 1740-60

20. Unusual eagle. 1740-60

21. Crowned cherub head. 1745-65

22. Large (Lancashire) cherub head. 1750-75

23. Cockle-shell, 1750-65

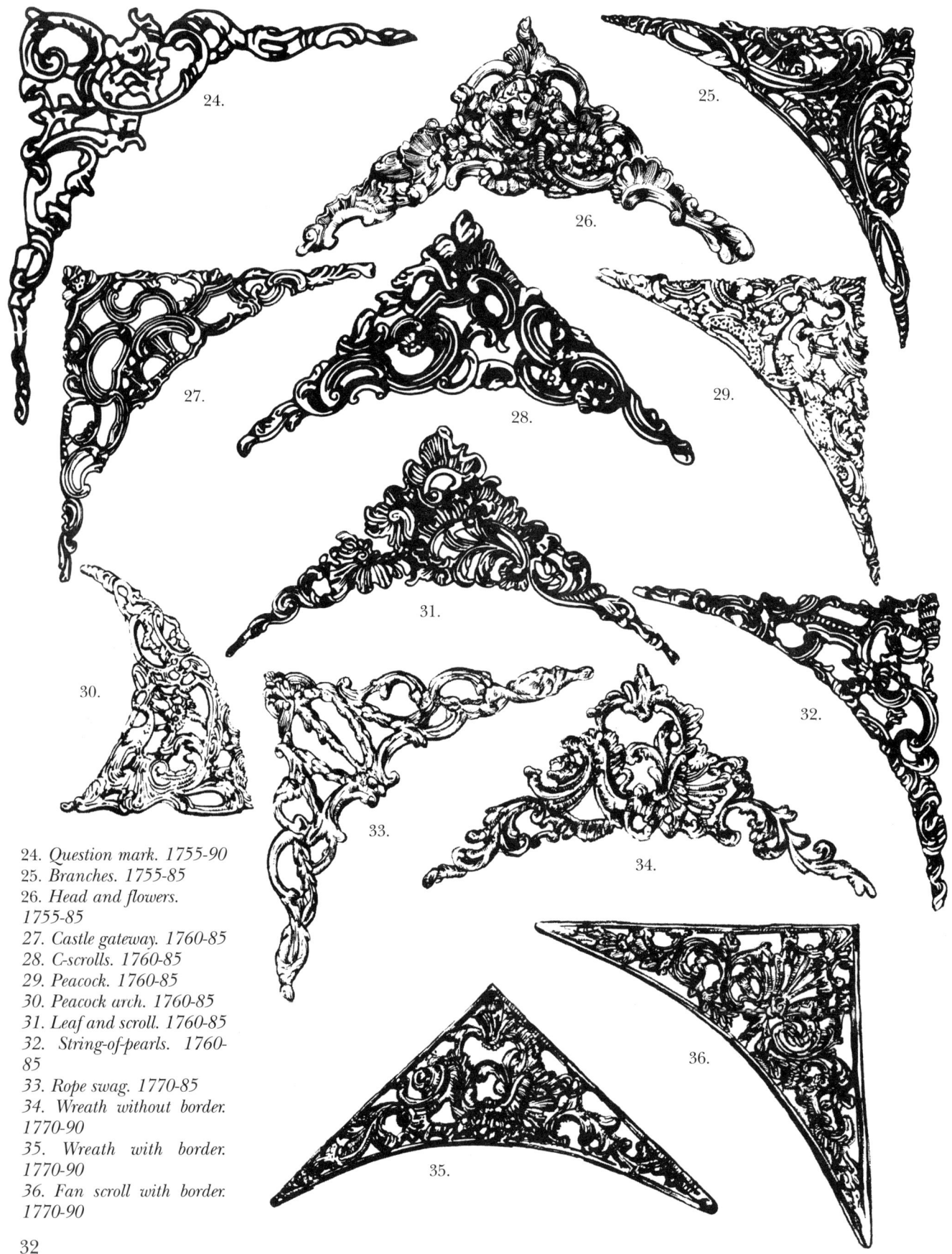

24. *Question mark. 1755-90*
25. *Branches. 1755-85*
26. *Head and flowers. 1755-85*
27. *Castle gateway. 1760-85*
28. *C-scrolls. 1760-85*
29. *Peacock. 1760-85*
30. *Peacock arch. 1760-85*
31. *Leaf and scroll. 1760-85*
32. *String-of-pearls. 1760-85*
33. *Rope swag. 1770-85*
34. *Wreath without border. 1770-90*
35. *Wreath with border. 1770-90*
36. *Fan scroll with border. 1770-90*

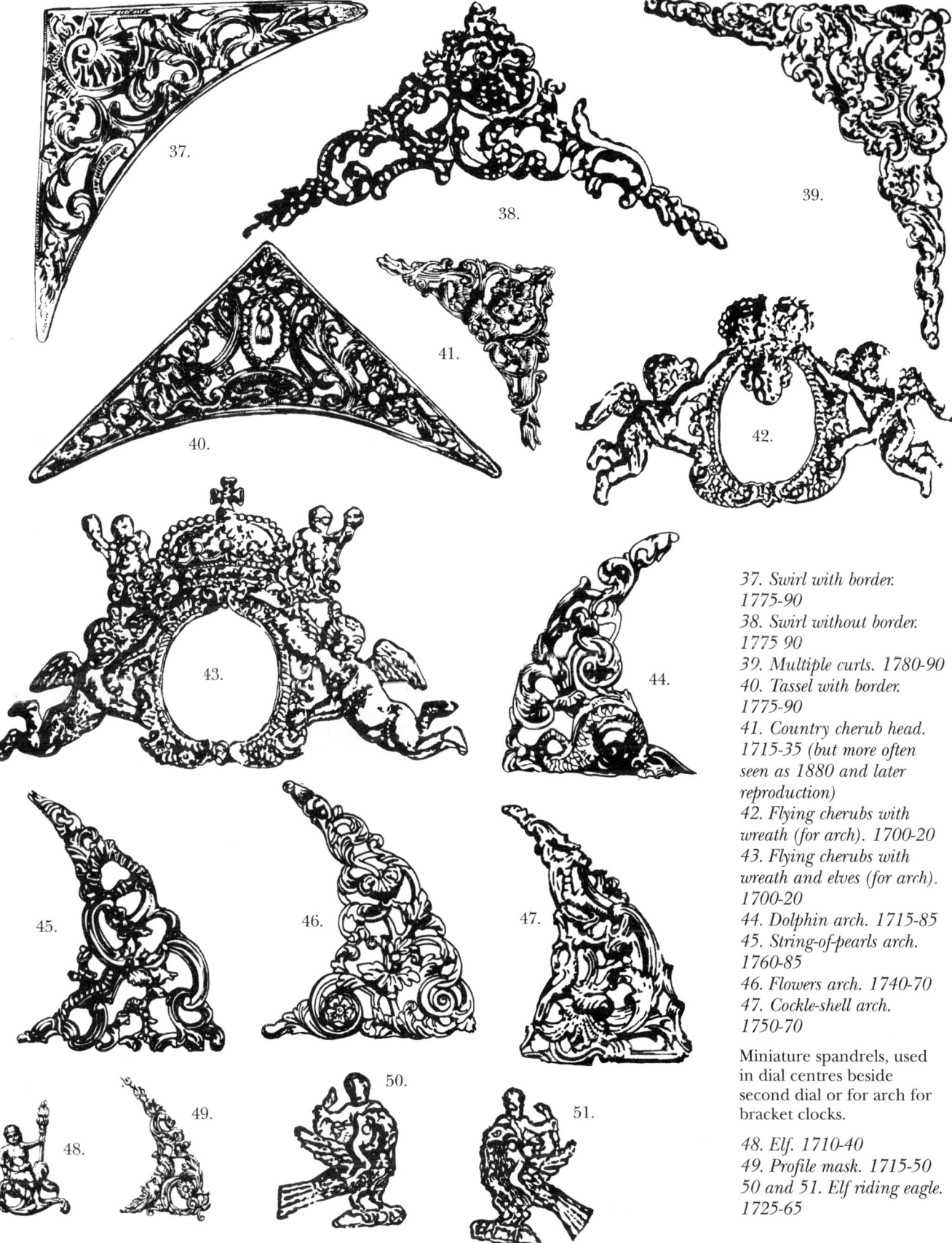

37. Swirl with border.
1775-90
38. Swirl without border.
1775 90
39. Multiple curls. 1780-90
40. Tassel with border.
1775-90
41. Country cherub head.
1715-35 (but more often
seen as 1880 and later
reproduction)
42. Flying cherubs with
wreath (for arch). 1700-20
43. Flying cherubs with
wreath and elves (for arch).
1700-20
44. Dolphin arch. 1715-85
45. String-of-pearls arch.
1760-85
46. Flowers arch. 1740-70
47. Cockle-shell arch.
1750-70

Miniature spandrels, used
in dial centres beside
second dial or for arch for
bracket clocks.

48. Elf. 1710-40
49. Profile mask. 1715-50
50 and 51. Elf riding eagle.
1725-65

33

CHAPTER TWO

EARLY LONGCASE CLOCKS IN LONDON

The earliest London clocks will be examined here briefly. Traditional lantern clocks continued to be made with the new escapements until about 1700, but in this chapter we are concerned with other weight-driven clocks, principally longcase. Those features first developed in London were for the most part continued when eight-day clocks began to be made in the provinces. This was not the case, however, with provincial thirty-hour clocks, which largely had a different origin (see Chapter Five).

The great majority of London-made longcase clocks at all periods from their inception in 1658 were of eight day duration or longer. There was a taste towards the end of the century for clocks of month duration, a fashion which soon seems to have passed. Clocks running for three months or six months or twelve months *were* made but were exceptionally uncommon.

A few thirty-hour clocks were made in London in the seventeenth century, perhaps to cater for the lower price market. They were made either as single-handers or two-handers, and either with plated or posted movements (square-pillar or lantern-pillar form – see page 87 ff.). Some were longcase examples,

PLATE 12. *Ten-inch dial of an eight-day longcase clock of about 1680 by John Wise of London, having six pillars and four latched dial feet. Minute numbers are within the track. Half-hour markers are of the early trident form. Fine-grain centre matting. Signature below the chapter ring. All these are early features.*

34

PLATE 13. *Eight-day longcase movement of the 1670s by Ahasuerus Fromanteel of London, showing latched dial feet and typical finned pillars (also latched). The strike detent passes to the countwheel through a slot in the backplate at this time. The butterfly suspension regulation was limited to this maker and one or two other early ones.*

others were hooded clocks, the latter usually recognisable, even if today caseless, by their exceptionally small dials – normally less than nine inches and frequently less than eight. A surprising number of these hooded clocks today seem to be caseless or to have been re-cased later (some to become 'grandmother' clocks). This may have been because their cases were of ebonised pine, or of walnut veneered on to pine, and have perished. Thirty-hour longcase clocks of any period represent no more than a tiny fraction of London work, to the degree that we automatically think of a London longcase

PLATE 14. *Month duration longcase movement of the 1690s by Joseph Knibb of London, showing latched dial feet, butterfly suspension regulation, outside skeletonised countwheel. The two bells are for roman striking, the top bell being of the typical Knibb 'pork pie' shape.*

clock as being an eight-day.

By the 1680s the London longcase dial had reached ten inches in size, developing gradually with time through eleven to the twelve-inch norm by the end of the century. Any of these three sizes may be met during the late seventeenth century, but by the end of the century and thereafter twelve inches was the almost invariable size. London dial sheets were normally solid castings and therefore relatively heavy, not having the 'cartwheel' spokes found on many provincial, principally northern, dials, which were often cast in that form – see Plate 23.

Later movements of this period normally had five pillars riveted to the backplate and fastened at the frontplate end by latches or taper pins. Early in this period six pillars were not unusual and occasionally more than six. The

PLATE 15. *Twelve-inch eight-day longcase dial by Christopher Gould of London, c.1690. Note ringing of winding and seconds holes, floral engraving between the spandrels, scrollwork around the calendar box, meeting arrowhead half-hour markers – all typical period features. Half-quarter markers are often shown at this period, as here. Fine original blued steel hands.*

high number of pillars was to give extra rigidity to the plates and six pillars (sometimes more) were also used at later times for extra strength in longer duration clocks, which would have to carry larger weights. By the end of the century it was found that more than five pillars were unnecessary and London work settled down to a normal five pillar construction. In the provinces fewer pillars were used at almost all periods than in London, other than with occasional contrary makers.

Pillars at this time normally had a central knop and were finned for decorative effect (Plate 13). Earlier clocks of this period often had their pillars latched, a latch being a small swivel catch which pressed into a notch in the pillar end to hold the plate fast. Some of the earliest clocks also have their (normally four) dial feet latched, but dial foot latching fell from use somewhat sooner than pillar latching. Not every maker used latches and those who did gradually ceased to use them on their movement pillars later so that they fell from general use towards the end of the century. Later in the century and for some years thereafter some makers kept a single latch on the centre pillar and pinned the other four. The point was that the single latch enabled the movement to be clipped easily together thus leaving the workman with both hands free to tap home his four taper pins. Latches are often regarded as a sign of high quality, but this is not an infallible guide. This probably arises from the fact that at this early period when latches were in use the quality of most clocks was very high anyway.

On eight-day clocks striking (and chiming too) was controlled by the countwheel (sometimes called a locking plate). In the earlier part of this period the countwheel was positioned outside the backplate, but by the 1690s the countwheel had been moved inside the plates, attached to the striking greatwheel, where it remained throughout its period of use. The two forms are known as 'outside countwheel' and 'inside countwheel'. Although rack

PLATE 16. *Twelve-inch month-duration longcase dial by Thomas Speakman of London dating from about 1690-95. Almost all the dial features are typical of the period. The Tudor rose centre engraving was used by some but not all makers.*

PLATE 17. *Twelve-inch eight-day longcase dial by John Berry of London. The herringbone border is one of several patterns used. Half-quarters were used by some, but not all, makers.*

striking was introduced about 1676, it was slow to come into use on longcase clocks, where it was seldom used before 1700.

Thirty-hour clocks almost always had countwheel strikework at all periods and on these the countwheel was always positioned outside the backplate. Occasional thirty-hour clocks had rack striking, but rarely before 1720. For more details on countwheel and rack striking see page 24, Chapter One.

London eight-day dials have a distinctive character about them, which is soon learned. Early examples from the 1680s have a matted centre, plain winding holes, often with no dial decoration in the way of engraving, sometimes without even a seconds dial or calendar, making overall a very plain appearance. Some early examples have maintaining power (see page 23, Chapter One). At this time the chapter-ring half-hour marker might typically be a trident or fleur-de-lis, like a refined form of that found on many lantern clocks, soon developing into the meeting arrowheads type.

Minute numerals initially are marked *within* the double engraved minute track. Signatures are often on the dial sheet lower edge and are frequently Latinised, often incorporating the word *Fecit* (Latin for 'made it'). Some dials of the early part of this period had a little decorative engraving on the dial sheet between the spandrels at each side and at the top, to balance the engraved name on the lower edge. Occasionally this between-spandrel engraving was used later, after the signature had ceased to be in that position.

By the 1690s the minute numerals move *outside* the minute track, in the slightly wider space left there for them by the now wider chapter ring, making the numbers somewhat larger than formerly. At this period a small marker indicates the half-quarters on many, but not all, dials (i.e. between 5 and 10, 20 and 25, 35 and 40, and 50 and 55) - see Plates 15, 17, 19 and 20.

Most dials by 1690 have a seconds dial below XII, the seconds calibrated on the *inner* edge of the seconds ring and numbered every fifth unit – 5, 10, 15,

PLATE 18. *Detail of the signature from a thirty-hour longcase of 1682 by John Williamson of London to show the nature of early engraved script and numerals. The half-hour markers are of the early trident form. Signing on a drape (sometimes called a lambrequin) was confined almost entirely to thirty-hour London work of this period.*

etc. Most also by now have a square calendar box above VI, the latter normally with a bevelled edge (a round box is more often found on provincial clocks). The calendar box is often surrounded by a little engraved scrollwork set into the matted centre. The calendar ring had teeth on its inside edge and was knocked on once in twenty-four hours by a pin on an intermediate wheel set on the movement frontplate. The calendar disc rested in two rollers (sometimes two lugs) set into the back of the dial.

Winding holes, plain at first, are normally ringed by the 1690s, as often too is the seconds hole. Signatures now begin to drop the Latinisation and are usually engraved on the chapter ring itself between V and VII. Some dials have an engraved border in herringbone pattern (alternatively called wheat-ear pattern), a feature found in London clocks roughly for about ten years either side of 1700.

The matted centre was that normally used on London eight-day clocks at this period and later too, exceptions being very unusual. Occasionally on some London thirty-hour clocks the engraved centre was used, that is a polished ground with all-over decorative engraving on the same principles as the old lantern clock (Plates 18 and 22). Some London thirty-hour clocks had the matted centre, as used on eight-day versions. However, it seems that the all-over engraved centre was not used on London eight-day work, though it was in provincial clocks.

Half-hour markers on the earliest London clocks were often a form of trident rather like those used on lantern clocks (Plates 12 and 18) and the stem of the trident was joined to the quarter-hour track. By the 1690s a different form of half-hour marker appeared increasingly often, being composed of joined arrowheads in varying configuration, the central stem still attached to the quarter-hour track, a style first used by Tompion in the 1680s followed by others later (Plates 15, 16 and 17). By the time of the first arched dials (about 1700) a fleur-de-lis form of half-hour mark appeared, initially joined to the quarter-hour track (Plate 24) but before long becoming a 'floating' half-hour marker no longer attached to the track (Plates 25 and 26). London dials of the early eighteenth century typically have floating half-hour markers, an increasingly popular style until the half-hour marker eventually ceased to be used at all. This was not necessarily so in the provinces, however.

By the late 1680s increasing use was made on two-handed clocks of the half-quarter marker positioned between the minute numerals at each 7½-minute

PLATE 19. *Twelve-inch eight-day longcase dial by Thomas Bradford of London dating from about 1700. Fine herringbone edging is repeated around the name zone. Here the diamond (or lozenge) half-hour markers are very prominent, used again for the half-quarter points. The movement has six pillars.*

PLATE 20. *Eleven-inch eight-day longcase dial of the late 1690s by John Haydon of Rotherhithe, which is strictly in Surrey, but the style is pure London. Meeting arrowhead markers are used for half-hours and half-quarters.*

PLATE 21. *The movement of the John Haydon dial shows the solid-sheet dial casting, finely finned pillars typical of this period, tiny cheese-head wheel collets, the calendar ring on its rollers. The backcock here has two alternative pendulum hanging positions, a feature offered on some longcases at any time or period.*

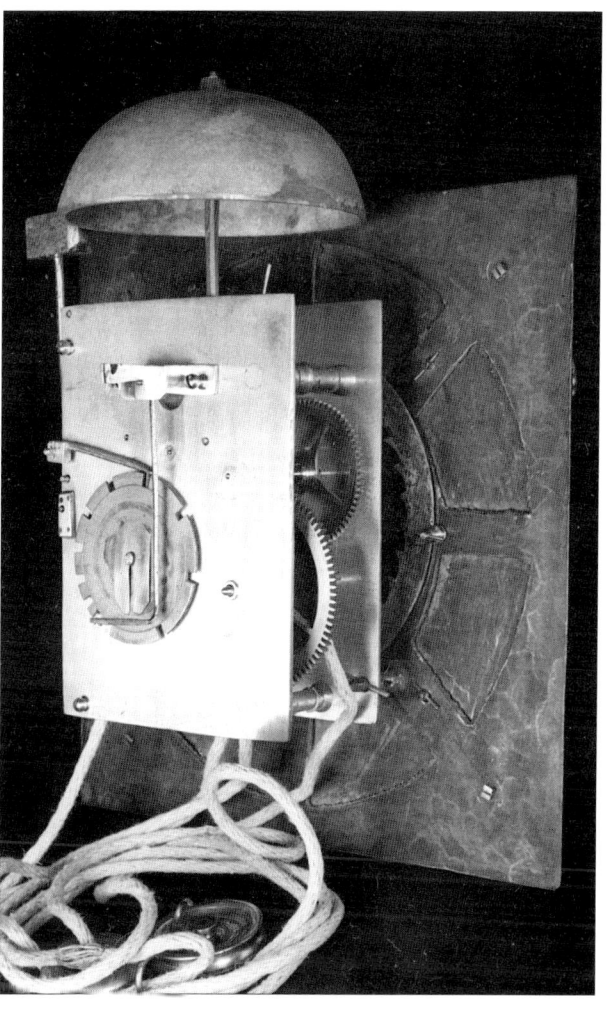

PLATE 22. *Ten-inch dial from a thirty-hour single-handed longcase clock of the 1690s signed 'Tho.Bridg Londini Fecit'. This is a most unusual clock and dial for London work, as is the original cast brass hand with iron backplate. This may have been a provincial maker who worked for a short time in London.*

PLATE 23. *Movement of the Thomas Bridge thirty-hour clock, showing cartwheel type of dial casting, most unusual for London work. Countwheel striking outside the backplate was normal for all thirty-hour clocks.*

point which was not a numbered quarter (see page 16 for an explanation of the function of this). This might be a cross, joined arrowheads, or, somewhat later, a diamond.

Several spandrel patterns were in use at this period, most of them also used in the provinces as well as in London. Patterns kept on changing as time went by, influenced not only by fashionable taste but by dial size. In London the small winged cherub head (Plate 12) was used on dials of ten inches and less. The larger winged cherub head pattern was used on dials of eleven and twelve inches (Plates 15 and 17). Towards the end of the century appeared the twin cherub style, of which there were two forms. The earliest, appearing about 1690, was that with twin cherubs holding a crown surmounted by a cross (Plate 20), the cross sometimes being removed on smaller (eleven inch) dials. Slightly later, about 1710, came the version with twin cherubs holding crossed maces below a larger crown, this form being slightly less common in general and on London work in particular

Hand styles are very difficult to summarise, partly because many clocks have non-original hands and replacement hands may themselves have been copied from non-originals. Generally the earlier hands of this period are simpler and more open in design, whereas those later in the century become more complex and have more piercings. The only way to become familiar with hands styles is to study many illustrations.

PLATE 24. *Twelve-inch eight-day longcase dial by Anthony Hebert of London. The very high arch, lavishly engraved, is a feature of some early arched dial examples. Herringbone edging separates the arch from the square, another feature of some early arched dials. Ringed winding holes are a carry-over from square dial styling. Date about 1710-15.*

PLATE 25. *Early twelve-inch arched dial eight-day longcase clock of about 1715 by Brounker Watts of London, having calendar work in the arch. Ringing of the winding holes is now falling from fashion, but here the maker surrounds his with a little scroll engraving. Floating half-hour markers show the beginning of a new style soon to become universal in London.*

THE FIRST ARCHED DIALS

Throughout the seventeenth century clock dials had been square, but towards the very end of the century the first arched dials appeared, sometimes called a break-arch dial. Tompion is believed to have been the first to produce arched dials in the 1690s, but the arched dial is generally accepted as first appearing on occasional London clocks just after 1700. For the first few years it was an experimental form in so far as the arch had not yet reached a settled height, width and proportion, though it did so within the next decade so that by about 1720 the standard arch dial had become a regular style.

London dials at this period were normally twelve inches wide and, although occasional eleven-inch arch dials are known, twelve inches is the regular London size from the time of their first appearance and thereafter. So set was this principle that when we speak of a London arched dial we mean a twelve-inch dial. A regular (twelve-inch) London arched dial will have an arch of approximately four and a half inches, making the overall dial height sixteen and a half inches. The spread of the arch will cover the central ten inches of the square, leaving a shoulder of an inch at each side. An early arch is instantly recognisable by varying from this either in the height of the arch or the span of the arch or both. In provincial examples of early arched dials this variance is usually even more extreme.

PLATE 26. *Twelve-inch eight-day longcase dial by John Hindmore of London dating from about 1730-40 with a late use of herringbone banding. The sunburst arch boss is not typical but is an echo of the engraved arch of the Hebert dial (Plate 24). Ringing of winding holes has now ceased in London.*

Other recognition factors apply to early arched dials. Often the dial was still seen as being square in design with the arch as an added portion at the top – all part of the one dial sheet but often having the arch section separated in its design. (Some early provincial arched dials have the arch actually constructed as a separate piece, then fastened on to the square.) These early arched dials appeared when the herringbone-banded style was in fashion and when this was continued into the arched period the herringboning often separated the square section from the arch.

Early arched dials are recognisable too by the fact that they continue in these first few years other features of style which belong more correctly to the square dial – for example, by the use of spandrel patterns which were strictly square dial patterns such as the two styles of twin cherub spandrel or the large cherub head one.

These transitional arched dials were often inclined to use the arch for a purpose rather than just for decoration. Some have calendar work in the arch, occasionally moon work or tidal work or both. Later examples commonly used the arch as a place to set the maker's name on an engraved boss, a feature not used in the early transitional period. By about 1720 the London arch dial was of a settled size and proportion, which it remained thereafter. Moreover, in London work the arch became instantly popular to the extent that it replaced the square dial almost totally within a very few years. After about 1720 square dials seldom appear on London clocks other than on clocks made for specialist use, such as regulators.

PLATE 27. *Ebony-veneered longcase clock of the 1670s by Ahasuerus Fromanteel of London showing the spoon and locking catch used on a rising hood. Note the iron door hinges and the lateral section at the top and bottom of the door, used on many early cases to prevent warping. The carcase wood is oak.*

PLATE 28. *Ebony-veneered longcase of an eleven-inch eight-day clock of the late 1680s by John Ebsworth of London. The lenticle glass and barley sugar pillars came into use about this time. This case has an architectural pediment, which soon fell from fashion, but the overall proportion and styling is typical of many early anchor escape-ment clocks.*

CHAPTER THREE

EARLY LONDON CASEWORK

1. GENERAL CONSIDERATIONS

The clockmaker was a worker in metals and did not make the wooden cases for his clocks. This applied from the very beginning to the very end of clockmaking, though there are known to have been very occasional exceptions of provincial clockmakers who also happened to be woodworkers and who probably made their own cases, examples of which are probably unidentifiable anyway. Most casemakers of all periods and regions remain unidentified as very seldom did they 'sign' their work.

The case might have been provided in one of several ways. Sometimes the clockmaker would have ordered a case for a particular clock, fitted his clock into the case and offered it for sale, hoping a customer would come along who liked it and would buy it. This method would tend to be confined more to city or town clockmakers who had a sales shop, and probably particularly to London makers. A clockmaker ordering cases in this manner may have ordered his cases in batches of several at a time of the same or varying styles, fitting each clock into whichever case he chose.

In fact it seems more often to have been the practice for a customer to order his clock in advance as a bespoke item and in those instances he would undoubtedly also specify what casework features he wanted and in what woods, perhaps being advised by the clockmaker on the most suitable or latest style.

A third method, however, of which we have some documented examples, was that the customer bought the clock caseless and either ordered his own case from a local joiner or cabinetmaker, or in some instances even made his own case, the latter applying more often in rural areas, where the customer might be more of a handyman. This latter factor may well explain some of the eccentric and even inaccurate features sometimes found in early provincial casework, some of which are discussed in Chapter Six.

An extension of this principle was that emigrants sometimes took with them a caseless clock and had a case built in the new country, so that we sometimes find English or Scottish clocks in their original, yet American-made, cases. Thomas Tompion's clock No. 30, dating from about 1686, is now in Australia and is housed in a case known to have been made there for it in 1834 in Australian cedarwood – though it is not known whether this was taken caseless or had its original case damaged or lost.

2. EARLY LONDON EIGHT-DAY CASES

The first longcase clocks and the first pendulum spring clocks were made in London from 1658 by Ahasuerus Fromanteel, whose workshops are believed to have had a virtual monopoly for the first ten years or so. Very few clocks survive from this first, ten-year period, even fewer in their original cases, and of those some have been restored or have undergone alterations and may therefore mislead us in certain aspects of style. Clocks of this very early period are examined in detail in *Early English Clocks* by Dawson, Drover and Parkes, but for our purposes we can summarise the situation as follows.

The first long cases (and the first bracket clock cases) were black, made from ebony veneer on an oak carcase, it being thought that black best set off the gold and silver of the clock dial. The pendulum clock was of course conceived

PLATE 29. *Twelve-inch eight-day longcase clock of about 1700 by Isaac Papavoine of London. This case is in pine, stained brown to resemble fruitwood, though most were ebonised. Proportion, styling and mouldings are typical of the turn of the century. Height 6ft. 9in.*

in a Puritan era. Before long ebony supplies proved inadequate and as an alternative 'black' cases were veneered in a fine-grain wood such as pearwood and polished with a black polish, referred to today as being 'ebonised'. The quality of workmanship and design of these cases was exceptionally high and represented the best that money could buy. They were often trimmed with ornamental metal fittings (mounts, escutcheons, pillar capitals, etc) in gilded brass or even sometimes in solid silver. Considerably later, commencing towards the end of the century, ebonising was done directly on to longcase carcases of oak or pine, the grain first being filled with a plaster filler, but this was more common in provincial work than in London.

At first the top was shaped like a rooftop, known as a pitched pediment or an architectural pediment, though before long variations appeared such as the flat top, the flat top surmounted by a single-tier caddy, the flat top with a double caddy, or the flat top with a carved cresting which swept up towards the centre and often had the embryonic shape of a swan-neck pediment, a style later to become widespread.

The first hoods had normally a fixed glass to protect the dial. To enable this to be reached the hood slid upwards along a channel at each side of its back, running on a protruding tongue of the backboard. This is known as a rising hood. It was held in the raised position by a sprung clip on its backboard. When lowered it was held in a 'locked' position by a spoon-shaped lever operated by pressure from the closure of the trunk door. A hinged glass door to the hood proved more convenient, as did a removable forward-sliding hood (needing less headroom), and both became the normal construction within a very few years. Some fixed glasses were later converted to opening and most rising hoods were later converted to sliding, both changes often being made at the same time. Many hoods had large glass side windows offering sight of the movement, probably because proud owners were interested to view the mechanics. Most hoods had three-quarter pillars attached to the door and quarter pillars at the back.

The rising type of hood rested, in its lowered position, on the top-of-trunk moulding, which at this period was convex (quarter-round) in shape. The hood itself, therefore, needed no lower mould either for strength or decoration or even to disguise the joining of two separate parts, as it sat on a mould which already had a decorative edge and formed part of the top-of-trunk mould itself. Most hoods, however, were finished round the bottom with a moulding, often in such a way that the mould had a downward-protruding lip, which made for a neat job of concealing the joint between hood and trunk and also helped locate it to its true position. A few hoods, however, had no lower mould at all, a feature occasionally found on provincial cases too.

As the forward-sliding hood increasingly became the norm, most hoods now did have the lower lipped moulding. This was a regular feature of London cases from this time forward and was to become the accepted method on provincial casework too, though the point in time when the latter became universal varied, as is explained later

The Fromanteel style of longcase was very slender – the short (verge) pendulum did not require a wide case to contain its swing and the dials were small (eight or nine inches). A few went straight-sided to the ground, coffin-clock fashion, but others had a base which was wider than the trunk, and this soon became the normal style. Perhaps the wider stance of the true base

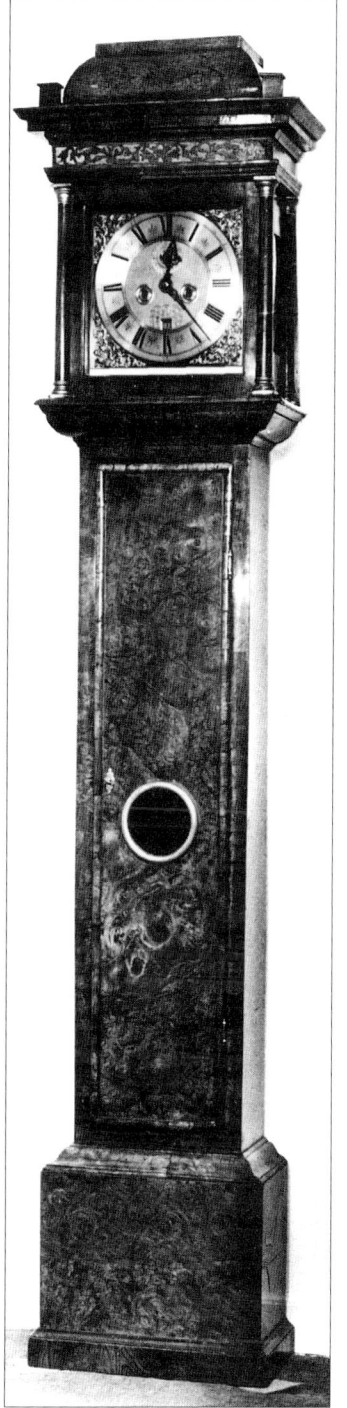

PLATE 30. *Fruitwood eight-day longcase clock of the 1680s by John Wise of London, still using the early panelled form of case. Here the lenticle glass rim is of brass, though many were of wood.*

PLATE 31. *Detail of the John Wise case showing three-quarter barley sugar pillars to the door, quarter pillars at the back. The D-mould edge to the door was typical of many cases from now on. Note the very large hood side windows.*

PLATE 32. *Caddy top longcase clock by John Haydon of Rotherhithe c.1690, the case in burr elm but of a typical London style more usually found in burr walnut. Large hood side windows. Height 7ft.2in.*

helped with stability. Raised panels of veneer broke up the plainness of the moulded-edged door, sometimes the base too and occasionally the case sides. The convex top-of-trunk moulding remained till the 1690s, when this mould changed increasingly often to a concave shape, though convex moulds still appear on occasion for another twenty years or more.

By the 1680s walnut veneer and other strongly figured woods such as olive-wood and laburnum were being used to produce a more spectacularly figured case, the veneers quartered or halved vertically, mirror-image fashion, to

PLATE 33. *Lacquer longcase (with green ground) by Thomas Bradford of London c.1700 with original double caddy top. The overall style is very much the same as on a walnut case of the day. Height just under 8ft.*

PLATE 34. *Dome-top longcase in quartered burr walnut by Anthony Hebert of London about 1710-15. Many hoods now have sound frets in the side rather than glass windows. Note the shallow base. The bun feet are later replacements. Height 7ft.4in.*

PLATE 35. *Longcase in quartered burr walnut by Samuel Townson of London c.1720. Some such cases had a caddy above the flat. Height about 7ft.6in.*

produce 'book-matched' patterns. Very soon strongly con-trasting woods were used set one into the other in complex and intricate patterns, often incorporating flowers and birds, known as marquetry (parquetry being used to describe more simple, geometrically-patterned marquetry).

The progression of style in London cases during the last quarter of the seventeenth century can be seen more easily than described. A little familiarity with a few examples soon makes dating a relatively straightforward affair.

Lacquer cases are known from the 1680s in London work, but more often date from the 1690s and the main period for lacquer was after 1700 (up to about 1760). Lacquering (sometimes known as japanning) was performed by a complicated application of lacs, gums and varnishes to produce a high-gloss coloured ground on to which were worked raised giltwork panels, usually of oriental themes. For a more detailed description of the processes see Tom Robinson's *The Longcase Clock*. Black was the commonest ground colour, followed by blue, then green, then less common shades of red and even yellow. Lacquerwork was usually applied to an oak carcase, though occasional later examples may be of pine with oak for the doors.

Oak was not normally used as a case wood in its own right on eight-day London clocks and examples occasionally seen can often prove to have been cases originally lacquered or ebonised and stripped down later, perhaps when the surface condition deteriorated. Some lacquer cases may have been made in the provinces, but the majority were probably London made, even those examples which are original cases on provincial clocks (see the Holloway clock in Plate 45).

London longcase dials grew wider over the next thirty to forty years, ranging from a ten-inch norm in the 1680s to eleven or twelve inches by 1700. This, together with the anchor escapement and wider-swinging long pendulum, in regular use by the later 1670s, resulted in a gradual widening of cases. A balance of slender proportions was retained by an increase in height, which also tended to make later clocks more imposing. The caddy top sometimes now became a higher caddy or a two-tier caddy, often ornamented by either two or three finials of ball or ball-and-spire or 'flambeau' (flaming torch) pattern. Case height, which had ranged from a little over six feet in Fromanteel examples, increased on occasion to as much as eight feet by the end of the century, even on square dial clocks.

The additional height of some clocks sometimes meant that later owners felt the need to remove their caddies in order to accommodate them under lower ceilings than those they were designed for, so that today many hoods which originally had superstructures are flat-topped. This is something which happened to many clocks of all periods but especially to once-tall examples from the end of the seventeenth century.

A small glass window, known as a lenticle glass, first appeared in the lower door of some longcase clocks in the late 1670s but was not widely used until a decade later and not many of the earliest clocks (i.e. of the 1670s and 1680s) have this feature. By the 1690s, however, it was present in most London cases and its use lasted in London clocks till about 1710, later in provincial examples.

PLATE 36. *Hood of a burr walnut longcase of about 1725 by James Stevens of London. Note the very fine frets to the hood front and sides. Many such clocks now have chevron walnut crossbanding inset as on the door of this example.*

49

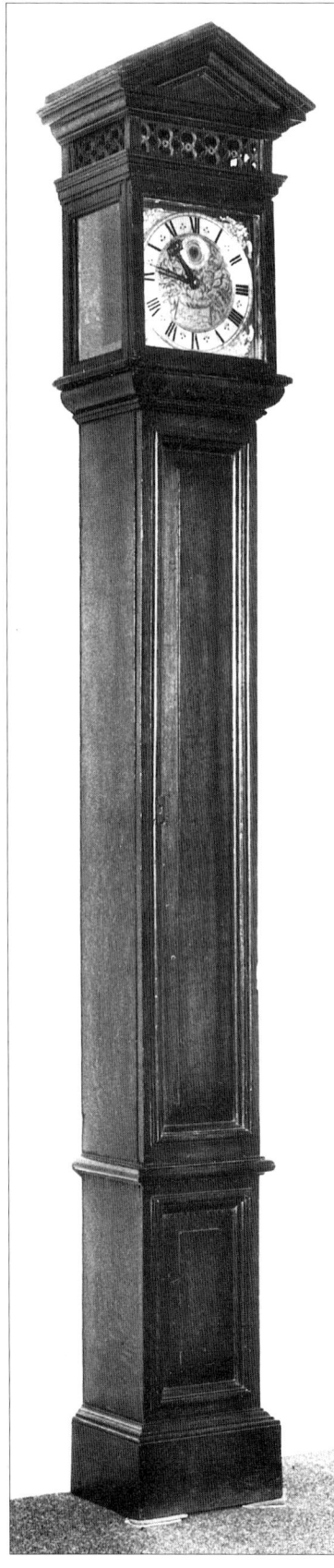

By 1710 the arched dial began to establish itself in London and, although arched dial clocks with a lenticle are known, they are unusual, as the lenticle glass faded from use about the same time as the square dial. The cut-out shape for the lenticle was normally circular or oval, edged with a half-round wooden moulding, sometimes gilded, but some examples have a brass mould. The glass was sometimes plain but many clocks have an apparently original glass of the 'bull's eye' type, despite unsubstantiated statements sometimes made that these are all later replacements. In fact these bull's eye glasses are so commonplace that it is inconceivable that they could all be replacements, and what in any case would have been the point in replacing them? The purpose of the lenticle glass is not known but is believed to have been to show the swing of the brass pendulum bob.

Early arched cases, like early arched dials, often retain features more typical of the square dial form. The lenticle glass is one such feature. Barley sugar pillars and marquetry are both features more often found on square dial clocks, but examples of both occur in some early arched dial cases, particularly during the period of transition from square to arched styles. The flat-top door is also a feature of square dial times, sometimes carried on into early arched dial casework, which soon developed into the standard London arched-top door.

3. EARLY LONDON THIRTY-HOUR CASES

The thirty-hour longcase clock was never a common type amongst London makers, perhaps because the lantern clock was still produced to cater for the thirty-hour market. Some thirty-hour clocks do exist, however, as hooded clocks, often with posted movements either of true lantern-pillar type or of the more usual square-posted birdcage form. Others exist in long cases, though few seem to survive from this very early period.

The thirty-hour London longcase at this time might be ebonised (pine) or might be of oak. The oak examples seem usually to be very dark and though this might be the natural result of the oak darkening with age it may also be that oak cases were stained very dark originally to give them some resemblance to the black of an ebonised case.

These early thirty-hour cases seem to resemble in style the eight-day cases of the day but are usually simpler and without the applied metalwork, probably to keep the price lower. Some, however, are upright 'coffin' cases more in the tradition of the lantern clock case (see page 106). They do not generally seem to have been made in fancier woods such as walnut, but occasional walnut examples are seen with clocks of a more unusual or special nature, for example musical or chiming clocks. It would be most unusual to see an early London thirty-hour clock in arched dial form.

PLATE 37. *Thirty-hour quarter-chiming longcase clock by John Williamson of London, c.1682. The contemporary case is in oak and incorporates many early features – architectural top, sound frets all round (on account of the chiming), full glass sides to the hood, early convex top-of-trunk moulding. No true base but instead the base section is a continuation of the case sides. The door covers the case sides fully leaving no visible door frame.*

CHAPTER FOUR
EARLY PROVINCIAL EIGHT-DAY CLOCKS AND CASES

1. GENERAL BACKGROUND

The concept, design, mechanics, layout and style of the eight-day longcase clock was firmly established in London by 1680. It was in this decade that the eight-day first began to be made in the provinces, though examples are very uncommon before the 1690s. I know of no pre-1680 provincial example, though it is not inconceivable that some may exist.

The eight-day longcase arrived in the provinces in two distinct ways. There were some clockmakers who worked in London (whether trained there or not) and who moved out into the provinces taking with them their fully-developed skills and knowledge of methods. Their work was therefore not simply *like* London work but *was* London work. John Williamson was one such, who moved to work in Leeds in 1683, as was Thomas Cruttenden, who moved to work in York in 1679.

This type of clockmaker had been schooled in the eight-day market, which predominated in London and was patronised by the upper income groups. The likelihood is that these men would stay with that type of clock in the provinces, the cheaper thirty-hour form making up only a very tiny proportion of their work, if any, and those tending in the earliest years to be lantern clocks rather than thirty-hour longcases. Such clockmakers moved to the principal towns and cities of the provinces, where they were accessible to a wealthier clientele. They were usually few in numbers as even a large town would hardly have provided a living for more than one clockmaker at a time at this early period. Most of the clocks we see by these few ex-London clockmakers will be eight-day examples (or occasionally of longer duration, of course). Their eight-day clocks, however, will be seen more frequently than eight-day examples by the other group described below.

The other way in which eight-day longcase clocks reached the provinces was through locally-schooled clockmakers, often of the 'clocksmith' type (see page 99), many being sons and even grandsons of blacksmiths or other metalworkers, carrying on a traditional family craft developed over several generations and now including clocks as a new product. Some of these were town-based but others lived in rural areas. Their principal product was the thirty-hour clock (lantern or longcase) and eight-day examples made up only a tiny proportion of their output. Such makers included men like John Ogden of Askrigg, Walter Archer of Stow-on-the-Wold, Jonas Barber of Bowland Bridge (see Chapter Sixteen) and in fact the great majority of the early provincial makers by whom eight-day clocks are known (as well, of course, as all those whose output was limited purely to thirty-hour clocks and who are discussed in Chapter Seven).

The fact that a provincial clockmaker of clocksmith origins may have concentrated mostly on thirty-hour clocks for the cheaper end of the market does not mean that he was not capable of fine quality eight-day work, frequently much finer than his thirty-hour work, when he could get orders for it. The great difference of quality between the humble, sometimes crude thirty-hour work and the high grade, well-finished eight-day work of such an early provincial clockmaker proved puzzling to past students of the subject to the point where it was often concluded that both could not have been made by the same crude hand. The inference was drawn that he himself made only rough clocksmith work and that he must have bought such fine eight-day clocks from some unidentified London clockmaker for the occasional well-to-do client and had his name engraved on it as maker, though really acting only as a retailer. In fact no documentary evidence has ever come to light to support this view, which as far as I can see is totally false, born of a conceit that fine clockmakers worked only in London and that rustic makers were capable of nothing better than smithy work. All the visual evidence indicates that such men *did* actually make their own eight-day clocks as well as their own thirty-hour ones.

These clocksmiths must have seen and examined and learned from examples of London eight-day clocks, but they frequently added their own personal touches of craftsmanship or idiosyncrasy of design, which may not necessarily have been better or worse than the London model but were essentially *different*. These little differences set them apart from the London school, where the set model was drilled into clockmakers from apprenticeship onwards as the 'right' way of doing things and where innovation and deviation were positively discouraged and were seldom practised (other than in such cases where some individual and usually outstanding maker might invent some new mechanical method of achieving a certain result or might produce an unusually complex clock for a specific purpose).

Provincial clockmakers at this very early period well knew the London style and principles but followed them only as far as their personal inclination extended. Furthermore, many added little personal touches of style or construction which were sometimes sufficiently whimsical to form what was virtually a trade mark. Examples of this can be seen from some of the clocks illustrated. It is that very individuality and eccentricity which gives provincial clocks an exciting and endlessly varying character, and especially these earliest ones which form the very beginnings of the craft nationwide.

Turning for a moment to the cases of early provincial eight-day clocks, those makers who moved from London probably retained contacts with the casemaking trade there and it seems that they bought some of their cases from London (for example the Williamson cases in Colour Plates 39 and 40 and Plates 586 to 588???) and where they had them locally made they would no doubt be based closely on London models. Provincial makers of the clocksmith school mostly had their cases made locally, but these too would have been styled to a greater or lesser degree on London principles. This London influence was sometimes also apparent in the cases of the thirty-hour clocks made by the clocksmith school, but often the thirty-hour case was of a much more rustic nature – see Chapter Six.

The question of dial engraving is a very difficult one to answer for any period and especially at this early time when it might be expected that very few engravers lived in the provinces, especially those rural areas of sparse population. This might seem to suggest that many provincial dials are likely to have been engraved by London engravers, but that is a conclusion all too easily jumped to. It is a question we could only answer on an individual basis, one clockmaker at a time, if we had a number of his clocks to examine, which is rarely the case. Some clockmakers undoubtedly could and did engrave their own dials. We can identify a few of these but for the majority we are very hard pressed to identify who engraved his own dials and who did not. The idiosyncratic nature and 'non-London' style of the engraving of many provincial dials often indicates that they were engraved locally, either by the clockmaker himself or by a local engraver.

Whether a clockmaker engraved his own dials, sent them elsewhere locally to engrave or had them engraved in London is of little more than academic interest. The London clockmakers, especially the most famous names, had much of their 'own' work done out of house. The Knibbs, Tompions, Quares, Windmills and other first rank London clockmakers put out to others such a high proportion of their work that some became little more than employers of other workers, in and out of house. It is not uncommon to find the name of a lesser London clockmaker engraved inside the work of one of the most famous ones. So, if the provincial clockmaker *did* send some of his work such as engraving out to others, we should no more think less of him for that than we do of Tompion or Knibb for doing the same. In fact we *do* feel far more pride and pleasure in the work of a man who did it all personally and if that reality exists anywhere at all it is in the work of the provincial maker in these very early times which represent the birth of clockmaking.

As we have seen, the cases of early provincial eight-day clocks might have been ordered from casemaking specialists in London or might have been made locally. We have seen from surviving examples that those provincial makers who came to the provinces from London (or their customers) did sometimes house their clocks in fine London-made cases in walnut or marquetry (for example the Williamson cases in Colour Plates 39 and 40 and Plates 586 to 587). So too on occasion did the early locally-trained clocksmiths (see the Holloway case in Plate 45). The picture is distorted. however. by the fact that some such early clocks have been re-cased later, their fine cases having probably been taken in the past to house 'better' clocks with London names.

Those cases which were made locally can usually be identified by the degree to which they vary from the London norms, even when based on London styles – for example, the case of the Stretch clock (Plates 56 and 57), John Hough clock (Plate 60), anonymous clock (Plates 41 and 42).

2. SOME EXAMPLES

The anonymous clock in Plates 38, 39 and 40 has a ten-inch dial and probably dates from the 1680s, but could perhaps be earlier. It is of an experimental mechanical nature and was probably a clock built purely for experiment, there being numerous spare holes in the plates where certain features have been tried and re-positioned in an ultimately successful attempt at what is an otherwise unknown type of repeating strikework. The reason the clock is unsigned may even be because it was an experimental piece rather than a commercial clock, but we shall never know.

It could conceivably be London made but is probably provincial. The dial is very much of the London design, the engraving resembling that on clocks by Robert Seigniour and/or Henry Jones. It seems likely that the engraving is by a London engraver, but we cannot be certain. Distinctive in character on this dial are the dots for quarter-hours alternating with four meeting arrowheads for half-hours on the inner circle. The same arrowheads are used for half-quarter markers. Half-quarter markers were known from the earliest longcase clocks but were in fact seldom used until after the minute numerals became positioned outside the minute track. The numbering is very early – note the curly-tailed ones and the curly-topped fives, a style of numbering fading out of use by the 1690s and gone by 1700.

The spandrels are of a pattern seldom seen and at first sight their style might seem to suggest a later period. In fact these are original to the dial. As they are

PLATE 38. *Anonymous ten-inch dial of an experimental longcase clock with unique strikework dating from the 1680s. Note the shape of the early numbers, especially the 1 and 5. Uncommon spandrel pattern.*

PLATE 39. *Detail of the striking barrel of the experimental clock. The trigger shunts back and forth to pick up the required number of pins at each hour. Five finned pillars.*

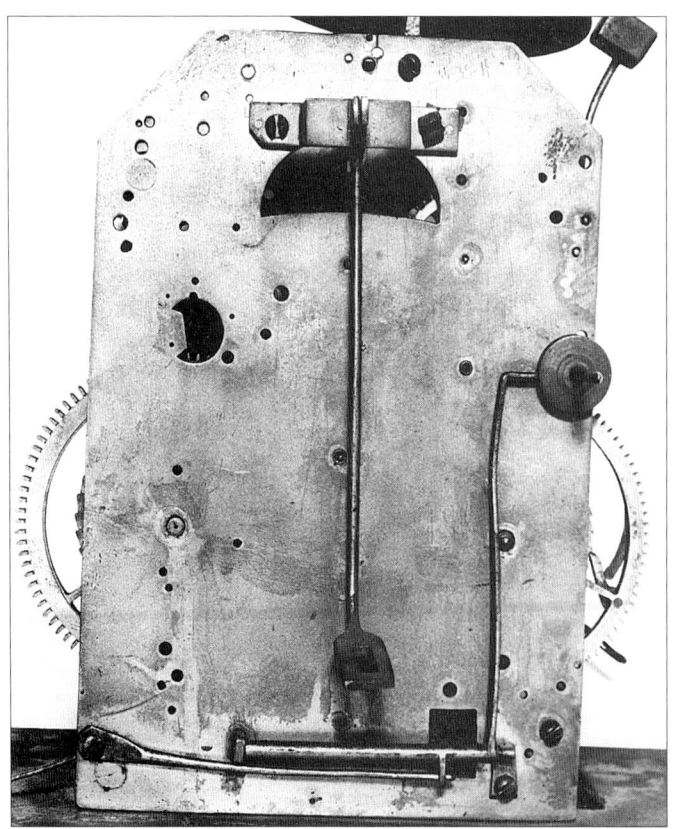

PLATE 40. *View of the experimental movement from the back showing shaped plates, striking shuntwork, escapement removal slot, and numerous holes where modifications to the system have been tried.*

PLATE 41. *Detail of the hood of the experimental clock. The hood is not removable and the movement has to be inserted through the opened hood door. Extraordinary backsplats run down from the hood on to the upper trunk. Access doors at each side of the hood to assist with assembly.*

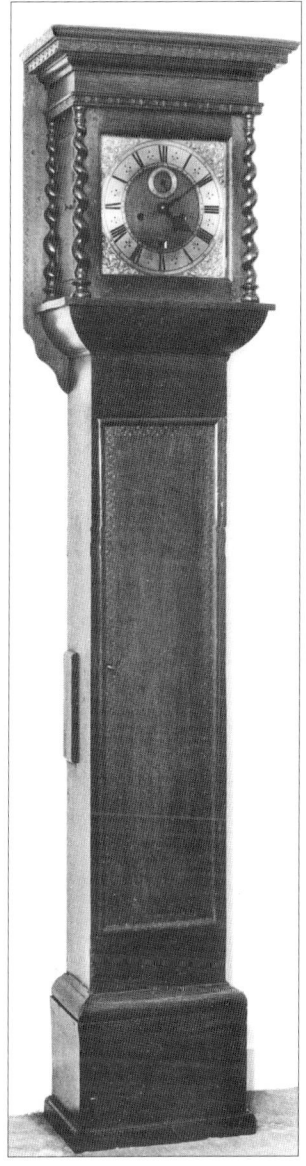

PLATE 42. *The oak case of the experimental clock showing many early features. Exceptionally heavy convex top-of-trunk mould. The door is hinged between the case side-members, there being no true door frame. Height about 6ft.6in.*

unusual I asked several experts for an opinion on them. One of the older school of restorers, very experienced in early London clocks, told me he often used to come across them but always replaced them with the 'nicer' cherub head style – which may account for the reason we seldom see them today!

The movement has a type of counting barrel apparently unique in British work, whereby the barrel makes one rotation at each hour. It has wedges across its width reaching from one to twelve 'teeth' wide, set so that the strikework is triggered on a varying number of wedges depending on how far back or forwards the pick-up trigger is shunted by means of a sort of rack tail set by a sort of snail. The effect is the same as that of rack striking, but is a more involved way of achieving the same result, which allows for repeating if desired. The system is wasteful of power, however, as the barrel completes a full revolution for any number of blows, doing the work of twelve to produce even a single blow, which is perhaps why it did not catch on. It also involves a spring-loaded shunt system more complicated than that in rack striking and therefore probably more fault prone.

It seems to me that anyone who had seen the rack striking system would not have troubled to invent this less perfect and retrograde method, and to my mind this would suggest that this clock pre-dated the invention of rack striking, which is usually set at 1676. However, those who know these things

PLATE 43. *Ten-inch dial of an eight-day clock of about 1685 by William Holloway of Stroud. Early form of numerals, especially obvious in the curling 1s. Superb original hands in blued steel.*

PLATE 44. *Movement of the Holloway clock showing five finely finned pillars pinning, most unusually, at the back, the centre one latched. Note high outside countwheel. Solid dial sheet.*

better than I do tell me the clock is not quite that early, so I date it tentatively in the 1680s. Perhaps it was made a few years after 1676 by someone who had not yet seen the rack striking system (which was after all not much used on longcase clocks till after 1700). This again suggests provincial work, as the rack strike system was certainly known to London makers by the late 1670s. The five finned pillars and neatly finished wheelwork suggest competent work, apart from the inventiveness of the strike system.

The oak case of the anonymous clock in Plates 38 to 40 may not be original to, but is contemporary with, the clock (Plates 41 and 42). It shows many interesting features of early country work. The large hood wider than the small and narrow base and therefore giving a top-heavy appearance characteristic of many early provincial cases, the full length door with an unnecessary and 'unused' space above and below it, the half-round moulding to the door sometimes called a D-mould, the barley sugar pillars, the extraordinarily heavy convex top-of-trunk moulding, are all very early period features. A most unusual feature, seldom seen and then only at this early period, is that the hood is a fixture to the case and does not remove. The clock is inserted by opening the hood door. Another feature indicative of this very early time is that the door hangs on the case sides, which themselves form the door 'frame' with no true framework. This is found only on a few very early longcases and sometimes on lantern clock cases.

A very unusual feature is that the backsplats to the hood section are unduly wide and extend well down into the upper trunk area. The flat 'mould' above the convex top-of-trunk moulding is in fact a solid piece of timber which also forms the seatboard, as with the lantern clock case in Plate 121, and is a feature of a few very early provincial cases. The 'carving' around the trunk door, backsplats and hood frieze is original and seems to have been produced by shaped punches used to form a decorative border pattern.

The clock by William Holloway of Stroud in Plates 43 and 44 was made about 1685. This maker, by whom lantern and longcase clocks are known, married in 1664 and died in 1694. The ten-inch dial shows by its formal 'London' style that he was familiar with what went on in the capital, or perhaps that his engraver was. The curly-tailed figure one, chapter ring layout, ringed winding holes and fine original steel hands are characteristic of the period and of high quality. The movement, however, has several features of a highly individual nature, most obvious of which is that the five finned pillars are riveted into the movement frontplate and the outer four pin behind the backplate, the centre pillar being latched at the back. This is exactly the opposite of London practice, where pillars pin or latch at the front. The clock has maintaining power of the bolt-and-shutter type (see page 23) but in this instance never had the usual shutters, again indicating individual preference by the maker. Striking is controlled by an outside countwheel, normal at this period. These factors demonstrate that this clock, whilst having the latest stylistic features, was *made* by Holloway, not bought by him from a London maker.

The blue lacquer case of this clock (Plate 45) is typical of the caddy-top London style and proportion (here with a two-tier caddy), but lacquer is very unusual for this early period even in London and much more so in the provinces. It seems likely that this case was bought from a London casemaking specialist as it is doubtful whether any local cabinetmaker could have produced lacquerwork of this quality at this very early date. Being a London case then, this has nothing to teach us about provincial casemaking.

Ninyan Burleigh worked in London from 1692, having trained under some unknown, probably provincial, clockmaker, but he left London again within the year. The clock in Colour Plate 1 (page 114) probably dates from the mid- to late 1690s and shows that he then worked at Pontefract, though within ten years or so he had moved to Durham. It is just possible that this Pontefract clock was made before he went to London, but the strong London styling suggests not. The movement has five finned pillars and inside countwheel striking and the ten-inch dial too is very much in the London style, suggesting he took his London skills with him to Pontefract. The chapter ring still has the minute numerals within the double circle of minutes in the earlier manner, but uses the newer style of meeting arrowheads for half-hour markers, a style first used by Tompion in the early 1680s and which came gradually into more widespread use in the 1690s. The original blued steel hands are especially fine and deeply chiselled and bevelled.

The clock by Edward Norton of Warwick (Plates 46 and 47) dates from about 1690, give or take five years, and that too has a mixture in the twelve-inch dial of older and newer features indicative of the transitional style of the end of the century. Here the minutes have moved outside the track, a newer feature of the day, yet Norton still uses a variation of the trident motif for half-hour markers, a feature by now becoming old-fashioned. Not much is known about this maker except that he repaired the church clock at Berrington in Shropshire in 1680. He is believed to have worked in Warwick, though he puts no place name on his dial. He is the earliest Warwickshire clockmaker by whom a longcase clock is known to survive, and this is the oldest known Warwickshire longcase clock.

His dial signature shows unusual floral sprigs under the name, which is on the chapter ring in the new style position for it. The spandrels are the cherub head type typical of most provincial clocks of this century, here having a little engraving between each spandrel. That in itself is a bit unusual as normally we find such between spandrel engraving on three sides only, leaving the space between the lower ones for the name (as on the Holloway dial in Plate 43),that

PLATE 45. *The double-caddy case of the Holloway clock is in blue lacquer and almost certainly London made. Large hood side windows.*

PLATE 46. *Twelve-inch eight-day longcase clock dating from about 1690 by Edward Norton of Warwick. Engraving between all four spandrels is an unusual variation. Minutes are now marked* outside *the minute track in the newest style of the day.*

being the earlier position for the name. So again we have a mixture of old and new styling, in this respect similar to Ninyan Burleigh's dial in Colour Plate 1. The square box calendar in the matted dial centre has a little engraving around it typical of the day, here showing the early form of flat-topped eight numeral. This feature too is similar in style to the Burleigh dial. The mixture of old and new style features on both these dials may be indicative of provincial engraving work, but whether of the makers personally is impossible to say. The movement has five finned pillars and is of high quality. Striking is by inside countwheel and the upper backplate has a slot for removing the escapement without dismantling the plates.

The Norton clock is in its original oak case, standing about 6ft.8in., vaguely based on London styling and typical of many provincial cases of this time. Now we have a true door frame holding the long flat-topped door with iron blacksmith-type hinges – at this early period virtually all doors are flat-topped. The door has a lenticle glass, as do many of this period, here with an apparently-original bottle glass or 'bull's eye' glass rather than the plain one sometimes found. The upper trunk corners have a little incised cut for decoration, not typical of cases in general but just a small touch by this particular (unknown) casemaker to individualise his work. Now too we have the newer type of concave top-of-trunk moulding to support the hood, a style which usually begins just before the end of the century and continues in concept thereafter, though with occasional variations and occasional late examples of the convex shape.

The hood pillars have a three-quarter section to the door and a quarter section set at the back inside a backsplat, which is now of the normal type, unlike the strange example in the anonymous clock in Plate 41. The pillars are turned in an interesting way sometimes called 'bobbin-turned' and quite different from the straight-section London pillar. This bobbin-turned pillar fashion lasted for a good few years and probably originated as a provincial derivation from and continuation of the earlier London-based barley sugar twist-turned style. Pillar shapes appear in a great variety of turning patterns at this time on provincial cases and are one of the most interesting features determining the character and appeal of the case.

The hood has its original caddy top along the London principle but simpler; the caddy on many such cases was removed later to reduce height. Across the hood front are a series of holes, designed to let out the sound of the bell. Many clocks have a fret here to hide the sound holes, but this case shows no sign of having had one.

Now we come to an eight-day clock (Plates 48 to 50) by a little known clockmaker, John Waklin, who signed his clocks without any place name and whose work place is therefore unknown. I have seen only two clocks by this maker, the other being a thirty-hour longcase (Plate 186) which is actually dated 1707. Judging region by style alone is very difficult at this early date, but my feeling is that this maker worked in central southern England. His twelve-inch dial shows an altogether different treatment from any eight-day clock yet examined. The dial centre is not matted but polished and engraved most beautifully with all-over engraving of the type formerly used on lantern clocks, being ornate floral work. In this example, as often in others of this nature, the symmetrical pattern stems from the calendar box as a central source, like flowers from a vase. At this period this form of all-over engraved centre work was very largely provincial, was used by a minority of clockmakers and was more often used on thirty-hour work than eight-day. In London it was just occasionally used on thirty-hour clocks, but not, as far as I am aware, on eight-day work at all.

The chapter ring uses the old form of minute numbering inside the track as well as the old form of trident (verging on fleur-de-lis) half-hour markers with unusual heavy dots marking the hours. These dotted type markings were used in London by Henry Jones and Robert Seigniour but in a much more delicate manner than here, where the whole dial is of a very bold and strong character.

Waklin's name is engraved in the older fashion along the lower edge of the dial sheet within the decorative border, which is of a very bold type of herringbone engraving, sometimes called wheat-ear engraving. The twin cherub spandrels of the earlier type have their crosses overlapping the border, which was not unusual. Ringed winding holes and ringed seconds complete this very busy dial. The original blued steel hands are of a very eccentric nature, the broad and heavy hour hand being almost two-dimensional. The engraving very strongly suggests the work of a provincial engraver.

The movement shows much of fine quality in, for instance, the four finely finned pillars. It was by now realised that four pillars were quite adequate and this clock illustrates the beginning of an increasingly strong provincial trend to four pillar work, whereas London makers stayed with the five pillar layout (or occasionally more, of course). The strikework still uses the older form of outside countwheel.

Features indicative of Waklin's own personal work are the extraordinary hammer spring which is shaped like a giant musical note (a straight piece of

PLATE 47. *The oak case of the Edward Norton clock stands about 7ft. The integral baluster pillars are a variation known as 'bobbin-turned' and are found on some provincial work at this period.*

PLATE 48. *Twelve-inch longcase clock by the little known maker John Waklin, using the all-over engraved centre style on to a polished ground deriving from the old lantern clock dial centre. Highly distinctive and bold numerals and borderwork full of individual character and a considerable departure from London practice. The original blued steel hands are also very individualistic.*

plain spring steel would have functioned equally well), the cut-out in the top centre of the backplate (done to facilitate easy removal of the anchor), the large heavy wheel collets and the large square-headed screws. The S-shaped trip lever can be seen top right for re-setting the strike in the event of its coming out of sequence. No such movement ever came out of London and everything about it indicates the personal craftsmanship of the maker. Dating is a little difficult as its traditional features combine with some up-to-date ones, but perhaps c.1705.

The bell stand is attached to the frontplate, a feature which is often made much of as being an early indicator, whereas attachment to the backplate is thought of as being the later norm, whatever 'later' might mean. Personally I don't regard this as very helpful and a glance at a few movements will soon indicate that it is almost a random affair as to which plate it attached to. The anonymous clock in Plate 40 has it on the backplate and so does the Holloway in Plate 44. It is probably true to say that most early clocks use frontplate fixing and most later ones use backplate, but there are so many exceptions as to make this of little dating value.

The Waklin case (Plate 50) is of pine with an ebonised finish and stands about 6ft.10in. Its basic style and features are based on the London model but in simplified country form. The long door with a bull's eye lenticle glass, concave top-of-trunk mould, and hood pillars attached to the door (three-quarter front pillars, quarter rear ones within the slanted backsplat) are all typical. Such provincial cases at this time appear also in oak, painted pine,

PLATE 49. *The John Waklin movement also shows signs of individual thought characteristic of better provincial work. The large musical note hammer spring is probably unique. Outside countwheel with outside detent. Solid dial sheet. Square-headed spandrel screws.*

fruitwood and walnut, the latter both in solid and veneered versions. The glass side windows to the hood appear on many, though not all, cases of this period. The case was probably made locally, but clearly with knowledge of contemporary London styling.

The eleven and three-quarter inch dial of the eight-day clock of about 1690 in Plates 51 and 52 is signed 'John Worth, Humberstone'. Humberstone is a village in Leicestershire and this clock is the oldest Leicestershire eight-day longcase clock so far recorded. The maker was completely unknown until this clock came to light in the outbuildings of a Leicestershire farm just two or three years ago. Subsequent research proved that John Worth was born in 1656, married in 1685 and was still alive in 1702 when his wife died.

This (solid sheet) dial has the same type of all-over engraved floral centre as the Waklin clock (Plate 48), engraved even inside the seconds circle, this pattern, like the Waklin one, springing from the calendar box. Minutes are numbered outside the track. Half-hour markers are of the meeting arrowhead type. This is a fine dial of highly distinctive nature and purely provincial, still having traces of gilding on the spandrels.

The movement has four finned pillars, fastened to the frontplate by latches and pins at alternate corners. Jonas Barber also used this method (see page

PLATE 50. *The case of the Waklin clock is in ebonised pine and stands 6ft.10in., a country version of the London case of the day.*

PLATE 51. *Dial of an eight-day clock of about 1690 by John Worth of Humberstone, Leicestershire, a maker as yet known only through this one clock. This dial also uses the engraved-centre principal, beautifully executed.*

PLATE 52. *The movement of the John Worth clock has the frontplate attached by alternate pins and latches. Note the planishing (hammering) and filing marks on the frontplate. Inside countwheel striking.*

436) which clockmakers occasionally hit upon independently. Striking is by inside countwheel. The winding squares are filed decoratively into a four leaf clover pattern, a charming feature occasionally found on provincial clocks. The wheel spokes are left unfinished from the casting, still having casting rag attached, and no attempt seems to have been made to clean off excess brass from those areas which would not affect the running of the clock. A small adjustment screw is fitted to control the hammer spring tension, an extremely unusual feature. The top left pillar has a screw-adjustable stop wedge to act as an anti-rattle device, a feature found on some London clocks but here done differently. Many of these features are indicative of individualistic handcraftwork.

The detail in Plate 53 is from the dial of an eight-day clock by George Battersby (spelled Battersbee) of Manchester, Lancashire, a maker whose work is known as yet only by this one clock. The record has been traced of his marriage in 1694, which is about the date of this clock, making it one of the very earliest Lancashire longcase clocks yet discovered. Even from this small section it can be seen to be of the all-over engraved style with minutes numbered inside the track and a variant of meeting arrowheads for half hour markers.

Peter Stretch of Leek, Staffordshire, was born in 1670, the son of blacksmith John Stretch of Rostherne, near Knutsford, Cheshire. Peter was a Quaker, working at Leek at least from the date of his marriage in 1693 till he emigrated

PLATE 53. *Detail from an eight-day longcase dial by George Battersbee of Manchester dating from the 1690s. Another example of the all-over engraved dial centre style.*

to Philadelphia in 1701, where he became the first clockmaker. This pins down very closely his working period here to 1693-1701. The clock illustrated in Plates 54 to 57 is one of very few known by him during what was only an eight year working span in England. It dates about 1695.

The twelve-inch dial is very much in the London fashion of the day, certainly as far as its up-to-date style goes with very fine grained matting, the minutes marked outside the track, and half-quarter markers, a feature becoming increasingly popular from this time forward – of the earlier dials illustrated only the anonymous clock (Plate 38) has them. The movement still has outside countwheel striking and tapered arbors, both features by now a little old-fashioned in London work. Five finned pillars, as used here, are still common. A crescent cut-out in the backplate is to allow easy removal of the anchor without dismantling the plates and was not standard practice (though a number of London makers did this). It is, however, very rare for there not be be some method of taking the anchor assembly out without dismantling the clock. The solid dial sheet is sometimes found on northern clocks of this early period, as with the earlier examples illustrated – perhaps he bought his dial sheet ready cast from a southern supplier or simply had not yet turned to the cartwheel casting method. The winding square ends are shaped into four leaf clover shape, a nice touch performed by a few provincial makers.

The walnut-veneered case was made locally, following the general London concept but more heavily proportioned, with heavier moulds (still using the convex trunk-top mould), the base taller than a London base, the lenticle glass aperture a bit large. The interesting cushion mould on the upper hood was not much used in London work but before long became a feature particularly popular in later north-western casework. The barley sugar pillars swell in the centres and stop short as a narrow spindle or 'neck' some distance from the top in a much more exaggerated way than in London work – compare the hood of the Webster lantern case in Plate 124. The hood lacks the side windows normally seen in London work and also in some northern work of the period. The hood door was apparently cut later into the surround, indicating that it was initially a hood with a fixed (non-opening) front.

The clock by John Hough of High Leigh (Plates 58 to 60) has an eleven and three-quarter inch dial with solid dial sheet, the movement having four latched pillars. Four rather than five in number was a very up-to-date feature for the time, yet latching was by now a somewhat old-fashioned feature of London style. Quite exceptionally for a provincial maker, the four dial feet are also latched, again by now an old-fashioned feature of quality found only in

PLATE 54. *Twelve-inch eight-day longcase dial by Peter Stretch of Leek dating from the later 1690s. The dial centre is very finely matted. Original blued steel hands.*

PLATE 55. *Movement of the Peter Stretch clock showing outside countwheel strike with outside detent. Solid sheet dial.*

the work of the very finest of the earliest London makers such as Fromanteel. The clock dates from about 1695 and has the earlier form of countwheel striking with the locking wheel *outside* the plates. The bell stand is on the backplate, sometimes taken as a sign of 'later' date, though personally I can never see any period indicator in this.

This is a recently-discovered clockmaker, known so far only through this longcase clock, a thirty-hour longcase (Plates 176 to 178) and a bracket clock. High Leigh was a tiny hamlet in the parish of Rostherne, near Knutsford in Cheshire, the parish of origin of the Stretch clockmaking family, who were Quakers, and there might be some connection (though that is only supposition). Hough was clearly a clockmaker of exceptional talent, obviously familiar with the most fashionable London dial style of the day, yet incorporating high quality technical features by now discontinued in London.

Herringbone edging, arrowhead half-hour markers, tiny fleur-de-lis half-quarter markers, twin cherub spandrels of the commoner (earlier) type, are all features indicative of the end of the century. This clockmaker (or his engraver) clearly knew of London styles and practices, though my feeling is that the engraving is probably local. Note the mistake in that the engraver has positioned one half-quarter marker between 45 and 50 instead of 50 and 55! The blued iron hands are original and are of an individualistic styling.

PLATE 57. *Detail of the Peter Stretch hood, showing absence of side windows, cushion mould, individual styling of pillars, flush finish to lower hood without the normal stopmould. The door is cut into the walnut surround, suggesting alteration from a fixed-front (i.e. rising) hood.*

The case is of oak ebonised on the surface. Again this is based on the current London style – barley sugar pillars, hood side windows, the newest fashion of top-of-trunk moulding (i.e. concave) – yet shows provincial signs. For example, the lenticle glass is rather larger than usual; the bottom hood moulding extends much lower to overlap the trunk topmould more deeply than on a London case, giving an overall ogee effect; the original iron strap hinges are obviously purpose-made or purpose-bought, but are fitted *the wrong way round,* with the long strap on the case side and the short one on the door, quite defeating their purpose which was to provide a better grip on the moveable part and to strengthen the door against warping. This is obviously the work of a provincial cabinetmaker doing his best to copy London by working from a sketch or memory or both. He is still very new to the recently required provincial skill of clock case making. It is another of those very rare glimpses of the beginnings of provincial casemaking, a highly skilled craftsman learning to make an object he has not been called upon to make before, a case for one of the very first longcase clocks ever made in the region.

Henry Simcock is known to have worked at Daventry in Northamptonshire (then sometimes written as Daintree) between at least 1693 and 1727, being one of the earliest makers in the county. I have come across only two examples of his work, this twelve-inch dial example (Plate 61) dating from the early 1690s being the earliest. The absence of a seconds dial is unusual and the bald look of the dial centre is simply because the matting does not show in the photograph. The hands are replacements.

The case (Plate 62) stands about 6ft.10in. and was probably made in London or, if not, then closely followed the London proportion. It is veneered most

PLATE 56. *Walnut-veneered case of the Peter Stretch clock, based on London concepts but with many provincial features, including narrow necked barley sugar pillars, heavy moulds, high base, large lenticle.*

PLATE 58. *Fine quality longcase dial of about 1695 by John Hough of High Leigh, Cheshire, being one of only three clocks yet known by this interesting maker. Fine engraving work but with the 52½ minute half-quarter marker positioned wrongly in error at 47½ minutes. Original blued steel hands of distinctive style.*

PLATE 61. *Eight-day clock of the 1690s by Henry Simcock of Daventry, then written Daintree. The three crown motif around the calendar box was popular with makers in London and elsewhere at this time.*

PLATE 59. *The Hough movement has latched movement pillars and latched dial feet, most unusual for a provincial maker. The trip lever on the right is to aid in re-setting the countwheel strike.*

PLATE 60. *The case of the John Hough clock is of ebonised oak and has many features indicative of provincial work copying the London principle. Height 7ft. Original pierced fret.*

PLATE 62. *The case of the Henry Simcock clock is veneered in laburnum wood using oystershell pieces to make patterns, producing a very bold effect. Overall styling is based on the London concept of the day.*

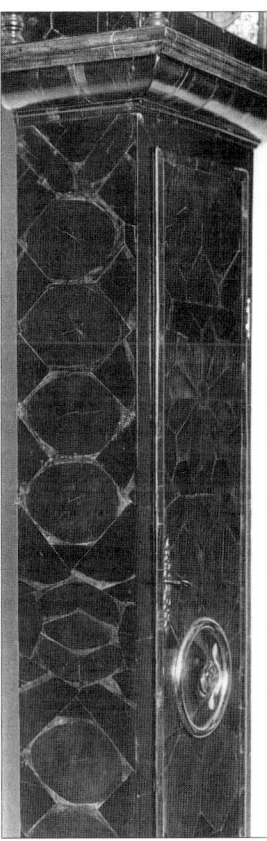

PLATE 63. *Detail of the Simcock case showing ornate oystershell patterns to the sides as well as the front. This is a form of parquetry.*

unusually in parquetry pieces of laburnum wood, cut into what is known as oystershells, done by sawing consecutive slices at an angle through complete smaller branches. These oystershells are then assembled into patterns, in this example even down the case sides (Plate 63), the pale outer wood of each circle making a contrasting border to produce a very striking effect. Even the mouldings are veneered in the same oystershells. Sound frets in the hood sides may be replacements for earlier glass windows.

Richard Washington of Kendal in Westmorland is the earliest clockmaker working in that county by whom work is known today, and that only a single longcase (Plate 64) and a lantern clock. This ten-inch dial, signed 'Rich[d]

PLATE 64. *Ten-inch eight-day longcase clock of the mid 1690s by Richard Washington of Kendal. Very much a London style of dial, perhaps even bought from London. (Photograph by courtesy of Tom Robinson from his book* The Longcase Clock.*)*

Washington Kendall. fecit', is the oldest known longcase clock from that county, dating from about 1690-95. One or two other clocks are known signed with the monogram RW and these may be by him. Its features are very much in the London style and it is impossible to say whether this dial was made there or locally. This maker was the son of a whitesmith and was working by the 1660s, but may not have taken up clockmaking till later. He was mayor of Kendal in 1685 and died in 1698.

The movement was made with five finned pillars (one later removed), outside countwheel strikework, a solid dial sheet, heavy ironwork including the square-section hammer shaft, and its winding squares have been shaped into four leaf clover shape, demonstrating a touch of individuality and suggestive of being made by Washington, not simply bought in from London.

The twelve-inch dial by Samuel Sugden of Selby dates from about 1695-1700 and is the only clock so far to come to light by this maker (Plates 65 and 66). It is very conventional for the period, incorporating all the usual typically London features with the exception of the circular calendar box, a shape more often found in provincial work and little used in London. The walnut-veneered case stands about 7ft. and is in the general London style, though probably locally made.

John Green is the earliest Skipton clockmaker by whom work is known to

PLATE 65. *Twelve-inch longcase dial by Samuel Sugden of Selby dating from the late 1690s. London styling but the heavy matting and unusual number shapes (especially the 5s and 4s) suggest local work.*

PLATE 66. *Walnut-veneered case of the Samuel Sugden clock based on London principles.*

survive today, working there from at least 1696 till his death in 1742. Oddly enough, only two longcase clocks by him seem to have been so far recorded, one of which is illustrated in Plates 67 and 68. Perhaps he worked at other skills besides clockmaking.

This clock might date from about 1700 or a little after. Several features of its style indicate provincial clockmaking of a whimsical nature. The eleven-inch dial here makes late use of the cherub head spandrel. The signature as 'John Green *of* Skipton' is usually an early indicator. The ringing of the winding holes and seconds hole is of a busier and bolder nature than usual. The chapter ring is riveted (at XII and VI) rather than fixed by the usual feet and pins. The maker has left a small 'hump' inside the ring at XII, but at VI his low-cut calendar box, itself of eccentric design, has forced him to omit the hump and instead to rivet right through the quarter-hour track. Meeting arrowheads still mark the half-hours but now his minute numerals are not only outside the track but are quite large.

Plate 68 shows that he used the northern 'cartwheel' type of dial sheet, cast in that form. The movement plates are on the tall side, showing the barrel arbors protruding unusually far at the back, the barrels themselves a little unusual in not being grooved for the line, a time-saving feature used by some provincial makers. The top two of the four movement pillars are finely worked

PLATE 67. *Eleven-inch longcase dial of about 1700 or a little later by John Green of Skipton. Very busy ringing and unusual ring-decorations to the square date box. Chapter ring riveted at XII and VI.*

PLATE 68. *Movement of the John Green clock showing unusual construction with top two finned and knopped pillars of brass and the lower two of plain iron rod. Cartwheel cast dial sheet.*

and finned in brass, but the lower two are of plain iron rod. This was a feature used by a number of northern makers, probably for cheapness and perhaps saving money on the lower two pillars as they showed less than the top two. The strikework has an inside locking plate, a feature often used unusually late in the North. Altogether a clock showing numerous signs of clocksmith work.

EARLY ARCHED DIALS

The arched dial began in the provinces about 1700 or very shortly after. Early provincial examples tend to have very varied shapes and sizes of arch, much more varied than in London, and this eccentricity in the arch proportion continued for longer than in London, where a set shape rapidly evolved. As in London, some early arched dials can be seen to continue features which originally belonged to the square dial style – cherub-head spandrels and herringbone borders. In the cases too there was often a continuation of square dial features, for example barley twist pillars and lenticle glasses. London clocks were, as we know, almost always of eight-day duration, whereas provincial ones of this period might be eight-day or thirty-hour examples. However, it is safe to say that the arched dial, being a more costly version, was almost entirely confined to eight-day work in the provinces too during this early period, arched dial clocks of the thirty-hour type being seldom met with before mid-century. These early arched dial examples form one of the most interesting of all groups of early clocks as they include excitingly unexpected variations as clockmakers experimented with the new form.

The clock by Edmund Bullock of Ellesmere, Shropshire, dates from about 1715-20 (Colour Plate 2, page 114, and Plate 69). The maker was born about

1683 and died in 1734. The arch is especially small in its span and here carries a very unusual form of twin cherub arch spandrel found only in early examples. This sort of arch spandrel would usually carry a boss type name-plate or legend, but here it is used with an open centre to frame a slot in the arch which reveals a swaying brass ball, rocking with the pendulum on an extension of the anchor arbor. This is probably intended to represent the sun, echoed as it is by a moon dial below the XII. The moon is of the 'penny moon' type, a disc part black waxed and part silvered, which shows the moon's shape in the sky. A box below the moon disc shows the lunar date. The penny moon is more commonly a feature of thirty-hour clocks and is unusual in eight-day work except at this early period, when it is more commonly found in the arch.

Bullock numbered some of his clocks, this one being No. 246. The number often seems to be added to the chapter ring in an awkward position as an afterthought, which might be an indication that the ring was engraved elsewhere and the number added later by Bullock. The chapter ring rather unusually marks half-hours and not quarters. Half-quarter markers appear outside the minute ring. The blued steel hands are believed to be original.

The original case (Plate 69) is of oak and represents the beginnings of arched dial casework, retaining several features of the square dial style – the lenticle glass, the barley twist pillars. What is exceptionally interesting is that here in this very unusual prototype example the 'new' arched dial shape is treated in exactly the same way as the square dial had been, that is by fitting it into a 'square' door, the square being actually pushed into an upright rectangle. The alternative, shown in prototype form in Plate 72 (the Bridge case), is to shape the door top to accommodate the arch of the dial, producing an arched door, which became the standard method of accommodating arched dials hereafter. The flat-topped (hood) door example kept the flat top to the case; the arched top (hood) door example had an arched top to the case.

Although this is a small flat-topped cottage style of clock standing about 6ft.6in., the 'new' arch shape is now echoed in the arched trunk door top and the arched hood side windows. The bun feet are modern replacements.

The twelve-inch dial by Thomas Bridge 'de' Wigan (Plate 70) shows another type of early arched dial, here with the arch much wider than the later normal span. The signature using 'de' is sometimes found at this early period (on square or arched dials) where the maker would have found it difficult to Latinise the town name or perhaps where no Roman name for the town existed. Examples are Henry Hindley de Wigan, Thomas Ogden de Halifax, Timothy Goodman de Towcester, where Wiganii or Halifaxii or Towcesterii would have looked most odd. Some used the English 'of' and others left out the preposition altogether, but it was probably felt that the Latinised form, where possible, or the (Latin or French derived?) 'de' form, looked more impressive and imitated the fashion of the London makers of the day.

The Bridge dial has the arch as a separate section joined by riveted straps on to the square. Some makers did this regularly, but it is most often found at this early period where the clockmaker was probably using square dial castings for most of his clocks and made the occasional added arch as an optional extra rather than keeping a stock of arched castings too. Here the arch runs almost to the very edge of the square.

The dial here uses the cherub head corner spandrel, more usual on square dials and only used on early arched versions, and for the arch a different version of the twin cherub spandrel as used by Bullock in Colour Plate 2, page

PLATE 69. *The oak case of the Bullock clock (see Colour Plate 2, page 114) incorporates some square dial carry-over features such as the lenticle glass and the full height hood door with barley sugar pillars. Charming work, purely provincial.*

PLATE 70. *Twelve-inch eight-day longcase dial of about 1715 by Thomas Bridge of Wigan with a different type of unusual arch, running almost the full width of the dial. Superb original hands of brass, engraved on the surface. The arch and square sections are two separate castings riveted together, as some early arches were. Heavy half-hour markers are very northern.*

114, but here carrying a plaque bearing the coat of arms of the local Chadwick family for whom the clock was made. The lavish engraving around the arch (Plate 71) is a feature of some such early arches and the general boldness of the engraving is more often found in the North-west than elsewhere. The maker worked in Wigan from 1712 till at least 1745. This clock is about 1715. The hands are superb original examples of brass, an unusual metal for hands at this time but used by occasional north-western makers, principally in Lancashire. They are very finely engraved on the surface. The half-hour marker, a form of fleur-de-lis, is now 'floating', i.e. not attached to the quarter-hour circle on the inner chapter ring edge. The mysterious black dots by II and VIII are rivet holes where the chapter ring was pinned down to a turntable during engraving.

The oak case of the Bridge clock (Plate 72) stands about 6ft.9in. and is of the domed-top style. The domes of such early clocks (when not surmounted by a further pediment) are quite deliberately heavy in their overhang and often, as here, give the hood a much bulkier appearance that to the modern eye can seem out of proportion to the small base which can be relatively slim and shallow (see small bases in the Kirk case in Plate 75 and the Clark case in Plate

PLATE 71. *Detail of the superb engraving of the Bridge arch, the spandrel cherubs supporting the coat of arms of the first owner, a family named Chadwick. Note the ruling out lines for the engraving, which a more careful engraver would have polished out.*

76). This is quite intentional and is a feature often found in these early arched dials. The proportion is very different from the occasional later example of the dome top, as in the clock in Plate 383 by John Dison of St. Ives. It is also very different from those clocks which now stand dome-topped but once had a superstructure above the dome, principally being originally pagoda-topped London walnut and London mahogany cases, cut down to dome tops to reduce the height..

At the rear of the hood are half-round pillars facing forward – some face sideways, but half-round rear pillars are often an early feature, usually in north-western England. In most areas hood front pillars are still attached to the hood doors as three-quarter pilasters (usually termed 'integral' pillars), but in the North-west separate pillars begin to appear from this time, as here.

This case has features which indicate that the casemaker was not yet familiar with casemaking, or at least with arched dial casemaking, as this was still a very new skill in the area. He has left his hood bottom mould too narrow to take his hood pillar brass bases which therefore sit uncomfortably in an overhanging position. More seriously, his hood door is too deeply recessed so that on opening it fouls against the right-hand pillar, even though he has fitted an exceptionally deep hinge flange to the door. Both these are constructional oversights which would not be repeated on the next clock case this joiner made, so again we have an illustration of the very beginnings of casemaking as the craft is still being learned. It is this sort of mistake which makes these early clocks so fascinating.

Note the eccentric shape of the trunk door top and the oddity of the D-mould lip running not only around the door edge but across the top shoulder of the door too. The bracket feet are original, which is instructive as so many are later replacements.

Joseph Kirk is a little known clockmaker, though the earliest maker in the county by whom longcase clocks are known. He was born about 1670-80, the son of London-apprenticed clockmaker, John Kirk, who moved out from the capital to work at Epperstone, Nottinghamshire. Several longcase clocks and one bracket clock are known by Joseph, signed at various nearby villages including Harstoft, Clay Cross, and Skegby, but he eventually moved to work in Nottingham town about 1715, the age of the clock illustrated in Plate 73.

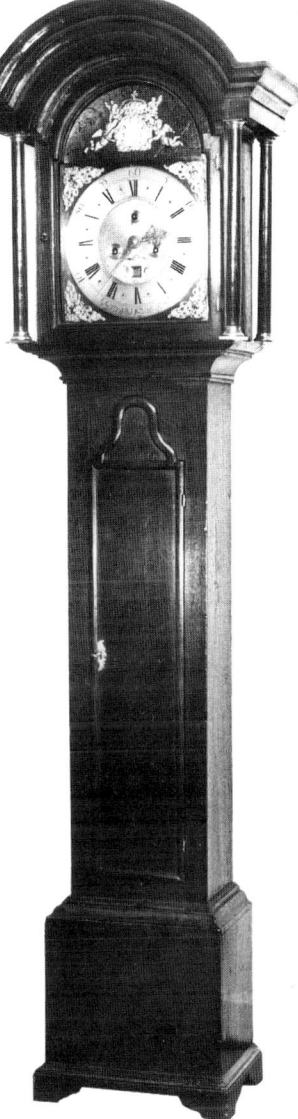

PLATE 72. *The oak case of the Bridge clock has a dome top with heavy topmoulds, indicative that it did not once have a caddy superstructure. Free-standing half-pillars at the back of the hood are found on some northern clocks from now on. Eccentric shape to the door top. Original bracket feet. Height 6ft.10in.*

PLATE 73. *Longcase dial of about 1715 by Joseph Kirk of Nottingham with an early wide-span arch almost full width of the dial. Herringboning only around the arch with scroll engraving to separate the arch from the square as was the early principle. Very decorative centre with extensive engraving into the matting.*

He is believed to have lived until after 1740.

This clock is clearly a very early arched dial (c.1715) by its transitional features – wide arch span, tall arch height, the arch used for a purpose (here a calendar), herringbone engraving in this example only round the arch area, decorative engraving to 'join' the arch to the square section, use of half-quarter markers. The matted centre features an odd but charming decorative border and also features two doves and four flies!

The engraved name 'Bocket' behind the dial of one of Kirk's Harstoft clocks is puzzling and might imply that he had (some of) his engraving done elsewhere, perhaps in London, where one Richard Bocket, clockmaker, is known to have worked in 1712. The name Boket *(sic)* was also found behind the dial of a Gabriel Smith longcase clock of the early eighteenth century – see page 365. On the other hand, the designs, patterns and boldness of engraved designs on clocks by both Joseph Kirk in Nottinghamshire and Gabriel Smith in Cheshire are not only very different from each other but are also different from any normal London work! The five-pillar movement has rack striking.

The oak case stands about 7ft.3in. and has the attached pillars we normally associate with early hood doors (Plate 75). The lenticle glass is a feature of earlier square-dial style, here continued into the arch dial form and therefore an early feature on an arched case. The small, shallow base is typical of the period, making the hood appear top-heavy compared to later styles. On this

PLATE 74. *Movement of the Kirk clock showing sturdy construction. Pillars drilled for central bolts from the seatboard, as many early clocks had. Seatboard runs full depth of case, a common early practice. Rack striking.*

PLATE 75. *Oak case of the Kirk clock with some light inlays to the upper hood. Lenticle glass in an arched dial clock is a carry-over from square dial styling. Ebonised beadings and moulding lips.*

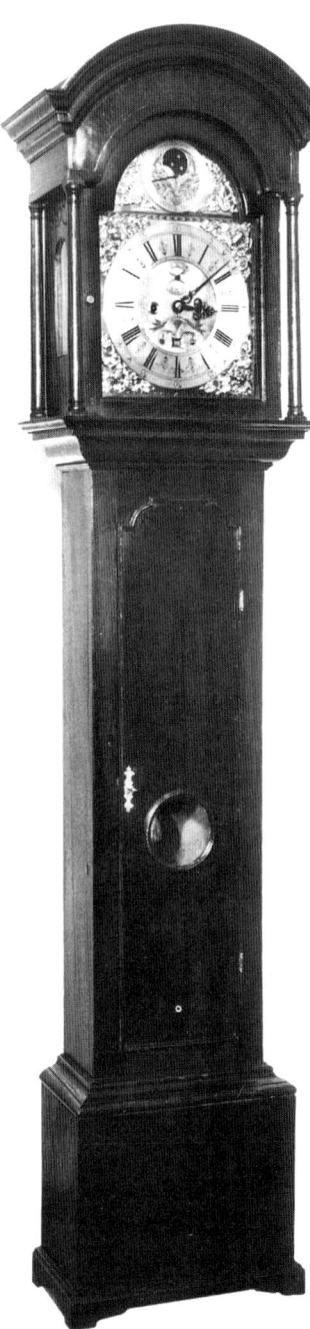

example all the raised beadings (for example on the door edge mould) are ebonised, an unusual treatment. The door top has a break-arch shape, as was often the case with early arched dial clocks, echoing the 'new' dial shape.

The dial of the clock by Thomas Clark of Warrington (Colour Plate 3) again illustrates the beginnings of the arched dial style and dates from about 1715. Lancashire was often in the forefront of stylistic development, and even the dial size at thirteen inches is an early innovation at this time, when most dials nationally were of the standard twelve-inch width. The arch is again put to a use, here for a moon dial of the 'penny moon' type. The dial edge is bordered with herringbone engraving, including the division joining (or separating, depending on how you see it) the arch section and the square. Cherub head spandrels, attached half-hour markers, half-quarter markers and the fine original steel hands are all features of this early period.

The dial centre features the two birds and a basket theme, a common subject in many regions and not especially associated with Lancashire. So common is this theme in widely separated areas, principally London and the Midlands, that it is tempting to think that it might be the work of a centralised group of engravers perhaps working in London. The name is cramped as if the dial was not planned for it – perhaps suggesting outside engraving with name added in house? On the other hand there is a strength and boldness to the engraving of this dial which is often a characteristic of north-western dials and is usually quite different from the more restrained London style.

The oak case of the Clark clock (Plate 76) is of the dome-top style, reminiscent of the Bridge of Wigan case in Plate 72. This one too has the up-to-date feature of separate pillars, but the lenticle glass and hood side windows are a carry-over from square dial times. The door top shape is an interesting variant, indicative of more varying shapes to come. The top-of-trunk moulding is here of ogee shape, in some ways harking back to earlier styles but this was a shape of mould which was to be retained in some north-western casework for many years. The height is about 7ft.4in.

John Burgess of Wigan began working there in 1711 and died in 1754. Like Thomas Bridge (and Henry Hindley) he signed 'de Wigan'. On the clock in Plate 77 his first name is abbreviated rather unusually to Jo. This is clearly one of his earliest, dating from between 1711 and 1720. The separately attached but original arch covers a narrow span and is used for a calendar feature. The twelve-inch dial has two engraved lines for a border, also bordering the arch section from the square; this division of the arch from the square is nearly always an early sign. The engraving is typically bold and strong and has the large half-hour markers we associate especially with the North-west, here attached to the chapter ring (later ones 'float'). These are a sort of fleur-de-lis almost becoming a flower, which later ones usually did.

A herringbone type of border is here used to surround the ringing around the winding squares and the penny moon dial which is positioned unusually above the VI numeral and has a distinctively strong 'face' with the lunar date showing through a small box above it. Herringboning in this position is unusual and was limited to the work of half a dozen or so makers (or their engravers?), located principally in the North-west. Another unusual feature is the presence of the two small spandrels beside the seconds dial, two small imps holding torches. These tiny spandrels were normally used as arch spandrels on bracket clock dials (and two similar ones sit atop the arch spandrel on the Thomas Bridge dial in Plate 71), but a few makers used them as here, again principally in the North-west.

PLATE 76. *The domed oak case of the Clark clock (see Colour Plate 3, page 115) has heavy topmoulds, free-standing half rear pillars, ogee top-of-trunk mould, shaped door top – all northern features.*

PLATE 77. *Bold eight-day dial with calendarwork in the arch, made about 1715 by John Burgess of Wigan. Penny moon positioned unusually at VI. Original hands are of brass, though here painted black.*

The half-quarter markers are retained from square dial times and are unusually large and bold. The hands are believed to be original and, though here painted black, are in fact of brass, a feature more common in Lancashire than elsewhere. Two rivet holes show at III and IX where the chapter ring was pinned during engraving. The dial is a good example of originality of thought and design, very different from the London norm of the day.

The original oak case (Plate 78) is trimmed with walnut crossbanding and stands 6ft.11in. The top is of the domed style and the base is typically small in relation to the top. The hood pillars are grooved (fluted), quite an early instance of this. Note the door top shape, which is now becoming more adventurous than the plain break-arch.

Randolph Bagnall worked at a village called Talk-on-the Hill (or Talk-o'-th'-Hill) in Staffordshire, the name in Plate 79 abbreviated to what at first seems a puzzling 'Talkothhill'. This clock dates from the 1720s and illustrates a few interesting new features. In particular the dial centre contains a superb tulip-based pattern, engraved on to a polished ground and reminiscent of the older lantern clock dial centres – the arched dial equivalent of the principle we saw on the square dials by John Waklin and John Worth (Plates 48 and 51). The arch features a penny moon which too is set within a similarly engraved floral centre. This very handsome centre engraving, combined with superb

PLATE 78. *Oak case of the Burgess clock with heavy dome top, unusual shaped door top. The crossbanding is in walnut. Simulated* verre eglomisé *panel in the upper hood is in fact painted wood.*

77

PLATE 79. *Eight-day longcase dial of the 1720s by Randolph Bagnall of Talk-on-the-hill, a village in Staffordshire. Fine engraved centre on a polished ground to the dial and to the arch penny moon dial. Herringbone engraving surrounds all and separates the arch from the square in the early manner.*

PLATE 80. *The oak case of the Bagnall clock has a starburst inlay to the door centre, a feature sometimes found at this period. Crossbanding is in walnut. Height 7ft.6in.*

PLATE 81. *Arched dial of about 1725 by John Holroyd of Wakefield. The calendar box is simply engraved on instead of being cut through, so is for appearance only. The unusual arch spandrel is found only on early arched dials.*

herringbone borders, bold half-hour markers and the now somewhat old-fashioned half-quarter markers, make this a very eye-catching dial and totally different from anything that ever came out of London.

The case (Plate 80) is of oak and stands about 7ft.6in. Blind fretwork in those otherwise unused areas above the arch is reminiscent of the treatment on London walnut cases of this period. Such London cases (see Plate 35), and perhaps this one too, often originally had a caddy superstructure above the flat. This case is crossbanded in walnut and has a starburst effect of varied inlaid woods to the door centre, this being a feature more often found in the North than the South. The case has little about it to indicate region, but the separate pillars and full rear hood pillars are generally northern pointers.

Little is known about John Holroyd of Wakefield, Yorkshire, though his clocks crop up not infrequently, apparently dating from the late 1720s to 1775 or so, the twelve-inch dial shown in Plate 81 being one of his earliest. The arch spandrel is a form of twin cherubs holding a name plaque, normally a very early sign but here used a little later than usual. This dial is not easy to date but could be from about 1725 or a decade or more later.

The 'calendar' box is in fact just an engraved outline, not cut through or ever fitted with a calendar, but simply done to give the impression of one, perhaps for economy, though it seems a bit of a penny-pinching method. The engraving around the box is not well done, being a bit feeble and crooked, and I get the impression from the several Holroyd clocks I have seen that he was not amongst the best of the clockmakers of the day. On the other hand, the chapter ring is well engraved, perhaps implying that he bought his chapter ring work from a professional engraver but made do with his own, less skilled, engraving for the dial sheet itself. Half-quarter marks are retained. The steel hands are believed original. The spandrels are one version of several forms of

PLATE 82. *Early arched rolling moon dial by Charles Butcher of Bedford and dating from about 1730. Small, floating half-hour markers are a southern trait. The moon sky is engraved to leave the stars proud, the background filled with black wax.*

PLATE 83. *The walnut-veneered case of the Charles Butcher clock stands about 7ft.6in but may once have had a superstructure. Very much based on London casework of the day.*

the female head pattern used over quite a wide period.

Charles Butcher of Bedford was born in 1702 and died in 1738. The eight-day longcase clock in Plates 82 and 83 dates from the late 1720s or 1730s. The moon dial is of the type which completely fills the arch area and is known as a rolling moon, this being the earliest example we illustrate of that type. Such a moon disc carries two moons, only one showing at any one time, of course, and therefore the disc undertakes a complete revolution every 59 days, each lunar month being consistently 29½ days in duration. Early moon dials have a starry sky between moon faces; after about mid-century the moon faces have painted scenes between them, often a landscape alternating with a seascape. A very early example, such as this one, has the moon faces and the stars engraved and waxed, the stars being left as raised brass and the hollows between them filled with black (or occasionally blue) wax before silvering, exactly the same process as with an engraved chapter ring. Slightly later ones, even of this early type, have the stars painted – see Dollif Rollisson dial in Plate 470, which has a painted starry sky.

The arch band of an early moon dial is often of irregular width, as this one is, being narrower at the top than the sides, and is often engraved with tracery on its surface, as here. Such dial border engraving is a sort of extension of the principle of herringbone borders. Some early ones too, as with this one, have

PLATE 84. *Twelve-inch, three train longcase clock of about 1720-30 by Jacob Lovelace of Exeter. The plain southern styling (plain centre, small floating half-hour markers) is enlivened by lavish herringbone borders everywhere, including a band to separate the arch area. Chime change lever at III gives the option of two different chimes.*

an opening to show the moon's age. In this photograph the clock shows no moon (new moon). The two moon 'humps' here have fanciful engraved surfaces showing a rising sun and a planisphere, merely for some form of decoration, and this again is usually a sign of an early moon dial. Later moon humps conventionally have two engraved hemispheres, a principle continued later in white dial clocks.

The main dial area itself shows typically southern styling very much based on London principles. The half-hour markers are very restrained and 'float' (compare those of the Bagnall dial in Plate 79). The dial centre is finely matted with just a hint of engraving spreading from the date box and vestigial remains of hole ringing.

The case is very much along the London lines in veneered walnut, London derived features being arched door top, hood side windows, shallow base, integral hood pillars, simple moulds. The hood top has a dome finish, but may originally have had a further caddy or double-caddy pediment. The height is about 7ft.6in.

Jacob Lovelace of Exeter was a fine clockmaker, as can be seen from the twelve-inch dial of the clock in Plate 84. The dial is restrained in principle, following the London concept for his matted, undecorated dial centre, with restrained chapter ring and small, almost delicate, floating fleur-de-lis half-

PLATE 85. *Movement of the Lovelace clock showing the nest of eight bells and the musical pin barrel. The chime change lever pushes the barrel backwards or allows it to return forwards, pushed by the bladespring behind it.*

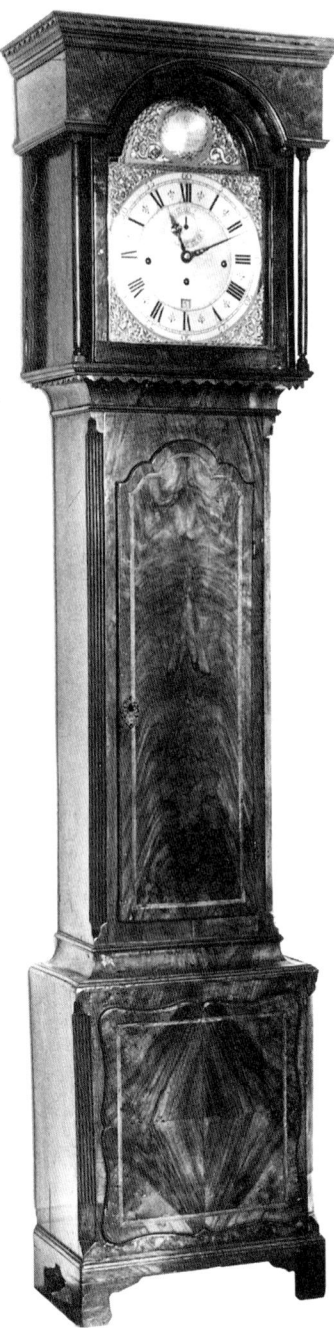

hour markers as well as the up-to-date stylistic absence of half-quarter markers. However, on the dial sheet edge he has run riot with strongly patterned herringbone edging, dividing off the arch section and also fringing the name boss in the arch, both being early signs.

The three winding holes, now *un*-ringed in the London fashion, are for a three-train movement, the third being a quarter-chiming train on eight bells. The chime change switch (choice of two) can be seen on the dial edge at III o'clock.

Lovelace worked at Exeter from about 1712 to 1750, dying at Aylesbeare in 1755. This clock could date from the 1720s, with the still acceptably fashionable herringboning and the up-to-the-minute London style chapter ring and dial centre, or it could date from the 1730s retaining what was by then rather old-fashioned herringboning. The blued steel hands are original.

The original case (Plate 86) is walnut, veneered on to oak, the major panels strung with yellow herringboned wood (boxwood?) and crossbanded outside that. The flat top may originally have had an upper pediment. The wavy-edged raised base panel is reminiscent of Bristol styling. The upper and lower trunk mouldings are of a very simple concave type, typical of many southern clocks. The frilly projecting lip below the hood, echoing that on the hood topmould, is an eccentric extra feature on this case and not typical. The height is about 7ft.6in.

William Stumbels was an outstanding clockmaker who moved from rural Devon to work in the country town of Totnes from about 1729 till his death in 1769. The eight-day longcase clock illustrated in Plate 87 dates from the 1730s, is based on the London (undecorated) matted centre principle with restrained 'floating' fleur-de-lis half hour markers, but has provincial flourishes such as the fine herringbone border. His arch contains a moon and tide dial registered by two hands, the 'minute' hand marking the lunar age (here the 27th) and the 'hour' hand marking high water times (here VII o'clock). A strike/silent switch is positioned above the 60 minute number. The blued steel hands are original throughout. The engraving is excellent, done by a master, perhaps Stumbels himself.

The eight-day clock illustrated in Plate 88 probably dates a little later, from the late 1730s or early 1740s. This is a very special clock, believed to have been made for Dartington Hall in Devon and, although it had been in the possession of the same family since new, it was found in a much neglected state in the 1980s, having been unused since about 1914.

This exquisite dial again shows the basic London influence but, unlike London work, is lavishly decorated, especially with the fine border engraving. The arch contains a ball moon, which rotates to show the moon phase, as well as the lunar age and high water times as shown on the previous clock.

The clock offers a choice of music or a chime on eight bells, selected by a switch positioned by the 15 minutes numeral. It plays at three, six, nine and twelve o'clock either three verses of 'Oh Worship the King' or three runs of the chimes, either of which can be repeated at will provided certain complicated rules are observed. Failure to observe correct operating procedures can prove disastrous to such complicated clocks. To avoid problems Stumbels provided instructions written in his own hand and pasted on to the trunk door, and these still survive today, reading:

Rules to be observed in keeping the Clock
You are to wind it up every eight Days,
the Day of the Month to be set by turning

PLATE 86. *The quartered walnut-veneered case is a provincial interpretation of the London concept of the day. Height 7ft.6in. Wavy-edged lipping is a whim of the casemaker.*

PLATE 87. *Fine twelve-inch dial from a longcase clock by William Stumbels of Totnes dating about 1730. London styling to the dial centre and chapter ring, but provincial departure in the herringboning. Strike/silent switch above 60. Moon and tides dial in the arch.*

PLATE 88. *Superb chiming/musical longcase by William Stumbels of Totnes dating from the 1730s. The three trains are wound through only two holes, a special double-ended key picks up the third ratchet in the right-hand aperture. The arch contains a ball moon and tidal dial.*

the Hand forward. The three Weights
are to be wound up by changing the Key

The Chymes may be repeated any Time
between twelve and two, and three and five,
and between six and eight and Nine and
Eleven, any other hours you cannot, The
Work being unlocked.—
The Chymes may be alter'd by Shifting
the Ditent up or down on the right hand.–

If by Chance you let it down, Set the
Ditent to Silent and turn the Minute hand
till it comes right to the Hour and all will
come right again.

What he is saying is that the clock will not repeat during the hour leading up
to the playing time.

The clock has three trains, the third being for the music/chimes, yet the dial
has only two winding holes. This was Stumbels' own design, used on two or three
known clocks, whereby the winding key is double-ended, the normal end being

PLATE 89. *Movement of the Stumbels chiming/musical clock seen in unrestored condition having lain unused for half a century. The lever on the right is to change the tune/chime. The third train winding ratchet can be seen around the right-hand winding square.*

used for going and strike trains but the 'negative' end being used in the right-hand winding hole, where it locks into a ratchet wheel, which turns a connecting gear and winds the third train located in the centre of the movement.

The two cases are similar in style, each standing a little over 8ft. and each having a dome top surmounted by a box top plus caddy and capped by a small cresting piece. So many clocks which once must have had similar tops have been cut down to a dome to reduce height, especially those London walnut cases which now stand baldly and unconvincingly dome-topped.

The first case (Plate 90) is in oak with walnut crossbanding. The Dartington case (Plate 91) is in mahogany (mostly on to a pine carcase), being a very early use of what was then a rare and costly timber.

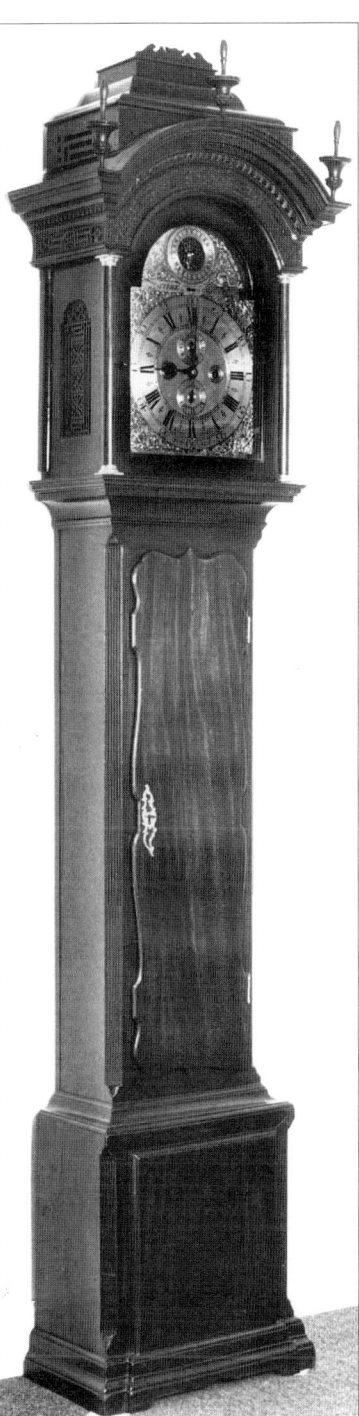

PLATE 90. *Oak case of the Stumbels clock in Plate 87, being a dome top with super-structure and finials. The crossbanding is in walnut. Height about 8ft.*

PLATE 91. *The case of the Stumbels musical clock (Plate 88) represents an early use of mahogany, a most unusual timber to be found in clock casework this early. This magnificent clock owes certain stylistic features to the Bristol area, most notably the serpentine door. Height about 8ft.*

Both have pillars attached to the hood door, hood side windows in the one, side frets to let out the music in the other. The raised panel base and wavy-edged trunk door in the Dartington case show Bristol influence, that being the city having most bearing on styles in South-west England. The 'breakfront' foot to the Dartington clock is not a regional feature but more an indicator of high quality and not often seen on a longcase clock. Both clocks have original wooden finials of the 'flambeau' (flaming torch) pattern, mahogany on the Dartington clock, gilded softwood on the oak clock.

CHAPTER FIVE

THE ORIGINS OF
THIRTY-HOUR CLOCKS IN THE PROVINCES

The earliest provincial longcase clocks were almost exclusively of thirty-hour type. The eight-day clock, which began in London with Fromanteel, was much more expensive than the thirty-hour longcase or the lantern clock and its clientele was initially amongst the most prosperous. There were some counties where eight-day clocks were being made in the 1680s, but in most counties longcase clocks of this great age tend to be exclusively of the thirty-hour type. Indeed, in some counties no longcase clocks of any kind are recorded until after 1700.

These earliest thirty-hour clocks are often prototypes of a very varied, interesting and exciting nature, and are almost totally unstudied – I know of no book which deals with them in any sort of detail. They are certainly very undervalued commercially, though this situation is likely to change shortly. They represent the very beginnings of provincial clockmaking yet have been almost totally ignored by writers on the subject.

So little is known about these clocks and their makers that the assumption is sometimes made that the earliest thirty-hour provincial clocks were not made by those whose names they carry, but were bought 'from London' and lettered to suggest local manufacture. We have already seen how many early eight-day clocks can be shown to have been locally made by their unusual personal

PLATE 92. *Anonymous nine-inch dial single-handed longcase clock of the 1690s with original iron hand. No obvious regional characteristics, though the clock was discovered in Leicestershire.*

PLATE 93. *The movement of the anonymous clock is of plated form yet most unusually has the trains set one behind the other, which was the normal layout for a posted movement. The same arbor carries the two mainwheels for going and striking.*

touches. The much more idiosyncratic nature of early thirty-hour clocks leaves no doubt that they were made locally – in any case, such clocks were not made in London.

POSTED AND PLATED MOVEMENTS

The eight-day clock introduced by Fromanteel in 1658 (and its grander associates of longer duration such as month, three month and twelve month) was a different breed of clock from the traditional thirty-hour (or shorter duration) lantern clock. The wheelwork of an eight-day movement was held between two upright brass plates, positioned front and back, and today we usually call this a plate movement or plated movement.

Lantern clocks were based on a different layout with four brass corner posts and plates (usually of brass but sometimes of iron) top and bottom, a principle which developed into a type of longcase clock we refer to as a post-framed or posted movement, the idea being the same as that of a four-poster bed. Sometimes this is called a birdcage movement. The posted movement construction in longcase clocks was confined (virtually always) to those of thirty-hour duration (and, of course, to lantern clocks and their derivatives, which we shall examine shortly); the plated construction might be used for eight-day or thirty-hour. The eight-day was (virtually) always built in plated form and an eight-day of posted construction is an eccentricity very seldom met with.

These were the two distinctly different and instantly recognisable types of movement. The plate movement had its two trains of wheels (one for going and one for striking) set side by side; the posted movement had the trains set

PLATE 95. *Detail of the Robert Page dial. The two winding holes are set eccentrically. The fine iron hand is original. (Photograph by courtesy of Tom Spittler.)*

PLATE 94. *Dial of a clock of about 1695 by Robert Page of Wivenhoe in Essex (here spelt Wevenhoo). This keywound single-hander has a finely-engraved dial centred by a Tudor rose. (Photograph by courtesy of Tom Spittler.)*

one behind the other, going train at the front, where it was closest to the hand(s) it drove. Any clock must fall into one category or the other, or so we would expect.

The normal and almost universal method of winding an eight-day clock (or one of longer duration) was through its dial by means of a key; that for a thirty-hour was by a pull rope or chain. Contradictions to these methods involved needless complications and are very rare but they do nevertheless exist.

PLATE 96. *The movement of the Page clock is an eight-day one, yet has two trains placed one behind the other, posted movement fashion though it has no corner posts and therefore would be categorised as being a plated type of movement, even though the plates are no more substantial than the typical cross bars of a posted clock. A most eccentric clock perhaps intended to be built into a cupboard or wall panel. (Photograph by courtesy of Tom Spittler.)*

It is always dangerous to generalise and in clocks there was the occasional eccentric who decided to do things quite contrary to the accepted method. I was recently shown the unsigned provincial thirty-hour longcase clock of the 1690s illustrated in Plates 92 and 93. This extraordinary and probably unique clock has a plated construction yet sets the two trains one behind the other, birdcage fashion! There is no good reason why plated movements should not have had this layout as a general rule, but they didn't, except (in my experience) in the case of this one clock by this one contrary and unidentified clockmaker.

I was recently also shown the extraordinary clock in Plates 94 to 96, made by Robert Page of Wivenhoe in Essex. It has a dial ten and a quarter inches square, a movement standing sixteen inches high, a form of anchor escapement which is almost a deadbeat and a pendulum fifty-four inches long. By its style we would date it to about 1695. The dial has a very fine all-over engraved centre akin in style to that of a lantern clock. The blued steel hand is original and unusually ornate, again akin to a lantern clock hand in style.

We are not sure of the purpose of this clock, which is a sort of domestic form of turret clock. It might have been intended to be built into a wall panel or cupboard, or to have stood on a wall bracket, or to have been housed in a wooden longcase. The movement construction is related to the plate frame form, though the back and front 'plates' are so very narrow as to hardly justify the term 'plate' at all. Its two trains are set one behind the other in the posted movement principle, yet extraordinarily it is wound by a key through the dial and runs for somewhat over eight days.

These two clocks are rare exceptions to the more general rules and an enthusiast might never come across such forms in a lifetime's interest in the subject. It is important for us to understand the normal methods and systems used by the clockmakers of the past and such eccentricities, thrilling though they may be to the dedicated, are of no more than passing relevance.

The 'frame' or 'cage' of a posted movement is a term used to describe the posts with top and bottom plates attached, within which are set the brass crossbars to hold the arbors, wheels, pinions and levers. In English lantern clocks this cage is of brass (just occasionally the top and bottom plates are of iron, rarer still are iron posts) and the round posts are normally of baluster shape – see Plate 118, Baxter clock. The cage of a post-framed thirty-hour longcase clock (which developed from the lantern clock) might have its posts and plates of iron or brass, iron being generally commoner on early examples and brass on later ones, but there was considerable variation, mostly on the whim of the individual clockmaker. The posts themselves were normally of square section (or rectangular section) rod. Occasionally, however, and principally on very early examples (those types which closest resembled lantern clock movements), these posts were decorative, baluster-shaped brass pillars of lantern clock type, a style which sometimes varied later to simpler round-rod pillars.

The plated movement construction was also used in making one type of thirty-hour clock, probably initially in London, but by the 1690s the plated thirty-hour longcase clock was known in the English provinces too. However, because of the already existing tradition of posted movement clockmaking in the provinces in the form of lantern clocks, it was the posted movement type of longcase which first emerged in most areas.

NORTHERN AND SOUTHERN DIFFERENCES

The posted movement tradition of clockmaking was strongest, at this very early period and much later too, in the southern part of England, perhaps because the lantern clock discipline was strongest there. Posted movement clocks are therefore less commonly found in northern areas where the plated thirty-hour school generally took hold at an earlier date than in the South. However even in northern England it does seem that the first longcase clocks were frequently of the posted type, though some of the first clockmakers there

PLATE 97. *Thirty-hour hanging wall clock by Walter Archer of Stow-on-the-Wold, Gloucestershire. The dial is that of a typical lantern clock with many idiosyncratic features typical of Archer's very individualistic work. The movement, however, is an iron posted hanging wall clock, making this a hybrid form between lantern and hook-and-spike clocks.*

PLATE 98. *Movement of Walter Archer's hanging wall clock, showing the mainly iron construction. Made as a simpler and therefore cheaper form of clock, looking from the front like a lantern clock yet saving on the cost by extended use of iron instead of the more expensive brass.*

very soon discontinued posted movement making and turned instead to the plated form as a preferred construction. This is evident in the very early work of certain makers such as Richard Midgley of Halifax, Samuel Ogden of Halifax, John Ogden of Askrigg, John Fletcher of Ripponden, Thomas Clarke of Warrington, and others, all of them very early makers known to have made posted movement clocks but whose later thirty-hour work is of the plated type.

In saying this I am aware that the informed opinion in recent years has been that northern thirty-hour clocks were largely thought to have *begun* in plated form, a view I shared until recently. We were perhaps misled in this by the fact that only seldom do very early (pre-1700) northern examples come to light, so seldom that this was a phase of clockmaking which has almost escaped being recorded, and also by the fact that northern makers turned early to the plated form. John Ogden, for example, was making plated thirty-hour clocks by the late 1690s, but those rare thirty-hour examples by him of earlier date are of posted form.

Whatever the varying rate of spread of the beginnings of domestic clock-

PLATE 99. *A seven-and-a-half-inch square dial lantern clock made about 1710 by John Fordham of Braintree in Essex. Still a true lantern clock but in its last form, its dial now resembling that of a longcase clock. Height 15in. This clock was later converted to a double-fusee spring-driven movement, when the minute hand was added.*

making into different parts of the provinces, it was the posted movement thirty-hour clock which was the first type to appear in all English counties, originating as the lantern clock and developing into the thirty-hour longcase. The posted movement construction ultimately changed to plated form, this change being rapid in most northern areas and more slow in the South.

A measure of this variation between North and South is seen to some degree in Wales (regarded for this purpose as part of the North) and more so in Scotland. In both of these countries where the lantern clock was virtually unknown in the seventeenth century (examples are known by only two makers in Scotland, none in Wales), the thirty-hour tradition scarcely existed at all – Welsh thirty-hour longcase clocks are uncommon and Scottish ones so exceptionally uncommon that I cannot recall ever having seen one! (The

PLATE 100. *Lantern clocks also developed into the arched dial form, mostly, as here, as small travelling alarm clocks. Height about 7in. This example is by Batty Storr of York and dates from the mid-18th century. Top finial replaced. Originality of hand doubtful.*

PLATE 101. *The movement of the Storr lantern clock with original verge escapement and bob pendulum (driving ropes here removed). The going train is behind the dial on the left of the picture; alarm train behind the backplate on the right. These are usually described as 'alarm timepieces' as the word clock refers strictly to one which strikes.*

Scottish attitude to thirty-hour clocks was admirably summed up by the late Felix Hudson of Dunfermline, a noted authority on Scottish clocks, who commented on the absence of thirty-hour clocks in Scotland by saying 'The Scots couldna thoil a poor thing!'.)

Ultimately the makers of thirty-hour clocks throughout the land switched from the posted movement to the plated form, even though in some regions this switch came very late. In my opinion they did this for the good reason that the plated form is stronger in construction and more trouble free in running over decades, even centuries, and is not prone to the loss of rigidity which can occur in the posted frame. The puzzle is why this change to plated construction took so long to happen in some conservative areas. This may have been because the posted construction had one considerable advantage

PLATE 102. *The six-and-a-half-inch square brass dial of a tiny hanging alarm timepiece, often referred to as a hook-and-spike wall clock with alarmwork. This is a very simple anonymous example, perhaps Quaker made, dating from the early 18th century. Original blued steel hand.*

PLATE 103. *Movement of the hook-and-spike clock in Plate 102, made largely of iron but using brass for the wheels. Such clocks were made as cheaper alternatives to a true lantern clock. The alarm hammer is double-headed.*

over the plated, namely that either of the two wheel trains could be dismantled for repair independently of the other, something which was impossible with the plated movement and which later posted movement makers perhaps clung to as long as they could. (In fact one or two very early London makers did make *plated* clocks with split plates, capable of independent train dismantling, but this practice fell from use in the late seventeenth century.)

LANTERN CLOCK HYBRIDS AND DERIVATIVES
The lantern clock had proved an adequate and reliable timekeeper and continued to be produced long after Fromanteel's introduction of the pendulum in 1658, both with verge escapement and later with the new, more accurate anchor escapement and long pendulum. Lantern clocks were still being made as late as the second half of the eighteenth century in some districts. So the progression from lantern to thirty-hour longcase clock was at times a gradual one in the provinces and some in-between types survive which illustrate the varying ways in which this transition evolved. These might be regarded as (ultimately) unsuccessful mutations in the evolution of domestic clocks, even though some of them escaped extinction for many years.

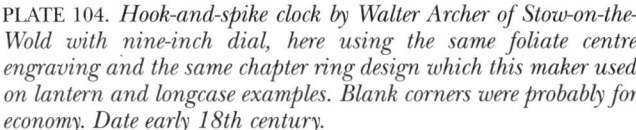

PLATE 104. *Hook-and-spike clock by Walter Archer of Stow-on-the-Wold with nine-inch dial, here using the same foliate centre engraving and the same chapter ring design which this maker used on lantern and longcase examples. Blank corners were probably for economy. Date early 18th century.*

PLATE 105. *The movement of the Archer hook-and-spike clock in Plate 104 is typical of many such clocks. This is a striking version with two trains, just the same mechanically as a posted movement thirty-hour longcase clock.*

For example, some lantern clocks took on the dial shape and form of the longcase, some having a square dial, some an arched dial. Some lantern clocks evolved into hanging alarm timepieces with simplified movements made principally in iron rather than brass (for cheapness, iron costing only one tenth of the price of brass), a sort of hybrid type. Some of these hanging wall clocks with a square brass dial and a posted movement (which might be of brass or of iron) barely resemble the lantern clock style from which they originated and to identify these as an interim type today we often call these hook-and-spike clocks, as they still used the lantern hoop and spurs for wall hanging.

Some hook-and-spike clocks, even when made as longcased versions, still hung by their hoops from the case backboard. Sometimes a hook-and-spike clock was bought as such originally and encased some years later, perhaps when the owner could afford the case or when it was felt that the exposed weight and lines had become a nuisance – they are a constant temptation to children and cats!

Another development which led from the lantern clock, perhaps via the hook-and-spike clock, was the hooded clock, which was often virtually the

PLATE 106. *Mid-18th century hook-and-spike clock hanging by its hook from the backboard of a long case. These clocks were sometimes housed in a case when first made but sometimes were cased many years later, when the owner could afford the case. The hook was sometimes used to hang it from, as in this example, instead of the normal longcase seatboard.*

same thing as an alarm lantern clock with varying shape of dial housed in a wooden hood, which hung on the wall. Almost all hooded clocks had alarmwork and therefore were timepieces (non-striking) and were of thirty-hour duration. Most had square or arched dials but round dial versions appeared later. Later examples often had plated movements but the earlier ones were usually post-framed, either in brass or iron. Hooded clocks were still made as late as the early nineteenth century, but by about 1780 had passed their main period of popularity.

The several different types of clock mentioned above, including the post-framed thirty-hour longcase clock, can be seen to have their constructional origins in the lantern clock. There are occasions when a longcase clock is in fact the same thing as a square-dial lantern clock in a case (see Plate 128). Hybrid forms deriving from lantern clocks, such as hook-and-spike clocks and hooded clocks, were occasional variants made by some clockmakers in some areas, but by no means everywhere. They appeared principally in the southern counties where the lantern tradition was strongest and lasted longest, and often those who made thirty-hour longcases also offered these derivatives as a cheaper option to the full longcase. Walter Archer of Stow-on-the-Wold is a typical example, making longcase clocks, very occasionally eight-day but more often thirty-hour, and at the same time making lantern clocks, hook-and-spike clocks and hybrid hanging wall timepieces (and perhaps also hooded clocks,

PLATE 107. *Anonymous nine-inch longcase dial (solid sheet casting) with cup-and-ring decoration. Date about 1700. The movement, however (Plate 108), comes as rather a surprise.*

PLATE 108. *The movement of the cup-and-ring dial in Plate 107 was originally an early 17th century balance wheel lantern clock 'modernised' about 1700 by having this replacement dial made for it to make it into a longcase clock (or hook-and-spike clock). The original steel hand was retained.*

though I don't know of any example of that type by him). See page 410 for a detailed look at Archer's work.

NORTHERN VARIANTS

In the northern English counties and in Wales and Scotland, these derivative types are largely unknown and in some counties entirely unknown. However, there are occasional northern examples of what have been regarded in the past as square-dial longcase clocks with lantern type movements with round brass posts, dating from the late seventeenth and early eighteenth centuries, often re-cased in long cases evidently later than the date of their making.

One prolific maker of such clocks was John Sanderson of Wigton, working there by 1691, whose clocks are so often found uncased or in later cases that it is a rarity to find one in its original, contemporary case. The late John Penfold, author of *The Clockmakers of Cumberland*, expressed the view that this was on account of the case destruction rate because of the sheer age of his clocks, although at that time Sanderson's dates were believed to be less ancient than recent research has proved. This is an opinion which I find difficult to accept as we do see clocks by other makers of equal age which *are* in their original cases and there seems to me no reason why Sanderson's cases should suffer a higher destruction rate than those of any other maker. It seems increasingly possible that some of Sanderson's posted movement clocks (and those by others of the 'Wigton school', which he began, such as John Ismay, Richard Sill and the early Barwises), rather than being longcase types may have been a northern equivalent of the hook-and-spike clock and may have

PLATE 109. *Nine-and-a-half-inch thirty-hour longcase dial by James De Launce of Froome (Frome), Somerset, c.1700, with original steel hand. The twin cherub spandrel sometimes projects at the corners of small early dials, as here. Spandrels are original though of two different patterns.*

PLATE 110. *The posted movement of the De Launce clock has cast brass pillars lantern-clock fashion. Note the strange spiked feet by means of which the movement sat on its seatboard (or perhaps a wall bracket). Two other clockmakers in this area used the same spiked feet. Solid sheet dial. Square-headed screws were used by many makers at this time.*

PLATE 111. *Early 18th century single-handed thirty-hour hook-and-spike clock signed Thomas Stanhope, believed to have worked at Preston, Lancashire. A most unusual type of clock for this region. The original hand is cast in brass, a material used by a few makers in this area but seldom elsewhere.*

PLATE 112. *Movement of the Thomas Stanhope clock showing posted construction with cast brass pillars, lantern clock style. Cartwheel cast dial. Tapered arbors. Anchor escapement with long pendulum.*

PLATE 113. *Anonymous hooded clock dating from the beginning of the 18th century, a timepiece alarm, which most were. The hood is in pine painted to simulate a richer wood. Dial six inches square. Most had anchor escapement and long pendulum (the latter removed in this photograph).*

been set on a wall shelf or bracket. This would account for their decorative lantern movements (pointlessly decorative if hidden inside a longcase) and the fact that most cased examples are today in later cases. Strength may be added to this view by the fact that occasional (longcased) clocks by such makers had more conventional, undecorated, square-pillar posted movements, more typical of those to be housed in a longcase.

CLOCKSMITHS
The men who made these first provincial clocks were for the most part descendants of the traditional, centuries-old metalworking trades. Some were blacksmiths or whitesmiths who probably entered clockmaking through having carried out repair work on public clocks. The word 'whitesmith' has fallen into disuse today, but was used to distinguish a metalworker who made smaller items of polished (white) iron from the blacksmith who made larger, heavier, unpolished ironwork such as gates, regarded as 'black' goods. Some such metalworkers were sons of blacksmiths, who turned increasingly to the new trade. At the time these men were often termed 'clocksmiths'. Many were self-taught (and proud of it) rather than apprenticed formally through the trade, as was the case with London-trained clockmakers. For example, the epitaph of clockmaker Thomas Peirce of Berkeley, Gloucestershire, who died in 1665, begins:

Here lyeth Thomas Peirce *whom no man taught,*
Yet he iron, brass, and silver wrought;
He jacks, and clocks, and watches (with art) made
And mended too, when others' work did fade… [my italics]

PLATE 114. *Hooded wall clock of the single sheet dial type made about 1760 by William Gill of Hastings with a ten-inch dial. Still a timepiece with alarm.*

PLATE 115. *Movement of the William Gill clock showing posted construction. The alarmwork is usually attached to the backplate, as here. Anchor escapement with long pendulum.*

A lantern clock which came to light recently (Plate 118) was signed 'John Baxter fecet 1670... of Conderton, blacksmith' and also was engraved with the following rhyme: 'John Baxter did me make And I will goe well for [hi]s sak'. This maker was previously unrecorded, interesting evidence of the relatively unidentified nature of many early blacksmith/clockmakers.

This origin in blacksmithing is probably the reason we find amazing variety of technique and skill in many early provincial clocks, often with a greater amount of ironwork than the more formal lantern clock carried. It is probably also the reason we find hanging posted movement wall clocks of quite primitive nature, sometimes with very crude attempts at dial engraving or even with no engraved dialwork at all except for numbers. Some clock dials have instead patterns made by tools such as drills, a decorative effect which could be done by someone incapable of freehand engraving – see for example the cup-and-ring type of dial (Plates 107, 160 and 161) and the well-known Oxfordshire 'zigzag' dials (Plate 149).

UNSIGNED CLOCKS AND QUAKER WORK
The background to such early clocksmiths may also account for the fact that a good number of early clocks were unsigned, or signed by initials only, or signed with a name and no place name, often leaving their makers difficult if not impossible to identify. A considerable number of early clockmakers were

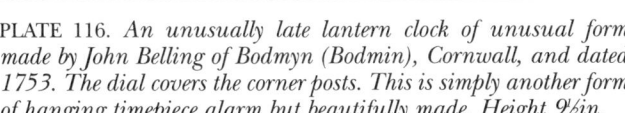

PLATE 116. *An unusually late lantern clock of unusual form made by John Belling of Bodmyn (Bodmin), Cornwall, and dated 1753. The dial covers the corner posts. This is simply another form of hanging timepiece alarm but beautifully made. Height 9½in.*

PLATE 117. *Movement of the Belling lantern clock with one door and bell removed to reveal the verge escapement with bob pendulum and double-headed alarm hammer. Alarmwork attached to the backplate. Half-round wheel collets.*

Quakers, especially in northern England where Quakerism was strongest. They seem to have taken to this kind of trade as being one suited to a self-sufficient life, perhaps in a remote location away from religious persecution, and some (though not all) Quaker clockmakers adhered to the principle that to sign a clock was a mark of vanity. Perhaps signing by initials only was a half-way house to anonymity. Many Quaker clocks were therefore unsigned, but there were also unsigned clocks which were not Quaker made. It can be difficult, at times impossible, to determine which unsigned clocks are Quaker made and which are not, but there is a general trend today on the part of the uninformed to ascribe all unsigned clocks to Quaker work, which is quite erroneous. There are several known features which can help to identify unsigned work as being that of a Quaker clockmaker and these are described in detail later (see page 132).

The reasons for a clock being unsigned are seldom clear to us today. It is sometimes said that the law obliged a clockmaker to sign his work. The Clockmakers' Company did occasionally try to enforce such a ruling, but their

PLATE 118. *Lantern clock dated 1670 and made by a country blacksmith, John Baxter of Conderton, Worcestershire, engraved with the rhyme: 'John Baxter did me make And I will goe well for [hi]s sak[e]'. No other clock is yet recorded by this maker. (Photograph by courtesy of Phillips.)*

PLATE 119. *Anonymous blacksmith-made country lantern clock of the 1670s, the iron posts hidden by the overlapping dial which is beautifully engraved with tulip motifs. Apart from the dial all parts were made in iron, including plates, wheels and pinions. Height 15in.*

jurisdiction held no sway outside London. Those early clockmakers who felt inclined left their work unsigned. This may sometimes have been to conceal their trading, where for example a clockmaker (or blacksmith) took clocks to sell at the market of a town where he was not a freeman and therefore prohibited on pain of fine from trading. Rural makers are known to have been prosecuted for doing this, when they could be identified as offenders, and the absence of a name on the dial would have made it harder to prove such a case.

John Sanderson, the Wigton (Cumberland) clockmaker who was a Quaker during part of his very chequered life, is known to have been fined for taking his clocks to sell in Edinburgh in 1715 contrary to local statute. Moreover, some clocks attributed (by me) to him are falsely signed 'Jeremiah Sanderson'

PLATE 120. *Movement of the iron lantern clock in Plate 119, constructed as balance wheel, 'modernised' in the 1690s to verge escapement. All wheels in iron except the two verge conversion wheels in brass. Note tapered arbors, iron bellstrap, strange screw-thread 'finials' to pillars, exceptionally heavy construction.*

(the name of an apparently non-existent clockmaker) and this may well have been a deliberate attempt at concealing his identity as the real maker in the light of his past fines.

Our understanding of the origins of provincial clockmaking is still very hazy. Much of the foregoing was totally unknown as little as thirty years ago. Most of it has come to light through the researches of a mere handful of people working in the last twenty years. The fact that so little is known for certain about the beginnings of provincial clockmaking makes the subject all the more interesting as new 'discoveries' constantly come to light. There is exactly the same opportunity for the new researcher to make exciting discoveries as for the old hand – if only he will document and photograph what he meets with!

PLATE 121. *Primitive oak lantern clock case of the 1670s modified in the 1690s to accommodate a thirty-hour longcase clock. It was almost certainly reduced in height at the same time as the thirty-hour clock would run with a much lower height. This is a primitive cupboard to house a clock. Height now about 6ft.6in.*

PLATE 122. *The same case showing front access door to insert movement and reset the hand (the hood being a fixture), and also one of the side access doors to permit pendulum attachment and strike sequence adjustment on the countwheel.*

CHAPTER SIX
THE ORIGINS OF PROVINCIAL CASEWORK

Some lantern clocks were housed in cases, though very few lantern cases survive today and the great majority of lantern clocks appear never to have had them, but hung from the wall instead. So, although the first provincial cases will nearly always be cases for longcase (thirty-hour) clocks, we can learn something about these from a consideration of early lantern cases.

It is doubtful whether any provincial longcase clock is known dating before 1680 and any earlier than the 1690s is exceptionally unusual. Clock case making had been an ongoing craft in London for some years by this time and by the 1680s a distinctive and regular style had evolved there which we can recognise today and which some provincial casemakers of the time also recognised and imitated. Some early provincial makers of eight-day clocks must have had their more sophisticated cases made in London. The case of the 1680s eight-day clock by William Holloway of Stroud (Plate 45) must have been made in London, as there would have been no local craftsman at that period with the stylistic knowledge to make a case of such sophisticated nature, especially in lacquerwork.

Most provincial clockmakers had their cases (lantern or longcase) made locally by the nearest joiner or carpenter, almost always in oak, but sometimes in pine. Some may have been in solid walnut or fruitwood or elm, but hardly any survive today in these softer woods, probably on account of woodworm damage and rot. The men who made the first (pre-1700) provincial longcases were men who for generations had made ordinary and simple household furniture, with which they were totally familiar – tables, chairs, cupboards. They knew how to handle wood, how to joint it, often using dowel pegs which typified the work of the country craftsman. These skilled woodworkers were now being asked for the first time ever to construct cupboards in which to house clocks and the fact that this was something entirely new to them is often evident when we examine such early prototype provincial cases.

For instance there are signs that the wooden case for the Henry Webster lantern clock (Plate 123) was originally built with a hinged door to the lower section in the manner of a double-height cupboard, even though this lower door served no purpose and was later fixed solid. A few other cases exist demonstrating this feature – they were just tall cupboards in which to house a clock, occasionally made to match panelling of the wall they were to stand against (for example the cup and ring case in Colour Plate 4 (page 118), the Ogden convent case in Plate 609 and the Leicestershire case in Plate 121). The first casemakers must have been working from sketches or from vague indications or dimensions (perhaps provided by the customer or the clockmaker) or perhaps from an example of a London-made case they had seen personally.

With the benefit of hindsight provided by our knowledge of nearly three centuries of casemaking we can often see how these early provincial case-makers were struggling not only with a style, shape and proportion they were as yet uneasy with, but also with the sheer functional mechanics of how the case 'worked' – which apertures needed to open as doors or could be fixed, how big some apertures needed to be for convenient access, and so on. When they got it wrong, as they sometimes did in these early years, the clock owner's task of assembling, winding or time-setting of the hands was made awkwardly inconvenient to the point where the casemakers would rapidly learn how to get it right next time in those respects. When we see such prototype cases we can sometimes conclude that the casemaker would set about building his next case differently and it is by these stylistic eccentricities and constructional mistakes that we can recognise the beginnings of casemaking in the provinces. Sometimes we can see that we are handling the very first case a particular joiner ever made, a unique and very exciting piece of horological history marking the beginning of clockmaking in a particular area. This kind of constructional error can be met with even into the second quarter of the eighteenth century as ever more woodworkers came new to a still young and specialised niche of the craft.

LANTERN CLOCK CASES AND COFFIN CASES

A few lantern clock cases survive from before 1700; many more must have been destroyed through obsolescence or because their primitive nature did not appeal to later owners with more sophisticated tastes. These early lantern clock cases tend to be straight up and down boxes with one or more access doors, rather like tall thin cupboards. A distinctive feature of their style is that often they lack the 'waist' normal with most later longcases or with later lantern clock cases or with more sophisticated London work. Instead the 'base' and hood sections are a continuation either of the actual timbers of the trunk sides or at least of the line of the sides. Often the base area is delineated merely by a separation moulding above the trunk 'base', whilst the trunk sides continue to ground level with a short plinth of an inch or two to form a foot, sometimes high enough to be regarded as a shallow base. Some such shallow-based clocks probably once had a very much higher base to bring the case as tall as maybe nine feet in order to match wall panelling, though often these have been cut down later.

The fact that the base is a downward extension of the case sides is a very important structural feature not only affecting the appearance of a lantern case or a coffin-type longcase, but also affecting the construction of many of those early (but post coffin-style) longcases of the more usual proportions which followed soon after and which *do* have a waist and a true base. Examination of many conventional longcases up to about 1730 in London work, and a little later in provincial work, will usually reveal that the base itself is built *on to* a frame which is essentially still a coffin-style frame, almost as if the base were an afterthought although it is part of the original construction. The base will be seen to be fitted around (and fixed by gluing and/or nailing on to) the lower case side timbers, which often extend right to the ground, or almost that far, *inside* the true (visual) base.

In other words, these early conventional cases often have a double-skinned base, originally probably devised to give rigidity to a tall thin object which might otherwise easily sway. In fact the true base probably originated as an attempt at giving longcases more stability in the form of a wider stance – it was probably as a continuation of that principal that mid-nineteenth century bases became very wide indeed. One aid to dating casework in the early eighteenth century is an examination of these internal side timber extensions, which gradually shorten as time goes by so that by the mid-eighteenth century they are no more than an overlap reaching just below the lower trunk moulding, sufficient for use as fastening supports for the base itself.

The hood of a lantern case or coffin case may be a continuation of the side and front timbers as a fixture with an opening access door, or may be a removable hood, with a fixed glass or with an opening glass door. The trunk door on such clocks will often cover the full width of the case front without any door frame as such and will instead be hinged on the case side upright piece. A search of the books on horology will reveal barely half a dozen such cases (see Plates 37 and 168).

The style of these early lantern cases is very interesting, few though they are. Plates 121 and 122 show such a case modified and modernised (in the 1690s?) to house a thirty-hour longcase clock of the 1690s. This suggests that the case may date from the 1680s or even earlier. The glass door can be seen to have been altered to square from an upright rectangle. The original 'seatboard' housing was for a lantern clock and still shows the four 'foot' marks; a later normal seatboard was then fitted above it for the later plated movement clock. The original 'seatboard' is in fact one piece of timber which protrudes to form a dividing moulding between hood and trunk, and this type of seatboard-cum-mould is a feature of quite a number of lantern cases and early longcases. The experimental clock (Plate 41) has this same feature (as do the Webster and anonymous clocks in Plates 124 and 121).

The hood of this anonymous Leicestershire clock is not removable but is a fixture, integral with the trunk. So too is the hood of the experimental clock (Plate 42). Plain three-quarter hood pillars were an alternative to barley sugar types, either form being an option at this very early period. The strange widely spaced dentil moulds under the hood cornice seem to have been an accepted feature of style with some early lantern cases, and similar ones are found on the Webster case (Plate 123), oddly enough from widely separated regions – here Leicestershire and Lancashire! It may be this was simply an architectural feature applied to furniture and a similar principle is seen on some cupboards and dressers of this period. This form of cornice dentilling is sometimes found on oak-panelled rooms and it may be that it was used on clock cases to match wall panelling. In fact there is some indication that the Leicestershire case may originally have been attached to (or built into) oak wall panelling rather than free standing (see Figure 14). Wall fixing would have avoided the tendency of such slim cases towards instability.

The oak case housing the lantern clock of about 1685 or perhaps earlier by Henry Webster of Aughton, near Ormskirk in Lancashire (Plate 123) has many structural and stylistic features in common with the Leicestershire one. This case might at first sight seem to have been made for a square-dial longcase clock, but the dial aperture is in fact an upright rectangle, the glass measuring 9½ inches high by 8¾ inches wide, and rectangular brass-dial longcase clocks are unknown (occasional ones are fractionally off-square, but not far off). It dates from about 1685, which is the same period as the Webster clock it contains and a time when precious few longcase clocks must have been made in Lancashire. In fact Webster, who died in 1697, is the earliest known Lancashire clockmaker by whom work survives, the next oldest not working till the mid-1690s. Moreover the topmost trunk mould, which also forms the 'seatboard', is cut for a posted (i.e. lantern type) movement, an unusual form of longcase mechanism for Lancashire. The case shows no sign of any other movement having been fitted and is believed to have always contained this clock. Many of its features are typical of early coffin cases, both of lantern and longcase type, and therefore informative as to the nature of early longcase work.

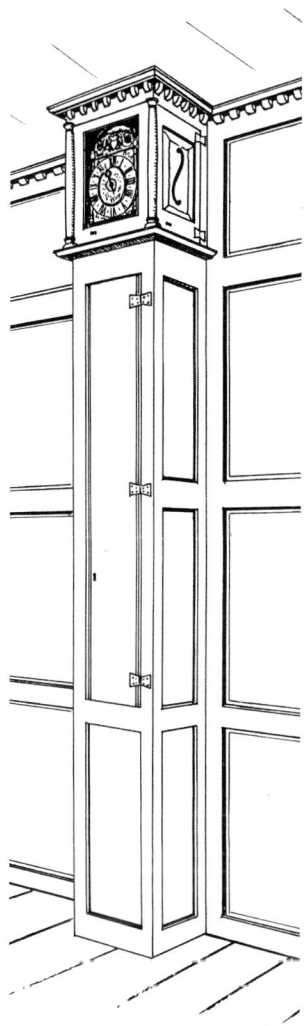

FIGURE 14. *An artist's impression of how the oak lantern clock case may have originally looked, designed to match in with wooden wall panelling.*

PLATE 123. *Lantern clock of about 1685 by Henry Webster of Aughton, Lancashire, in its original oak standing case. Height now 7ft. 3in. but this probably once stood taller. The initials and date are believed to have been added in 1703.*

The glass door of the Webster hood does open but the door does not run to the full width of the hood, as with the later pillars attached style, but hinges within the outer hood framework like a cupboard door. The hood slides forward for removal in the normal manner, with rebated safety restraints of softwood either side locking into the lower hood moulds. Some early provincial cases had rising hoods in the London manner, but by no means all of them and the fact that a hood slides forward rather than rises up does not in itself deny early age. At the left side of the hood is an opening wooden door; on the right a fixed wooden panel simulates a door, presumably just to match for appearance's sake. This opening wooden door may have been to assist access for fitting the long pendulum. This clock was constructed with anchor escapement and long pendulum and rack strikework, but for a countwheel-striking clock such a door would permit re-setting the strikework into sequence. These tasks might have been as easily performed by removing the hood and the precise function of such side doors remains conjectural.

The barley sugar pillars of early cases such as this are often of an interesting, almost experimental nature. They sometimes have unusual capitals/terminals – those on this example have an extraordinary long 'neck'. Quite often, as here, they do not reach fully to the base, but stop short as 'hanging' pillars. Of course not all of these very early cases have barley sugar pillars – some have plain pillars and some have no pillars at all.

The initials ISE and the date 1703 are carved on the door and seem quite genuinely of that age. The cross bar through the I suggests this is intended as a J, these two initials being used interchangeably at this time. This form of initialling is usually that of husband and wife and might, for example, refer to John and Ellen Smith, the central letter being that of the surname. The year would be one of significance to them, often a wedding year. The identity of the couple is unknown and the clock was around twenty years old at the time these initials were carved. It might be that the clock was inherited in 1703 or passed on as a wedding gift then and that event was commemorated by the initials and date, though this is only conjecture.

The lenticle glass in the door is formed by grooving the timber to leave a rounded lip-mould, rather than by adding a D-mould to the surface. This is amazingly like that in the cup and ring case in Colour Plate 4 (page 118) and is very unusual and apparently very early practice.

The case in Colour Plate 4 was made for a ten-inch square dial longcase clock, but has some amazingly similar features to the two lantern clock cases previously illustrated. Clearly it has been reduced in height, but originally had three sections to the coffin trunk. The top section is hinged as a D-mould edged door. Below this is a tiny D-mould edged fixed square section containing a small lenticle glass with the very same crude lip-mould as that on the Webster case. Below that is a further D-mould edged fixed panel, which is presumed to have originally been the same length as the door, now reduced in height. The present plinth may be a modification of what was originally a short bracing foot. The side extension pieces alongside the lenticle glass are where the case has been cut and widened to allow for over-wide pendulum swing (explained on page 430).

The hood has a fixed glass, not an opening door. As there are no safety restraints the hood can be lifted off in either an upward or forward direction, but clearly the intention was for it to slide forward and the lack of safety restraints was an oversight. The hood sides have neither doors nor windows but have sunken panels, as indeed most unusually does the backboard. It is

PLATE 124. *The hood of the Webster case in detail showing such interesting early features as the hanging barley sugar pillars, dentil moulds, hood-to-trunk meeting moulds and the rectangular glazed access door just high enough to permit a full view of the clock (not the bell).*

very difficult to date this case and the best we can say is c.1700. Of course, a case with a rising hood (and usually also one with a recessed 'cupboard' type of glass door) would have no internal 'mask' to frame the dial.

TRUE LONG CASES

Longcase clock making in the provinces appears to have begun no earlier than 1680 and in some regions not till considerably later. If there are pre-1680 examples, then we are unable to recognise them, as stylistic features at this very early time are imprecise and most very early thirty-hour clocks are unsigned. The style of early provincial longcases is something we can only judge by reference to the London equivalent and of course to early lantern cases, even though London examples were almost always built with more costly or exotic woods. London cases were made for a wealthy clientele and needed to conform to what was regarded as the refined taste and styling of the day. In the 1660s the cheapest eight-day London ebonised longcase would have been in excess of £20; a little later a simple provincial cupboard case of this nature cost about £1. For comparison an uncased London lantern clock could be bought for £3.10s.0d., a provincial lantern clock for £2.10s.0d., a provincial thirty-hour clock and oak case (of the 1690s) for £3.10s.0d. (Plate 125) and a provincial

109

PLATE 125. *Tiny early longcase thirty-hour clock in oak by Samuel Ogden of Halifax dating from the 1690s. Though loosely based on the London concept, this case has many provincial features including slender baluster pillars, exceptionally heavy hood topmould, peg fastening for the door. Height 6ft.6in.*

eight-day clock and oak case for £5.0s.0d. (Colour Plate 5, page 118).

The London casemakers had gone through their trial and error stages in the 1660s and 1670s and by the 1680s a recognised styling had evolved which was both functional and aesthetically pleasing. Proportion and balance were important and these were carefully judged, so that examples of London cases from the late seventeenth century represent classic design in the view of the casemakers, clockmakers and customers of the day and often too in the opinion of present-day collectors, some of whom feel that such elegance was never again achieved in later longcases.

The provincial joiners making their first longcases may or may not have been aware of the style current in London, or may have deliberately chosen to ignore it. Eventually most provincial cases did follow the London design, faithfully or otherwise, ultimately (by the late eighteenth century) departing from it intentionally to a style that was purely provincial, but the date at which these trends happened varied considerably with such factors as remoteness of the area from the capital and the whims of the individual joiner, customer or clockmaker. Some provincial cases as late as the 1730s still demonstrate archaic features which were met with as early as the 1690s in other provincial examples. The Henry Mason case (Plate 248) is one such later case still showing early prototype features nearly half a century later than some.

CASE DOORS

The London case of the 1680-1700 period had a long slender door within a balanced door frame. Some provincial examples have no door frame at all, so that the door hinges on and closes into the ends of the solid side uprights of the case (see Colour Plate 8 and Plates 37, 168, 169 and 248) This was a method of construction sometimes found on lantern cases and on some pre-1680 London coffin cases, but one that can cause some difficulty for the maker as he has precious little to hinge on or fasten to. The case of the Mason clock (Plate 248), though considerably later than 1700, still shows this 'frameless' door construction principle. In that example the unusually deep frontal sections above and below the door (as also in the anonymous case (Plate 42) and as sometimes seen even in later provincial cases) are probably to give strength to what otherwise might be a weak construction. These sections throw the door 'frame' area out of balance and are often a characteristic of very early casework. The Knibb case (Plates 168 and 169) also is of frameless door type; the fielded panel gives the impression of a door within a frame (looking therefore in proportion rather like a London trunk), though in fact the whole front panel swings open in one piece hinged between the two case side uprights.

HINGES

Wrought-iron hinges were usual on these early cases, the flanges either fitting inside the case or externally, and occasionally one half inside and one half outside (e.g. the Mason clock in Plate 248). Some primitive cases had iron staples interlocked, though today these have usually worn through and have been replaced by hinges (as on the Midgley case in Plate 190). The same hinging principles were often applied to hood doors as to the main door, though some hood doors swung on a hidden pivot pin in the pillar end. Brass cranked hinges were in use in London casework by the 1690s but iron blacksmith-made hinges continue to appear on some provincial cases well into the eighteenth century.

HOODS

London cases of this time had pillars to the hood in the form of three-quarter pillars attached to (and opening with) the hood door. They normally hinged by means of a pivot pin set into the top and bottom of the right-hand pillar and many provincial cases copied this pivot method. Occasional hood doors pivoted or hinged on the left, presumably at the request of the owner. Often the pillars were of the twist type we call 'barley sugar' pillars, though this style gave way increasingly to plain pillars towards the end of this period and in the early eighteenth century. The plain pillars were in fact often 'bellied', i.e. having a pronounced swell or entasis in their central areas, and the balance of this entasis can be very varied in provincial examples according to the whims of the individual carpenter. Some provincial cases imitated the London fashion but some had no hood pillars at all, a practice which continued with some country casemakers as late as the 1760s and is sometimes found in Scotland even in the mid-nineteenth century.

The hoods of early provincial cases sometimes betray their prototype nature in one or more of a variety of ways. It is often obvious that the casemaker (usually a country joiner or carpenter, but sometimes perhaps even the owner making his own case) was not experienced at the task, even though he may have had experience at making other kinds of furniture. A clock hood calls for several specialised features, the function of which may only become apparent during use and the absence of which can be a considerable inconvenience to the owner. When we see early cases lacking these necessary but perhaps unexpected features we can imagine that the casemaker would do things differently next time with the benefit of hindsight or a complaining owner. Sometimes we can feel sure that a certain case must be the first one this joiner ever made, for surely he would approach the problem differently next time to avoid a particular difficulty.

RISING, SLIDING OR FIXED HOODS

Everyone who owns a clock knows that the hood can be removed, most by sliding forwards. A few early ones had rising hoods, a feature originating with the earliest London-made clocks but only seldom used in early provincial work. The rising hood system ultimately proved unsuitable, probably on account of the need for a ceiling considerably higher than the clock, and most provincial casemakers never used that form. The forward-sliding hood was a more practical system and had become well established by 1700 in London and probably in most other regions too. Most of the older, original rising hoods were long ago converted for convenience to slide forward – though present-day owners sometimes convert them back again to the original rising form as an attempt at restoration to 'originality'.

A carpenter making his first clock case could be forgiven for not knowing that the hood should be removable, something which would not be apparent from a drawing. The case in Plate 41 (experimental clock) *looks* like a normal late seventeenth century country design apart from the unusually extended hood back splats (which we can put down to a quirk of the maker, and which is also seen in Plate 130). However, the hood is not removable but is a fixed part of the whole case, as with some early lantern clock cases (for example, the anonymous Leicestershire case in Plate 121). The only way to get the clock in or out of it is to open the hood door and feed the movement through complete with dangling lines and pulleys, a very inconvenient method and one never used in normal case design by anyone experienced in clock cases. A small access door

PLATE 126. *Hood of the Webster clock half-removed showing absence of hood stop mould and hinged side access door.*

at each side of the hood assists when fitting the pendulum into place; without these doors the owner would have to try to feed the pendulum in blind by reaching up from the trunk door, an awkward task even for the most experienced. A further inconvenience is that this case with its fixed hood is abnormally heavy (top-heavy, in fact) and awkward to move single-handed.

At this very early period (pre-1700) provincial clock hoods may have been fixed but were normally removable, may have had fixed glass panels but normally had opening glazed doors, and even exceptionally seem to have had a dial aperture without any glass at all.

HOOD TOPMOULDS

The topmost moulding of very early provincial hoods almost always finished flat. This cornice moulding often projected unusually far to give an unexpectedly wide overhang (for example, the Midgley case in Plate 190), but it tended to lessen as time went by to the more normal proportions we are accustomed to seeing in clocks of the mid-eighteenth century. This wide overhang is often present at this period in London cases too, though is less pronounced.

HOOD SIDE WINDOWS

Many early longcase clocks, both London and provincial, had glass side windows to the hood, the reason for which can only be surmised as there are no written records to explain this sometimes puzzling feature. These windows occur mostly on eight-day clocks though some thirty-hour clocks have them too. Casemakers must have assumed that the owner might wish to look into the movement, perhaps to watch the mechanics at work on what was after all the first machine (see Plate 215). Even today newcomers to clocks are often intrigued to watch the movement perform, especially the strikework. With an eight-day clock a glance through the hood side window might well have provided an indication as to whether the clock needed winding, though it might have been as easy to unlock the trunk door to establish this as to get a buffet to gain vision height. No such benefit would result on thirty-hour clocks, yet many have hood side windows, as do some lantern clock cases. It is difficult to see what point these could serve other than to give a view of the wheels turning. Whatever the reason, many hoods do have side windows, a practice which lasted longer in London and parts of the South than elsewhere. In the North side windows were less popular at any period and seldom occurred after about 1730. (Scotland is a special case where Edinburgh copied London – see page 298).

HOOD SIDE DOORS

Instead of windows some early provincial clocks have doors in the hood sides, sometimes in just one side and sometimes in both sides. The experimental clock (Plate 41) has a door at each side; the Webster lantern clock (Plate 126) has a door on the left and a door-like fixed panel on the right. These doors are confined mostly to thirty-hour clocks rather than eight-days and seem to appear during the same periods as the alternative side windows, though not as far as I am aware on London-made clocks. They seem to be a carry-over from lantern cases (which sometimes had fixed hoods) and may have been to assist when fitting the pendulum, but it is more likely that they were to give access for re-setting the strikework sequence on the thirty-hour countwheel striking system, which could come out of sequence when the clock ran down. The same would apply for many early eight-day clocks, most of which also had countwheel striking until well after 1730. This is only a suggestion on my part as nobody knows the reason these doors were sometimes fitted. Oddly enough, I do not know of any examples having one window and one door, nor any with just one window (though some do have just one door), but maybe this was to give the case either side uniformity of design.

HOOD STOP MOULDS

Those clocks which did have a removable (sliding) hood ultimately had the lower hood moulding so shaped that it stopped tight up against the lip of the top-of-trunk mould when it was slid into position. This was done by having an internal right-angled corner inside lip of the bottom hood mould, so that this fitted flush when in place. In other words, the hood fitted neatly into place in an established and recognisable position determined by the meet of the two moulds, and this was accepted practice in London casework by the 1680s if not earlier. It was simply a common sense method of fitting. However, a joiner unfamiliar with regular clock casemaking might not know this and some prototype cases reveal this ignorance in the maker. The absence of such a stop mould will therefore often be an indication of the beginnings of casemaking in the area or by that particular joiner. For instance, the Richard Midgley case in Plate 191 has a lower hood mould which is merely a flat strip of timber joined at the corners with no edge moulds at all. The hood can only be judged to be fully 'home' when the back fouls against the upper backboard. The Webster case (Plate 126) is made in the same way, i.e. lacking the lip, as is that by Thomas Clark (Plate 127) and that by Ogden (Plate 125), whereas the Mason case (Plate 248), despite its primitiveness, has such a lip.

DIAL MASKS

By the 1690s (or earlier) most London cases and some provincial ones had acquired a feature which was to become standard on virtually all regular-made longcases thereafter, and that was an inner framework to surround the dial in the way a mount surrounds a picture. This is only fully visible when the hood door is open, but can often be partially seen with the glass door closed.

On most later clocks this inner frame, which we usually call a 'mask', was made of softwood, often pine. Sometimes, however, the mask was in oak, especially on earlier clocks, say those before the mid-eighteenth century, though there is considerable variation in the wood used for the mask. The opening in this mask was normally the same size as the glass in the hood door, the principle usually being that the mask would overlap the dial edge by

PLATE 127. *Hood of the oak thirty-hour clock by Thomas Clark of Warrington (Plate 216) seen half-removed to show flat hood basemould with absence of stop mould.*

COLOUR PLATE 2. *Early arched dial of about 1715-20 with unusual shape of arch by Edmund Bullock of Ellesmere, No.246. Fine bold engraving. Early and unusual form of penny moon. Rocking 'sun' in the arch. See page 70.*

COLOUR PLATE 1. *Ten-inch dial from a longcase clock of the mid-1690s by Ninyan Burleigh of Pontefract. The engraved work is very fine. Note the early form of the numbers and the exceptionally fine original blued steel hands. See page 57.*

COLOUR PLATE 3. *Thirteen-inch arched dial longcase clock of about 1715 by Thomas Clark of Warrington. Fine engraving of very bold nature and with large half-hour and half-quarter markers, typical of some north-western work. The penny moon in the arch is an early feature. See page 76.*

perhaps a quarter of an inch all round, which would mean that the mask and door glass for an eleven-inch dial might measure 10½ inches, sometimes described as 'showing' 10½ inches.

Sometimes however the hood door and its glass might be made deliberately larger than the dial/mask size to produce the effect of a narrow border of mask showing around the dial through the glass when the hood door was closed. With some clockmakers, Will Snow for example, this was a regular stylistic feature – Snow's masks were painted a blue/green colour and showed intentionally through the hood glass, producing quite a pleasing effect.

The mask arrangement is rather variable on provincial clocks before about

1750. The mask did not always overlap the dial, particularly dials with a herringbone border where the dial might sit forward 'inside' the mask aperture in order that the engraved border would not be partly obscured. This same principle of the dial sitting inside the mask rather than behind it was sometimes used with ten-inch dials of the type with twin cherub spandrels, as these spandrels sometimes protruded at the dial corners and would therefore not permit a mask to overlap.

The purpose of describing this mask arrangement at this juncture is that the makers of some early prototype and primitive cases were not yet aware of the mask feature and built their cases without any mask at all! The hood of the Mason clock (Plate 248) was built without any mask, probably because the casemaker had not yet come across the feature. So too was the Ogden case (Plate 618) and the Midgley case (Plate 190). The hood of the clocks by Cooper and Stevenson (Plates 700 and 313) were built without masks in order to avoid hiding the decorative herringbone dial edges. Rising hoods had no masks as they would have fouled the hands. Masks were therefore a development in style, which, though known early, came very late to some country work.

The total absence of a mask on some of these early primitive cases sometimes resulted in the hood being especially weak, as all that held the two hood sides together might be the front mould and on some of these cases this could be a very flimsy affair, as with the flimsy front moulds on the Ogden and Midgley cases (Plates 618 and 190). To avoid this structural weakness some early casemakers fitted internally a lower horizontal front strut to link the two hood sides inside the hood door, looking rather like the lower member of a mask (with the top and two side pieces absent). This can be confusing to the inexperienced. The Ogden and Clark cases both have such a lateral strut, simply as strengthening pieces.

Past owners of such clocks have sometimes mistaken this lateral strut for the remains of a mask and have sometimes built new masks with the best of intentions in an attempt at 'restoration'. A case which never originally had a mask (or perhaps had a single lower strengthening strut looking to the inexperienced like a solitary surviving mask section) may therefore have acquired a full 'new' mask at some time during its chequered life. The case of the Midgley clock (Plate 190) was built with a single lower strengthening strut, but had later had the other (mistakenly believed missing) three mask sides added by some well-meaning owner, in mahogany – a wood, which was not available in England when the case was built! When I bought the clock I removed the incorrect mahogany mask pieces. Such a situation can prove very confusing to the beginner as this case (and others like it) will now appear to be missing the top three sides of what might erroneously be thought to have been a mask and the still visible glue marks will be apparent proof that they once were fitted!

It is important to understand the mask principle as a clock married into a wrong case has often had the mask replaced or altered to camouflage the marriage, especially to make a case accept a dial of the wrong size (see page 358 ff.).

HOOD SAFETY RESTRAINTS

Inexperienced makers of early primitive cases were sometimes caught out in yet another respect, which again would have come only through experience and use. Most clock hoods were fitted with a very essential yet simple safety device to prevent the hood accidentally falling off. When the hood door is opened, as for example when re-setting the hands or when winding an eight-day clock, the displaced weight of the open door will often cause the hood to tilt forward in a quite alarming manner and, unless a safety device is fitted, the hood may well crash to the ground or on to the head of the owner. A regular safety feature had become standard on virtually all longcases by the second quarter of the eighteenth century and on London cases by the 1690s. In fact the Webster lantern clock case of the 1680s (Plate 124) has such safety restraints, though it is impossible to say whether these were added at a later date.

The solution was just a thin strip of timber nailed on the outside of each of the seatboard cheeks in such a way that the inner framework of the lower hood mouldings would contact against both in the event of the hood tilting forward in the manner just described. Some early casemakers were not yet aware of this potential problem and failed to fit any device. The Ogden case (Plate 618) was built without any safety device and whenever the hood door was opened the hood threatened to fall off. The structure was such that the hood sides touched up flush against the seatboard cheek sides leaving no space for such safety strips, indicating that the casemaker had not thought of this potential problem (which was in fact solved by a present-day owner fitting two small pegs passing through the rear of the hood sides into the thick oak backboard).

The Mason case (Plate 248) also lacks any hood safety device. The Midgley case too (Plate 190) has its hood sides flush against the case sides with no room for such safety struts, but the casemaker evidently realised the problem late in the day and got round it by cutting a groove in each side of the seatboard support 'cheeks' (i.e. the upper case side extensions) to restrain the lower hood mould. This left the cheeks seriously weakened and this is probably the reason why the Midgley seatboard is nailed between the cheeks instead of resting on top of them as was usual with the great majority of clocks.

TOP-OF-TRUNK MOULDS

The topmost trunk moulding, the one which supports the hood, was usually convex in very early clocks and remained so till about the end of the century, by which time in London casework it had become a concave moulding. In provincial cases the date of this change from convex to concave moulds is rather variable and it is not unknown for a clock of the 1730s still to be using the early form. Some provincial cases of this period have a very small mould there (Plate 618), and some have what is known as an ogee mould, that is one starting as convex and turning to concave in its width. The ogee mould was used later in the eighteenth century in certain regions more than others, so an ogee mould in itself is not necessarily a dating factor. Some early provincial cases may have very eccentric moulds bearing little resemblance to convention (see the Shepley case in Plate 181).

BASES

A few very early cases had no true base but were of the coffin clock type, though today we seldom see this type. Many clocks of this great age have had their bases replaced in relatively recent times by well-meaning owners who have 'restored' bases like ones they have seen in photographs. Most restored examples now have a true base, perhaps faithfully copied from other clocks, but this may conceal the fact that some of these were once coffin based.

SEATBOARDS

Seatboard is the name given to the piece(s) of wood on which the movement sits. Some very early eight-day clocks (Fromanteels, for instance) had two

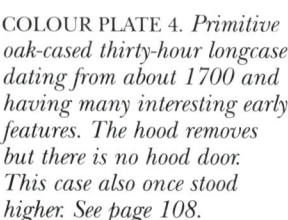

COLOUR PLATE 4. *Primitive oak-cased thirty-hour longcase dating from about 1700 and having many interesting early features. The hood removes but there is no hood door. This case also once stood higher. See page 108.*

COLOUR PLATE 5. *Interesting oak-cased eight-day clock of about 1700 by Joseph Hawthorn of Great Chell in Staffordshire, a previously unrecorded maker. Many early features including hanging barley sugar pillars, frets to three sides of the hood, heavy caddy top, ogee top-of-trunk mould. Height 7ft.*

sturdy strips of timber nailed front to back so that the movement plates sat astride them. Some posted-movement thirty-hour clocks had the same system but usually with the two timber strips fixed sideways across the gap between the cheeks. Most clocks, however, had a board, often as deep as the case would allow to the backboard, with holes to allow passage of the gutlines or rope/chain. This board was usually attached to a plated movement's lower two pillars by bolts screwed into the pillars themselves on many earlier examples, and later by seatboard hooks which lapped over the pillars and were held by a nut below the board. Posted movement thirty-hour clocks usually sat loose on the board, as did many plated thirty-hours. Occasionally posted movements were bolted down.

The seatboard itself might or might not originally have been nailed to the cheeks (which were normally the case side uprights). Sometimes later owners would nail them down for safety. For ease of handling and assembly modern practice is to have the board attached to the movement and unattached to the cheeks.

COLOUR PLATE 6. *Hood of an anonymous early thirty-hour longcase clock of about 1700 with heavy moulds, early heavy caddy and frets to three sides.*

The seatboard housing method was subject to considerable variation and is not particularly helpful as a guide to period. Sometimes, for instance, the seatboard would be nailed *between* the cheeks rather than *on top* of them, as with Richard Midgley's clock in Plate 191. Occasional makers (Thomas Lister of Halifax was a regular) fitted their movements to a sort of small 'table' that sat inside the upper cheeks. Variations in fitting methods were more of a personal whim than an indication of period, except that the separate strut system was for the most part superseded by the full seatboard by the beginning of the eighteenth century.

REGIONAL STYLES

In this very early stage of provincial clockmaking cases are very simple indeed, some being no more than a joiner-made cupboard in which to house a clock. Such styling as they have is usually modelled on the London equivalent and as yet no regional stylistic trends seem to have begun.

CHAPTER SEVEN

CLOCKSMITH WORK
(PRIMITIVE AND PROTOTYPE THIRTY-HOUR CLOCKS)

1. SOME VERY EARLY UNSIGNED AND MONOGRAMMED EXAMPLES

Many of the earliest provincial thirty-hour clocks were either unsigned or signed by monogram only with the result that their makers cannot be identified. We can date them only by assessing their period features as we have no facts to help us regarding the lives of their makers, who were probably blacksmiths, whitesmiths or clocksmiths.

The anonymous longcase thirty-hour clock pictured in Plates 128 to 130 has an eight-inch dial, a posted, lantern style movement and a primitive pine case of the coffin type. It is in effect the same thing as a square dial lantern clock with anchor escapement, having baluster brass posts, and fancily filed and shaped hammer stop. The dial pins on to lugs rather than feet, the dial itself being of the 'cartwheel' type of casting, which might suggest a northern origin – or might not! The circular grain to the matting shows quite well.

The chapter ring has an early form of trident half-hour marker and is riveted to the dial sheet at IX and III, a similar dot marker beneath each hour numeral attempting to disguise the rivet heads. Riveting of chapter rings is often indicative of rural clocksmith work. A slip of the engraver's tool shows through the head of the V. The imperfection in the chapter ring below XII is a casting fault. The dial has plain corners without any spandrels, which is a feature of some Quaker work but could just as well have been done for

PLATE 128. *Eight-inch dial of an anonymous late 17th century thirty-hour longcase clock of possible Quaker origins. Original steel hand in the manner of a lantern clock hand.*

PLATE 129. *The movement of the anonymous clock, being exactly the same thing as an anchor escapement lantern clock of the day. Note cartwheel cast dial.*

PLATE 130. *Primitive pine case of the anonymous clock. Removable hood with fixed door. Note the unusual backsplats which extend from the hood well down the upper trunk area of the case. Height about 7ft.*

economy as for intentional styling. The fine steel hand is original and is very much of the style used on lantern clocks. The rivets on the dial sheet below VI hold the lower pinning lug, and would have been concealed in more professional work.

The pine case, originally painted but now stripped, is charmingly crude and primitive and stands a little under 7ft. The hood has the most unusual and interesting extended backsplats like those on the experimental clock in Plate 41. Absence of hood pillars is to be expected on such a primitive case. The hood has no opening door but a fixed glass so that the hood must be removed for re-setting the hand. A needlessly high trunk space above and below the door is typical of some of these early cases. The wrought-iron blacksmith 'butterfly' hinges show plainly. Few such cases survive today as their crudeness has caused many to be destroyed.

Such a clock is very difficult to date accurately; it could be from the 1670s but it is more likely to date from the later 1680s or 1690s, perhaps even later.

The anonymous dial shown in Colour Plate 7 (page 122) is nine and a quarter inches square. The dial sheet is of copper, the chapter ring of brass. In fact some dials and pendulum bobs which appear to be of copper may simply be brass with an unusually high copper content, but the fact that they *look* like copper cannot have been accidental. The very finely engraved centre pattern is based on an open flower and is reminiscent of a lantern clock dial. The fine original iron hand has an extra long tail to assist leverage when setting and is also in the lantern clock manner. The strange blank centre zone serves no purpose other than for decorative effect, but initially suggests a missing alarm disc. In fact the clock never had alarmwork. The chapter ring is well engraved

COLOUR PLATE 7. *Nine-and-a-quarter-inch copper dial of an anonymous longcase of the late 17th century with superb flower and early trident half-hour markers. Very interesting original steel hand reminiscent of a lantern clock hand. See Plate 128.*

PLATE 131 (OPPOSITE). *Movement of the copper dial clock (see Colour Plates 7 and 8) with iron plates and posts. Fine filed work to the hammer spring and hammer stop. Tapered arbors, square-head screws. Solid dial sheet.*

with a trident half-hour marker and suggests more expert and experienced work than the example in Plate 128, the engraving being bold, crisp, more stylish and without slips. The tiny cherub head spandrels also give a more professional look to the whole. The dial sheet itself is of the solid type.

The movement (Plate 131) has iron posts and plates, unusually ornate filing to the hammer stop and hammer spring, tapered arbors, iron square-headed screws – many of these features being reminiscent of lantern clock work. The date could be 1680s or early 1690s.

The original oak case stands 7ft. high and is of relatively refined styling (Colour Plate 8). Although clearly provincial in execution, it is based on London concepts, which are: lenticle glass, hood side windows, convex top-of-trunk mould, integral hood pillars. The heavy topmould with wide overhang is very provincial, as is the double-D door lip mould. An extraordinary feature is that the case front which forms the door frame is all one piece of timber (the joints in the lower section are a repair). This is very uncommon practice on account of the likelihood of shrinkage causing splitting, though surprisingly this has not happened here.

The unsigned clock in Plate 134 has a ten-inch dial, floral (lantern-clock style) centre based on tulips. This clock too has this purposeless central blank zone, as used when alarmwork was fitted, though it never had alarmwork. It is almost as if some provincial makers at this time could not decide how to decorate the central zone which on a London eight-day clock would probably have Tudor rose engraving to it (see the Williamson clocks in Plates 582 to 584).

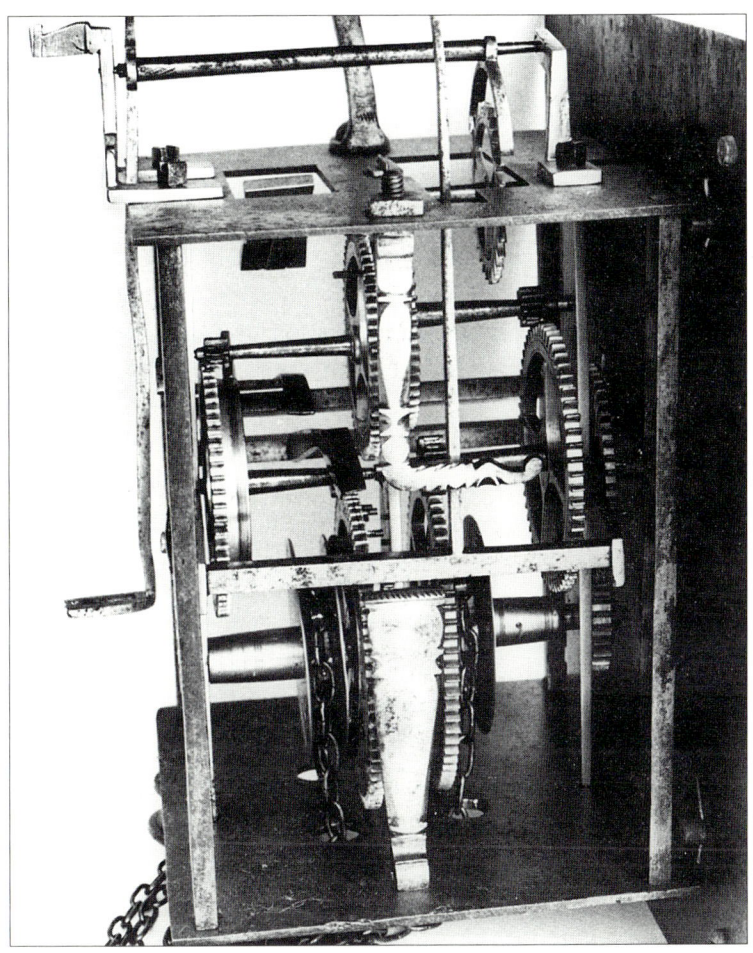

COLOUR PLATE 8. *Oak case of the copper dial clock standing about 7ft. Many early and primitive features including heavy topmould and double-D door lip mould. See page 122.*

The trident half-hours are strong and mostly well done, but notice the slip on the half-past eleven marker causing a crooked shape. Twin cherub spandrels of the later type suggest this might date from about 1710 but it could be earlier. The very fine iron hand is again reminiscent of lantern clocks.

The posted movement shows in its front posts that the dial fits by means of two feet rather than lugs. The hammer spring and stop are here very simple in style with just a hint of shaping. Note the massive square-headed iron screws and massive bellstand. I cannot guess whether this clock is southern or northern.

The charming clock by Thomas Whittaker (a maker whose working place and dates are unknown) has an eight-inch dial of very thin brass (note cracks top right and bottom right) and has blank corners (Plate 136). Both these features are typical of some Quaker work, but we cannot say whether this man was a Quaker. The dial sheet is solid – being so thin it would have had no strength if of the spoked type. Circles are used for decorative effect – these could be done by a clockmaker who could not engrave and might suggest that he had his chapter ring engraved elsewhere. Asterisk half-hour markers are made up of meeting arrowheads. The chapter ring rivets can be seen plainly. The original hand is of iron and of a marvellously primitive style akin to a lantern clock hand. The signature is naïvely engraved and positioned eccentrically. The movement (Plate 137) has brass posts and plates, with just a hint of decorative filing to the hammer stop and spring. The dial is held by two dial feet, top and bottom, pinning into the front crossbar.

The original oak case (Plate 138) is charmingly primitive and stands 6ft.2in.

123

PLATE 132. *Ten-inch anonymous dial of a longcase of about 1700 with superb central engraved work reminiscent of a lantern clock centre. Spandrels are of* lead. *Original iron hand with unusual screw-thread tail.*

PLATE 133. *Movement of the anonymous clock showing decorative filing to hammer stop, extraordinary double-hoop bellstand, pearl-headed terminals to all ironwork protruding beyond each nut, square-headed screws. Cartwheel cast dial.*

PLATE 134. *Ten-inch dial of an anonymous thirty-hour longcase of about 1710 with tulip centre and fine original steel hand. Bold engraving.*

PLATE 135. *Unrestored movement of the ten-inch tulip dial clock with the dial removed showing plain but very sturdy work. Holes in the flat front posts are for the two dial feet.*

PLATE 136. *Charming eight-inch dial from a longcase clock by the unidentified maker Thomas Whittaker, apparently Quaker work. Date about 1690-1700. Superb original steel hand.*

PLATE 137. *Movement of the Thomas Whittaker clock with brass plates and posts. Plain work but sturdy. Solid dial sheet (too thin to have been made cartwheel fashion).*

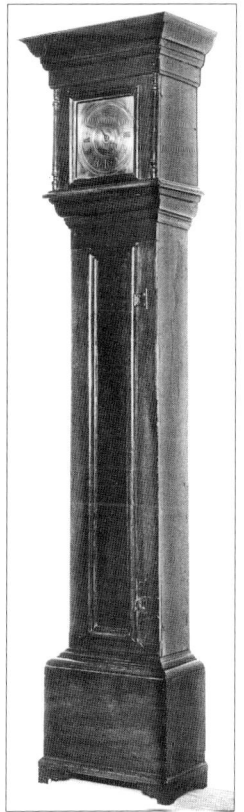

PLATE 138 (right). *The oak case of the Whittaker clock stands 6ft.2in. Many primitive characteristics.*

PLATE 139 (left). *Detail of the hood of the Whittaker clock with fine baluster pillars and interesting, heavy mouldings.*

COLOUR PLATE 9. *Dial of a tiny (less than six inches) and very primitive longcase timepiece with alarmwork. Original iron hand. See page 128.*

COLOUR PLATE 10. *The movement of the timepiece alarm has original verge escapement and the alarmwork side mounted on to a purpose-made strap. See page 128.*

The hood has half-round baluster pillars integral with the 'door', an alternative to the usual three-quarter types, and an odd little door-like panel, beaded on three sides only, in the upper left side (Plate 139). This never opened but just gives the impression of a door – perhaps the casemaker was copying from a drawing of a case with an opening side door, as this feature seems quite purposeless. The glazed dial aperture is a fixture with no opening door and the hood therefore needs to be removed to re-set the time. The clock is difficult to date – perhaps the 1690s.

The tiny clock pictured in Plates 140 to 143 is signed with the unidentified monogram TW. The solid dial sheet measures only six and a quarter inches square. The chapter ring, with meeting arrowhead half-hour markers, is competently engraved whereas the matting of the dial centre is uneven in its graining, perhaps suggesting the engraved work was done elsewhere. The tiny spandrels at first sight seem odd but in fact are the tips taken from the earlier twin-cherub pattern. The iron hand is a modern replacement based on an old pattern.

The movement is of the plated type in brass and was made as a simple going train only. A striking train was added as an afterthought (or perhaps slightly later) with iron plates! The date is about 1700.

The original oak case is tiny, standing only 6ft.6in., and is conventional for a

COLOUR PLATE 12. *Eleven-inch dial by John Knibb of Oxford, exquisitely engraved and matted with superb original hand. A top quality maker producing a cheaper clock. c.1690. See page 139.*

COLOUR PLATE 11. *Primitive pine case of the timepiece alarm. Height about 6ft. See page 128.*

cottage case of the day except in one respect, namely that its hood door is set within the overall hood front, opening like a cupboard door. Top-of-trunk moulds are concave. The beading around the edges of the trunk door, hood door and top-of-trunk mould are all chip-carved in an interesting manner, as are the two turnbuckle knobs, and these help give the clock a little more character.

The anonymous single-handed clock with alarmwork in Plate 144 is in fact a hook-and-spike clock, made with an iron posted movement and a tic-tac escapement which in effect is like an anchor escapement with a shorter anchor span covering usually only three or four teeth of the escape wheel. Tic-tacs usually have a wider swing than an anchor and therefore a considerably shorter pendulum, this one being eleven inches long.The dial measures five and five-eighths inches high by five and a quarter inches wide, and can be seen to be of very thin brass, with Quaker-style blank corners.

The alarm is set by rotating the required hour number on the disc to the hour hand tail, so is here set at 12.30. Like most alarm clocks, this is non-striking, the bell being purely for the alarm – known strictly as a timepiece alarm. The iron hand is original and resembles that of a lantern clock. The engraving is naïve with a sort of teardrop for half-hour markers. The date could be about 1690-1700, maybe a little later.

Another timepiece alarm illustrated in Colour Plates 9 to 11 (pages 126 and

PLATE 140. *Six-and-a-quarter-inch dial of a longcase clock signed TW, an unidentified maker. The spandrels are the tops of a twin-cherub set. Crude matting.*

PLATE 141. *The plated movement of the TW clock shows that it was made as a timepiece only, using brass plates, but that a strike train was added later using* iron *plates. The solid-sheet dial is too thin to have been made cartwheel fashion.*

127) has a tiny (solid) dial less than six inches square and is unsigned. Blank corners are a feature of many such primitive clocksmith clocks, not only those made by Quakers, and the reason was probably because it kept the price down and no spandrels exist of this small size. The engraving is naïve, bordering on crude, perhaps indicative of the maker attempting his own engraving. A simple dot marks the half hours, which was easily done with a drill though not standard practice. The primitive nature of this clock may mislead us into too early a dating, which is probably about 1700 to 1710.

The movement has a verge escapement with short (bob) pendulum, unusual for any longcase clock. The iron posts and plates carry the alarmwork side mounted, probably because rear mounting would have meant much deeper plates.

The primitive case is of pine and, rather surprisingly for the age and for such a tiny clock, it has full hood pillars.

The anonymous single-handed clock pictured in Plates 145 and 146 has naïve bird and flower engraving to its centre, suggesting the work of a non-professional engraver, probably the maker himself – compare the professional chapter ring. The dial is just over nine inches square. The well-made movement is of plated form and has the unusual feature of rack strikework. Rack striking was an option for thirty-hour as well as eight-day clocks and was sometimes provided at extra cost at the request of the customer to offer a repeating facility, but is uncommon on thirty-hour clocks. This is a difficult clock to date but I would guess about 1710.

The ten-inch single-handed clock signed TB (Plates 147 and 148) probably dates from about 1720, if we are to go by the latest features which are the uncommon spandrels based on an urn centre. These normally date between about 1720 and 1740 and, if we assume they are original to this dial, must be

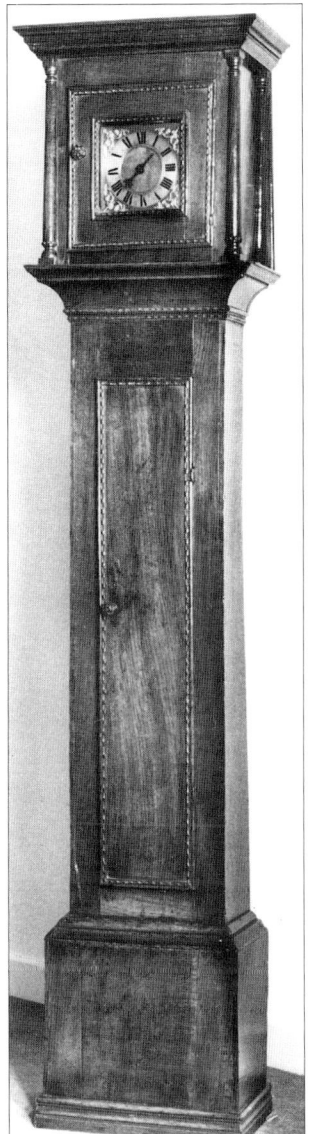

PLATE 142 (left). *Oak case of the TW clock standing 6ft.6in.*

PLATE 143 (above). *Close-up of the hood of the TW clock showing interesting chip-carved moulds. The hood door is set within the overall hood front.*

PLATE 144 (right). *Tiny and primitive hook-and-spike wall clock with alarmwork, dating from the late 17th century with tic-tac escapement. Interesting original steel hand.*

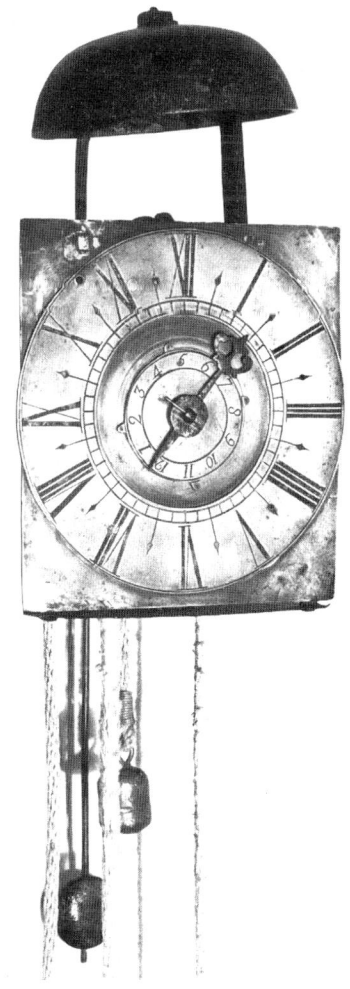

taken as the latest feature. Most other features of this clock suggest an earlier date – the trident half-hour markers, early, simple hand, cup-and-ring marked centre with radiating lines of 'matting' done with a matting tool. The decoration of the centre of this solid dial could be done by someone unable to engrave, so perhaps this maker bought his highly competently engraved chapter ring work elsewhere.

TB is thought by the owners to be Thomas Bacon of Tewkesbury, Gloucestershire and, although the clock has been in that area for many years, this attribution is far from certain. The posted movement has some highly decorative features – shaped hammer spring, hammer stop shaped like a crocodile mouth or a bird's open beak – but notice the initials IB on the spring. The I has a cross bar through it, suggesting it is used for a J. It is not known who JB was but perhaps TB's son or some other member of TB's family work-force.

The dial of this next clock is eight inches square and is made of copper (Plate 149). The dial centre is of a type we usually call a zigzag dial, a sort of wrigglework performed by using an engraving tool in a guider while turning

PLATE 145. *Nine-and-a-quarter-inch anonymous dial of a thirty-hour longcase clock dating from about 1710. Naïve centre engraving of flowers and birds. Original iron hand.*

PLATE 146. *The movement of the clock has rack striking to allow repeating facility. Solid sheet dial.*

PLATE 147. *Ten-inch longcase clock signed TB, the centre with a kind of wrigglework pattern done by someone incapable of engraving. The fine bold chapter ring is the work of an expert engraver.*

PLATE 148. *Movement of the TB clock showing a most interesting hammer stop filed as a bird beak or crocodile mouth. IB is monogrammed on the decorated hammer spring.*

PLATE 149. *Eight-inch copper dial of a hook-and-spike wall clock in a style known as a zigzag dial, characteristic of a group of Oxfordshire Quaker clockmakers. Original iron hand.*

PLATE 150. *Movement of the copper dial clock showing hanging hoop and spikes, the latter here riveted to the baseplate. Solid dial sheet.*

PLATE 151. *Nine-inch anonymous dial of an early 18th century hook-and-spike wall clock, probably also Oxfordshire Quaker work. Original hand. Starburst centre is an unusual variant.*

PLATE 152. *Movement of the starburst dial clock, here having the spikes as an integral part of the rear posts. Solid dial sheet, very thin.*

PLATE 153 (left). *The same starburst hook-and-spike clock before restoration. It had been made into a lantern clock with 19th century frets and bellstand, all of which were spurious later additions.*

PLATE 154. *Six-inch dial of an anonymous hooded alarm timepiece of about 1720-30. The spandrels are uncommon since not many dials are this small. Original steel hand.*

the dial sheet on a turntable. Though this is a form of engraving, such work could be performed by someone unable to engrave freehand. Here the dial has two circles of zigzags; on other clocks the number of circles may vary. The chapter rings of many of these zigzag clocks seem to have been done by professional engravers, as here.

It is not often at this early period that we can detect indications of regional styling in clockwork, but in this instance we can. This style is, so far as we know, confined to a group of Quaker clockmakers centred on North Oxfordshire, most of whom did not sign their work, as that would have been regarded as a display of vanity. The outer dial edge and corners are decorated by punchwork, again an indication that this clockmaker could not engrave. The chapter ring, however, was engraved by a skilled engraver in what would be the up-to-date style of about 1710-20, or perhaps a more old-fashioned style if the clock was made somewhat later, as the relatively plain and simple movement might suggest. Such clocks are very difficult to date accurately.

From the movement (Plate 150) it can be seen that this is not a longcase but a hook-and-spike wall clock, but its styling and construction would have been little different if it had been designed as a longcase. When such clocks do have

PLATE 155. *Unsigned nine-inch thirty-hour hook-and-spike wall clock of unmistakable origin being a product of the Oxfordshire Quaker zigzag school. Original (repaired) steel hand. c.1730-40.*

PLATES 156 AND 157. *The case now housing the zigzag clock was purpose made for the clock but some years later, about 1760. The age of the carving is uncertain. Height 6ft.1in.*

PLATE 158. *Plain oak case of another Oxfordshire zigzag thirty-hour clock of the second quarter of the 18th century. Height just over 6ft.*

spikes, they seem to have been positioned at random, either being attached (riveted) to the baseplate, as here, or as projections from the rear posts, as in the next example. This dial is quite thick, unlike most of the ones of this style which are often of brass so thin as to be flexible. Most zigzag dials are solid dial sheets – they would have had no strength if cast in cartwheel form. The naïve iron hand is original and reminiscent of lantern clock times.

This nine-inch solid sheet dial (Plate 151) has a starburst pattern engraved into the matted centre. The carefully worked but inexpert dial centre contrasts again with the immaculately engraved (meeting arrowhead half-hours) chapter ring, which was probably done out of house and might date the clock about 1710 or a decade or more later. The dial can be seen to be very thin and bent here and there, especially near the corners. The tiny female-head spandrels are rather like a sphinx head and were often used for small dials of the zigzag type, as few spandrels were small enough for the purpose. This unsigned clock is also probably Oxfordshire Quaker work, though the starburst centre is a variant from the normal zigzag. The blued iron hand is original.

The movement (Plate 152) shows this to be a hook-and-spike wall clock, not a longcase. Notice in this instance how the spikes are formed as an integral part of the posts, whereas in the last example they were riveted to the plate.

Plate 153 shows how this clock looked before being restored. It had been made to look like a lantern clock by being modified (maybe a century ago), involving the fitting of frets, finials, lantern style bellstrap and side doors, all in much more modern brass of a quite different colour from the original.

This unsigned dial (Plate 154) measures six inches square and is a plated alarm clock made as a hooded clock. It might date from about 1720-30. The diamond-shaped half-hour markers (sometimes called lozenge shape) here have a leaf-like pattern engraved within them, but are more often plain and waxed black. These markers were fashionable about this time, particularly with some Quaker clockmakers, probably because they were less ornate than some other types of the period and Quaker preference was often for sobriety.

The unusual spandrels show a male head wearing a sort of felt hat. The steel hand is original. Engraved work on the chapter ring and alarm disc is well done.

The zigzag style of dial was, it seems, limited to a group of Oxfordshire Quaker clockmakers, most of whom did not sign their work, though occasional ones did, especially towards the mid-eighteenth century and after. Some of these later clocks are very difficult to date, as is the one in Plate 155. This clock has a nine-inch dial and might date from about 1730-40, but could be as much as twenty years later. The zigzag centre style varies considerably in its make-up, here having three circles, but any number between three and seven might occur. The tiny spandrels of the female bust/sphinx head pattern were used for a long period as they were one of the very few patterns to fit small dials – these zigzag dials are usually ten inches or less. The spandrels are therefore not as reliable a dating guide on this type of clock as one might expect. The diamond- or 'lozenge-' shaped half-hour markers were also used over a long period and they too are little help in pinpointing the period.

A number of these zigzag clocks were made in the hook-and-spike form, so could be hung on a wall caseless if desired. Some were apparently used that way for some years and cases bought for them at a later date, sometimes a good few years later. This is the situation with this particular clock, the case of which is 'original' to it, but apparently dates to about 1760, some years later than the clock,

This means that such a clock seen today as a hook-and-spike clock may have been built that way originally, cased later at a time economically convenient

PLATE 159. *Anonymous longcase dial showing lavish use of cup-and-ring turning and dating from about 1700 or perhaps earlier. The cut-out engraved brass cherub spandrels are most unusual. Very bold engraving.*

for the owner, and may later still have lost its case and reverted to being a simple hook-and-spike clock.

When these clocks were cased, either at the time of the clock's making or later, the clock was often hung from a hook in the backboard rather than set on to a normal wooden seatboard. The problem with that form of hanging is that over the years the spikes tend to wear their sitting holes over deeply with the result that the clock can hang crookedly or tilt forward awkwardly. When that happened they were often fitted with a seatboard instead. This explains why on some clocks mysterious unused (half-depth) holes appear in the backboard and unused hooks and spikes behind the clock.

The case of this particular clock (Plates 156 and 157) is carved, and stained black, as was usual with carved examples. Whilst it is relatively easy to establish the age of a case, it is difficult if not impossible to establish the age of the carving on it – see page 189 for more about carved cases.

Another example of a tiny Oxfordshire zigzag Quaker clock in its case is shown in Plate 158, also unsigned, and dating from the second quarter of the eighteenth century. Some very simple cases such as this lack pillars and have simply-shaped moulds. Many are dowel-pegged, as this is.

2.CUP-AND-RING DIALS

Another form of clock, often early and often unsigned or monogrammed, is the type with what I call a cup-and-ring dial. Like the zigzag form, this sort of dial decoration could be done by someone incapable of engraving, usually by using a drill. Also as with the zigzag dial, the fashion for this decorative effect was not always confined to the earliest periods, though it did first appear then. Any clockmaker anywhere could have decided to use this decorative form and no doubt examples exist from various parts of the country. However, it was a style of decoration used by certain known early clockmakers in the Lake District and North Yorkshire, and I usually associate this style with those regions.

Pictured in Plate 159 is a very interesting but unidentified anonymous dial of the cup-and-ring type. The chapter ring has versions of trident half-hour

PLATES 160 AND 161. *Two ten-inch cup-and-ring longcase dials of the 1690s monogrammed RW and thought to be by Richard Washington of Kendal. The second clock has been converted to eight-day with added minute hand, but both retain the original identical (hour) hand. Both dials appear almost identical apart from spandrels, though the chapter rings are riveted in different positions.*

PLATES 162 AND 163. *The plated movement of the single-handed thirty-hour RW clock (front and back). It has* copper *latches on all four pillars, an early feature used by very few provincial makers. Round iron rod pillars seem incongruously simple for the relative sophistication of latches. The iron rod crutch support was much simpler to make than a true cast brass backcock.*

PLATE 164. *Ten-inch cup-and-ring dial of a thirty-hour longcase by the unidentified maker using the monogram HR. Fine original steel hand. First quarter 18th century.*

PLATE 165. *Primitive early 18th century dial with cup-and-ring decoration by John Ismay of Oulton, Cumberland. Several blow-holes and casting faults in the brass. Probably the maker's own attempts at engraving.*

markers, done, if a little crudely, by a true engraver. The cornerpieces are 'spandrels' made from flat cut-out brass pieces primitively engraved in the form of winged angel heads, each showing considerable variation from its partner. Presumably the clockmaker paid for the engraved work of these items and the chapter ring.

The dial sheet itself could have been made by the clockmaker as all its decoration is in the form of cup-and-ring turns of varying sizes and patterns. This dial is very hard to date and we have to rely principally on the chapter ring and the hand. The movement is of the birdcage type, though we have no photograph of that. It seems to date from the late seventeenth century, or perhaps the early eighteenth. Rivets in the chapter ring at III and IX are a clue to clocksmith work, as is the whole primitive style. The holes by XI and V were probably a rivet holes for holding the chapter ring to its bed during engraving. I would not like to guess where this clock was made, though it is believed to have surfaced in Wales.

The two ten-inch dials shown in Plates 160 and 161 are of the solid sheet type and are from single-handed clocks by the unidentified maker, RW. The second clock has been converted later to two hands by fitting an eight-day movement, but the dial shows that this began as a single-hander. These clocks came to light in the Lake District, where the cup-and-ring fashion was strongly popular, and I believe the maker was Richard Washington of Kendal, who was a whitesmith and died in 1698. Two other higher quality clocks are known by him signed with his full name and place (one eight-day longcase and one lantern clock) and it may be that he signed his better work fully, using a monogram for his cheaper clocks.

Both dials are identical, apart from the spandrels of Plate 160 (the single-hander), which are believed to be replacements. The single hands are also identical, the original hand being retained as an hour hand in the conversion example. If these clocks are by Richard Washington, and I don't see how we shall ever resolve that question, then they must date from the 1690s. If they are

137

PLATE 166. *Ten-inch longcase clock monogrammed by the unidentified maker GR and dated 1759. If it was not dated the clock could easily have been thought of as earlier. Original blued steel hand.*

PLATE 167. *Posted movement of the Knibb clock (see Colour Plate 12, page 127) with square iron pillars, the lower projecting threads forming 'feet'. Superior ratchet clickwork of the eight-day type.*

not, and we have to date them by relying on style alone, we might conservatively assume a slightly later date of about 1710.

The movement (Plates 162 and 163) is especially interesting as a very early plated example with features characteristic of whitesmith work, based on cheapness of materials and ease of making. The pillars are round iron rod instead of the more usual cast and knopped brass. The pendulum support shaft is a squared iron post hammered (riveted) into place, much easier to make and cheaper than a cast brass backcock. The crutch is supported direct from (squared on to) the anchor arbor and does not need backcock support. The bellstand is on the *front*plate, said often to be an early indicator. The pillars are held at the front by *copper* latches on all four, an unusually sophisticated feature for humble country work and one which perhaps had influence on other makers in this area later. The cone-shaped wheel collet is a whim of the maker.

The single-handed dial shown in Plate 164 is monogrammed HR, an unidentified maker. This clock is believed to have been made in the Lake District, though this is only an opinion. This too is a very difficult dial to date, but, judging principally by the original hand and chapter ring style, it should date from the first quarter of the eighteenth century, perhaps 1720-30.

The dial in Plate 165 is by John Ismay of Oulton, a hamlet near Wigton, Cumberland, and shows extremely primitive styling and execution, so primitive it is difficult to believe the work is that of a professional engraver, but rather that it was done by Ismay himself. The style suggests a date very early in the eighteenth century, but from the facts we know about him it is unlikely this dates before about 1720-25. Ismay was apprenticed in 1711 to John Ogden at Bowbridge, North Yorkshire (see page 383), but was working back in Cumberland by 1718 so may not have completed his training. He worked with

his relative, John Sanderson, as a member of what I call the Wigton Group of clockmakers (see page 392) and this dial is shown here as an example of how cup-and-ring drill turning was sometimes used conveniently to fill up a space which otherwise might call for professional engraving.

The casting faults, showing as gaps in the chapter ring between IX and X and as holes in the lower right dial corner, illustrate the crudeness of the casting and suggest this may have been done in-house. Ismay's work is closely akin to that of John Sanderson.

The maker of the ten-inch dial monogrammed GH and dated 1759 (Plate 166) is unidentified but it is thought he was also working in the Lake District. This clock has a plated movement. The dial is naïve in style with cup-and-ring turnings to surround the matted centre and spandrels of the twin cherub style, which are decidedly old-fashioned by now. The lower left spandrel is a non-matching replacement. Were it not dated we might be inclined to suppose this dial was twenty or thirty years earlier

The engraver of the chapter ring and attached 'name-plate' was clearly a professional.

3. SOME EARLY SIGNED EXAMPLES

The eleven-inch solid-sheet dial by John Knibb of Oxford (Colour Plate 12, page 127) dates from perhaps the late 1680s or 1690s – behind the dial is scratched 1689, which might be the date of its making. Knibb was a very famous clockmaker, though less so than his elder brother, Joseph Knibb of London, and was working in Oxford from about 1673. The dial is very formal (as befits the work of a man 'never known to laugh', said about him in 1716), having a finely matted centre (compare with some later crude rustic mattings), restrained trident half-hour markers and a superbly elegant

PLATES 168 AND 169.
Ebonised oak case of the John Knibb clock showing many early provincial features such as full-front door, hood side doors, heavy overhang to topmould.

139

PLATE 170. *Ten-inch dial of a longcase clock by John Pagett of Bridgnorth, c.1700. The grain of the matting can be seen clearly.*

PLATE 171. *The movement of the Pagett clock is a lantern movement, made with verge escapement and converted soon after to anchor. Iron collets integral with the tapered arbors. Finials, bellstrap and feet are all true lantern clock features. Iron plates. Solid sheet dial.*

original blued steel hand. The undecorated matted centre is based on London influence and at this time on thirty-hour dials it was more usual on those from the South-east than elsewhere. The signature is in the earliest (first) position on the dial sheet below the chapter ring.

The posted movement (Plate 167) has the iron posts, which unusually are threaded at their bases to take nuts to hold the baseplate in position, leaving the threaded extensions as 'feet'. The movement itself is sturdy yet plain, and has the superior type of ratchet click as used on an eight-day clock for smoother winding. Some early clocks, such as this one, have such finely engineered movements with such crisply cut wheelwork and arbors that to the novice they appear much more modern than expected.

The plain case (Plates 168 and 169) is of oak, so dark as to be almost black in colour, and probably ebonised originally – the Knibbs are known to have had a fondness for black cases, which they felt best showed off the beauty of the clock dials. It stands 7ft.4in and is of elegant yet primitive construction, showing some typical features of early country work. Joints are dowel-pegged. There is no framework to the trunk door or hood door, but instead these hinge by pivots attached to the case and hood sides. Both doors therefore swing open fully across their width. A large 'reed' run into the sides of the doors gives the effect of pillars. Both doors fasten by wooden turnbuckle knobs, each having an iron spike 'latch' on its inner part. The trunk door has

PLATE 173. *The Pagett movement sits on two blocks set front to back across the upper trunk top-board (which also forms part of the projecting top-of-trunk 'moulding'). This appears to have been the way early lantern clocks sat in their cases.*

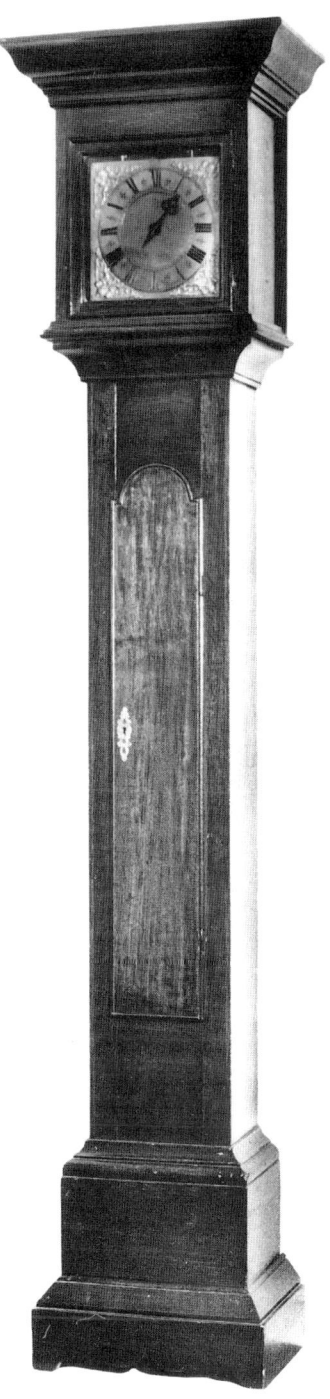

PLATE 172. *The oak case of the Pagett clock is exceptionally slender with early features such as absence of pillars, heavy topmould overhang, hood door inset within the framework, unused space above and below the door. Height about 7ft.3in.*

a fielded panel (i.e. a raised panel with chamfered edges). This is a feature only occasionally met with and it echoes the applied veneer panelling of some very early eight-day clocks by such makers as Fromanteel.

The hood topmould has a wide overhang, characteristic of many early country cases. Moulds are simple and small, the top-of-trunk moulding being a sort of ogee shape instead of the more usual convex or concave forms. The hood has a small door in the right-hand side, hinged on iron staples, probably intended for access to re-set the countwheel strikework without removing the hood. This is matched on the left-hand side by a fixed raised panel with a beaded edge. There is no inner hood mask surrounding the dial, which sits close up behind the hood door. The inner hood mask did not become a regular feature on all country cases for some years and primitive examples may be found without a mask as late as the mid-eighteenth century. The bun feet are a restoration and may be incorrect, based on grander clocks, as country cases more often had a small plinth.

This style of very simple case with exceptionally long door and hardly any 'waist' is sometimes called a 'coffin' case. There is no lenticle glass in the door of this clock and, although this became a very popular feature from the 1690s and during the following thirty years or so, it was not found on all examples and in particular it was usually absent from the very earliest.

The clock by John Pagett of Bridgnorth (Plates 170 to 173) appears to date

141

from about 1700-1710. Dating is a little difficult because this clock is of an exceptionally unusual type, having a true lantern movement originally made with verge escapement and converted not long after to anchor. The movement (Plate 171) has iron plates (most are brass but a few lantern clockmakers used iron). It has iron wheel 'collets' integral with the arbors – in other words there are no separate collets but the tapered arbors are shaped into a collet-like ending. This is a principle of some lantern clocks and some very early longcases. The clock has finials, bellstrap and feet (and once had doors and frets, now missing) as a true lantern clock, not simply a round-posted movement, which are relatively common. The maker made a lantern clock and then fitted a purpose-made ten-inch longcase dial, which fits by means of just two dial feet, attaching to the central stem of the front crossbar.

The dial has twin cherub spandrels of the earlier of the two types, original blued iron hand, matted centre without calendar or decoration, and an early form of floating fleur-de-lis half-hour marker.

The case (Plate 172) is of oak, dowel-pegged in country fashion, with the heavy overhang of the topmould characteristic of many early cases. The case sides pass right down inside the base, as with many early cases, to form a double skin. The hood has no pillars and the hood door opens within its framework, cupboard door fashion, again almost always a very early feature. Iron blacksmith hinges are used throughout. The case sides project forward beyond the door frame instead of being set behind the frame, which was the normal later method of most casemakers. The wide plinth is an ancient though later addition, probably to give better stability to what was a very slim case.

The movement sits on its feet on two wooden blocks set front to back (Plate 173) instead of a seatboard, this being the system used by Fromanteel and a few of the earliest makers and also by some early makers of cased lantern clocks (see Henry Webster on page 108). The top-of-trunk 'moulding' ends with a quarter round piece (on which the hood sits) which is not a moulding as such but the edge of a solid board having a central cut-out for the ropes. The hood has no stop-piece, but sits at whatever position it is pushed to. These are all features of prototype casework indicating a joiner not yet fully used to casemaking.

The question we cannot answer is whether this was a lantern clock modified (soon after making) into a longcase or whether the maker used a lantern clock movement for a customer wanting a longcase clock at the time of making. The date of movement, dial and case are so close together that this cannot be determined.

The eleven-inch thirty-hour longcase clock by Thomas Power of Wellingborough, Northamptonshire, dates from the 1690s. He is believed to have been working from the 1680s and is known to have died in 1709. The (solid) dial sheet itself (Plate 174) has a plain (polished) centre, which is very unusual. Undecorated dial centres are normally matted, decorated ones normally have an engraved design set on to a polished ground. Here the maker is ignoring the convention of the day, with which he must surely have been familiar, perhaps because a polished centre is easier to do than either of the normal methods and, of course, this would be an easy way out for a clockmaker who could not engrave his own dial centres. It may be there was an element of local styling here for Henry Simcock of nearby Daventry also made some clocks with plain polished dial centres about this time.

The chapter ring, however, is professionally engraved using the trident form

PLATE 174. *Eleven-inch two-handed longcase clock by Thomas Power of Wellingborough dating from the 1690s. Unusual polished ground to the dial centre.*

PLATE 175. *Plated movement of the Thomas Power clock with unusually tall plates and latches to all four pillars. The frontplate is finished (i.e. filed smooth) only around the edges, perhaps to give a smooth working surface for his latches.*

of half-hour marker with minutes numbered *inside* the minute band and very up-to-date styling of half-quarter markers, which here are dots. The town is quaintly spelled as 'Wellingborrow'. The two-handed form is also very up to date for such an early provincial thirty-hour clock, the great majority of which are single-handers. The spandrels are the standard late seventeenth century cherub head pattern. The hands look like modern replacements.

The plated movement (Plate 175) is conventional except that its pillars are latched, an unusual feature for any provincial work but especially so for thirty-hour work (but see the clocks by R.W. and John Hough in Plates 162, 163 and 177). The bellstand is attached to the front plate, supposedly an early indicator, though I think it is a mistake to make much of that point. Many clockmakers left their frontplates unfinished, i.e. raw from the casting. Power has taken the extraordinary step of finishing his plates around the outer edges only, as can be seen in Plate 175. This was probably to give the perimeter a smooth surface so that his dial feet ends and his latches would sit flush against the plate front. Both dial and movement show signs of fine quality work.

The ten-inch solid dial thirty-hour single-handed longcase clock by John Hough of High Leigh in Cheshire dates from about 1695 or perhaps just before. The dial style is very formal for a northern clock (Plate 176), the matted centre lacking any decoration or calendar feature, but then this man's work shows a strong southern (London?) influence. The interesting hand is of iron and is original. The chapter ring is professionally engraved and uses the arrowhead marker for half-hours. The late seventeenth century large cherub head is the spandrel pattern.

The movement (Plate 177) is finely made and has the unusual feature (for

PLATE 176. *Ten-inch longcase clock of about 1695 by John Hough of High Leigh, a hamlet in Cheshire. Original blued steel hand.*

PLATE 177. *The movement of the John Hough clock showing the solid dial sheet and finely finned movement pillars, which are latched. However, the dial feet are also latched, an exceptionally unusual feature on any provincial clock. Tapered arbors without collets, lantern clock fashion.*

PLATE 178. *Oak case of the John Hough clock standing 6ft.8in. Simple and elegant showing a little more refinement than many country cases of this age.*

a provincial maker) which we saw in the Thomas Power clock (Plate 175) of having latched movement pillars. However, this clock also has latched dial feet, something virtually unknown outside the work of a handful of early London clockmakers and quite exceptional in a provincial maker. The maker did not use wheel collets and used tapered arbors, both normally signs of very early work and a carry-over from lantern clock times.

The original oak case (Plate 178) stands 6ft.8in high and has side windows to the hood and a caddy top, which reaches only halfway back, a feature of some early north-western caddies. Three-quarter pillars are virtually always integral with the hood door at this early period. Moulds and overall styling are quite simple and graceful, when compared with other early northern casework such as the Shepley case in Plate 181.

The two clocks by John Shepley of Stockport in Plates 179 to 183 date from the beginning of the eighteenth century, about 1710. Of the two the nine-inch clock (Plates 179 to 181) is probably slightly earlier, this opinion based on the smaller dial, earlier-looking hand, earlier pattern spandrel and earlier, more primitive look to the case. Both chapter rings use the meeting arrowhead type of half-hour marker but that of the smaller dial incorporates a dot, possibly harking back to the dots of the trident marker style. Both dials are signed in Latinised form, as was the early custom – 'Johannes Shepley' and 'John's Shepley Fecit'. However, they are signed on the chapter ring itself, not the lower dial edge, indicating a move on from the first position of the Knibb clock period. The apostrophe in 'John's' indicates that it is an abbreviation for Johannes. Both have solid dial sheets rather than the cartwheel type of casting, and have plate-form

PLATE 179. *Nine-inch longcase dial by John Shepley of Stockport c.1710 with all-over floral centre.*

PLATE 180. *The Shepley (solid-sheet) dial with the chapter ring removed showing three holes at XII, IV and VIII for the chapter ring feet, and three riveted ends of the dial feet at XI, I and VI.*

PLATE 181. *Oak case of the Shepley clock with iron strap hinges fitted externally, tiny caddy, and fine baluster pillars. Height 7ft.*

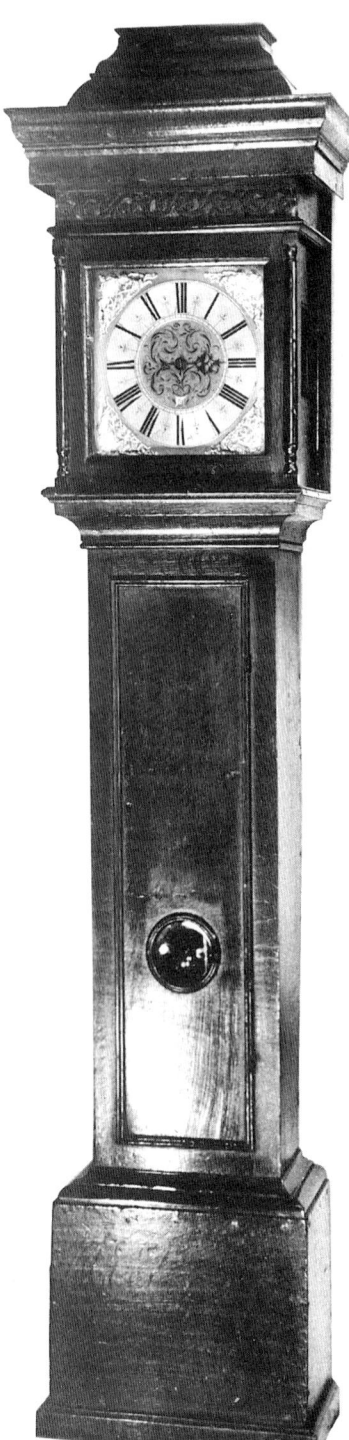

PLATE 182. *Ten-inch longcase dial also by John Shepley of Stockport, dating from a little after 1710, again with floral dial centre. Original blued steel hand.*

PLATE 183. *Oak case of the second Shepley clock standing about 7ft. This example is somewhat more sophisticated than the earlier one with good proportions helped by the caddy top.*

movements. The finely engraved dial centres are a development from the tulip theme (originating in the lantern clock). The spandrels are of the twin-cherub style, the crossed maces with large crown usually being popular a little after the earlier, small crown form of that same spandrel.

The engraving of both dial centres and chapter rings is superb, the work of a professional, whether Shepley himself or an outside engraver. The chapter rings attach to the dials in a professional manner by three small feet. The removed chapter ring in Plate 180 shows three holes (at XII, IV and VIII) where these feet fit. It also shows how the dial feet ends are riveted over in such a position as to be concealed by the chapter ring itself, at XI, II, and VI. Both of these features imply forethought and indicate far better workmanship than riveted chapter rings (Plate 128) and dial feet ends showing in the dial centre (Plate 230).

Turning to the cases, both are in oak which has darkened with age and has a good patina. Both are clearly early in style but the nine-inch case (Plate 181) is obviously much simpler and cruder with iron strap hinges outside the door as opposed to inside on the other case, and no opening door to the hood, whereas the larger case has a normal hood door. The ten-inch case (Plate 183) is based more on the lines of the London case of the day, a bit more refined in proportion, balance and overall style than the other case. The smaller case still has the convex top-of-trunk mouldings, which have changed to concave on the later case, indicating the more up-to-date styling of the larger case. The later case too has hood side windows, London-style.

The clock by unrecorded maker Jones of Abingdon is a hooded wall clock with a seven-inch dial. This is difficult to date and could be as early as 1690, but might more conservatively be estimated about 1710. This is a single-handed, posted-movement timepiece with alarmwork, the bell being removed in Plate 184 to show the double-headed alarm hammer. The Tudor rose motif alarm disc and its early numbering are exactly as found on a lantern clock of the day. The spandrels too are in fact dolphin-pattern frets made for a lantern clock with their edges clipped to make them fit this tiny dial.

The maker of this clock was clearly accustomed to making lantern clocks as we can see from the photograph of the baseplate (Plate 185), which is the *unused* dialplate for a lantern clock. We can see that this dial sheet was never

PLATE 184. *Seven-inch dial of a hooded clock, a timepiece alarm, dating from about 1710 by Jones of Abingdon in Berkshire, the bell removed to show the double-headed alarm hammer. The spandrels are dolphin pattern lantern clock frets cut down to size. Original steel hand.*

used for a lantern clock as it has no central hole for the hand, the holes presently there being off-centre and being the chain holes for the drive train and the alarmwork of this hooded clock – the alarm chain is missing in the photograph.

There is more to it than that, however, as by examination of the engraved pattern of the chapter ring with those strangely stylised ring-and-dot fleur-de-lis markers, and of the lantern dial sheet now used as a baseplate, we can recognise this very distinctive engraving style as being identical with that used at certain periods by Walter Archer of Stow on the-Wold (see page 410). This means that either Archer engraved for or supplied Jones, or vice versa, or that each used the same engraver. In fact the chapter ring and alarm disc are both lantern clock parts, used to make a hooded clock. Why the already-engraved lantern clock dial sheet was used as a baseplate I cannot imagine, unless perhaps there was some error or slip in the engraving.

The ten-inch dial from a thirty-hour longcase clock signed John Waklin Fecit, a maker whose dates and workplace are unidentified, shows a very

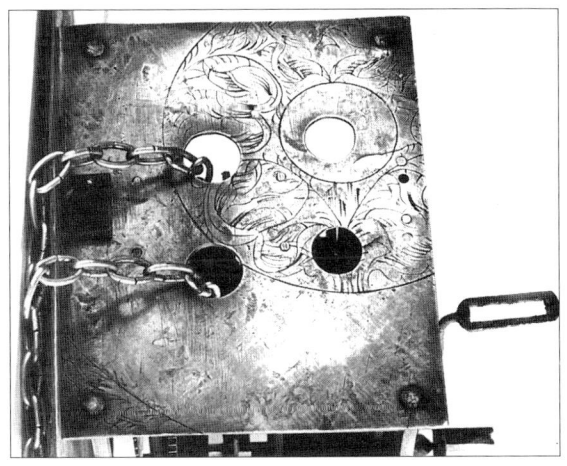

PLATE 185. *View of the baseplate of the posted movement of the Jones clock, which reveals this was the unused dial of a lantern clock. The engraving of this and the dial itself is recognisable as being by the same man who engraved Walter Archer's clocks (see page 410).*

PLATE 186. *Ten-inch longcase dial by John Waklin dated inside 1707. Superb engraved work with fine detail and Tudor rose centrepiece. Interesting original blued steel hand.*

PLATE 187. *Ten-inch two-handed longcase dial of about 1710 by Thomas Muddle of Rotherfield in Sussex with fine floral centre. Original steel hands.*

PLATE 188. *Ten-inch longcase dial of about 1710 by Richard Midgley of Halifax, with unusual floral centre based on the flowers-from-a-vase concept sometimes used in lantern clocks.*

PLATE 189. *Posted movement of the Midgley clock with brass posts and decorated hammer stop.*

PLATE 191. *Detail of the Midgley case showing seatboard nailed between the cheeks, which are weakened by being cut for a safety restraint.*

interestingly ornate centre busily engraved all-over in the manner of a lantern clock with a superb central Tudor rose. The dial (Plate 186) is reminiscent of the eight-day clock by John Worth of Humberstone in Plate 51. Attached fleur-de-lis form the half-hour markers with the earlier form of the twin cherub spandrel The style suggests a date of about 1700-1710, but the movement is actually dated 1707, a most unusual feature for any clock. The region of origin is unknown but the style suggests central or southern England.

Most early thirty-hour clocks we have examined so far were made in the provinces and the great bulk of them in rural areas, where clocks were still a novelty and the single-handed form was no doubt easiest to understand for those people previously unaccustomed to clocks of any sort. Two-handers were also made in these areas at this period but were very much in the minority.

The ten-inch dial shown in Plate 187 by Thomas Muddle of Rotherfield in Sussex is an early two-hander and dates from about 1710. The maker was born in 1670 and died in 1756. The chapter ring is very much akin to an eight-day ring of the day, here using meeting arrowhead half-hour markers and the same style for half-quarter markers. The all-over engraved centre is reminiscent of lantern clock centres and all the work is clearly by a professional engraver.

The movement is a conventional one of the birdcage type. The calendar is driven by a twenty-four hour wheel, as in an eight-day clock, a better system than the twelve-hourly knock-on method normally used on thirty-hour clocks, but more costly to produce and therefore not often found on thirty-hour examples. The original steel hands are based on London styles of the day.

The ten-inch solid-sheet dial of the longcase clock signed 'Richard Midgley Fecit' (Plate 188) is finely engraved in the floral centre style, originating from the lantern clock tulip theme, and dates from about 1700-1710. This maker worked at Ripponden near Halifax in West Yorkshire, though details of his life are vague. Half-hour markers are a variation on the meeting arrowhead motif. The steel hand appears original, though is unusual in having no 'tail'. The posted movement has brass corner posts and a decorated hammer spring stop-piece.

PLATE 190. *Oak case of the Midgley clock, unusually slim. Fine baluster pillars, heavy topmould. Height about 7ft.3in.*

PLATE 192. *Dial by Simon Worley of Staverton c.1710. The random matting grain shows.*

PLATE 193. *Posted movement of the Simon Worley clock, the iron posts fastened below with nuts, leaving shaped 'feet' projecting, along the lines of the Knibb clock in Plate 167.*

The oak case (Plate 190) is unusually long and slender, standing about 7ft.3in. Early characteristics include the heavy overhang of the hood top-mould, the D-mould round the door, superb baluster hood pillars attached to the door (as three-quarter pillars to the door and quarters to the rear). The concave top-of-trunk mould is a repair and might originally have been convex. The flat top to the hood door is normal in all areas and is not at this early stage an indicator of region.

The case incorporates one of those errors sometimes seen in prototype casework in that it was designed without any hood restraint, so that when the hood door was opened the hood was in danger of falling forward. The casemaker evidently had not thought of this problem and at the last minute cut a groove into the seatboard upstands (often called the cheeks) to serve as a restraint by accepting the inside lower hood mould. This left the seatboard upstands (here the case side extensions) unduly weak and it is probably for this reason that the seatboard is nailed between the uprights (Plate 191) rather than resting on top of them, as was usual practice. Such an interesting error is indicative of the fact that we are still in the infancy of provincial casemaking.

Simon Worley worked at Staverton in Wiltshire, though little is known of his working life. The clock featured in Plates 192 and 193 dates from about 1710, the dial being very conventional and signed 'Simon Worley Staverton fecit' (sometimes written as Starton). The interesting aspect is that the posted movement has nuts holding the baseplate to the square iron posts (in the manner of the John Knibb clock in Plate 167) and the lower threaded post extensions are turned to form 'feet'.

The eleven-inch dial (Plate 194) by James Brewer of Darlestone, near Stone in Staffordshire, dates from about 1710-20, in which latter year the maker died. The male maskhead spandrel has a sort of headdress and is sometimes known as the (Red) Indian Head pattern. There are several variations on this spandrel theme and it was used by Tompion, amongst others, from about 1700. The half-hour marker is a simple form of attached (to the quarters ring) fleur-de-lis. The dial centre has a background which is matted, fairly crudely

PLATE 194. *Eleven-inch dial of about 1710 by James Brewer of Darlestone with floral engraved design worked on to the matting, the beginnings of a north-western style.*

PLATE 195. *The oak case of the James Brewer clock has several early features indicative of the period: lenticle glass, baluster pillars, hood side windows, cushion mould below hood topmould, double-D mould to door.*

and irregularly, on to which is worked primitive floral engraving in a symmetrical pattern, the relatively inexpert execution of which suggests that this was done by the clockmaker himself. All-over floral engraving on a dial centre had so far normally been on to a polished ground, as with the lantern clock from which it developed. This dial breaks from that tradition by engraving florally on to a *matted* ground, and indicates the beginnings of a style which was to become particularly popular in north-west England. The movement is a conventional one of plated form, which became the normal type in the North from now on.

The original oak case (restored in the base) is of very slender and pleasing style, clearly provincial yet with a hint of the London taste about it (Plate 195). Hood pillars are of handsome baluster form and attached to the door, often known as 'integral' pillars. Hood side windows show a certain attempt at sophistication. A cushion mould below the topmost hood mould also has a hint of London influence to it. The lenticle glass in the trunk door is oval, a nice variation. The double-D mould around the door is an alternative sometimes used from now on instead of the earlier and more usual single D-mould, and therefore a sign of progression of style. The height is about 6ft.8in.

The ten-inch (solid) dial longcase by Joseph Norris of Abingdon, Berkshire, dates from about 1710-20. This maker was the brother of Edward Norris and went with him as part of the Fromanteel work-force to Amsterdam in the 1670s, returning to England to his birthplace, Abingdon, between 1692 and 1696. Under Fromanteel he worked on some of the finest clocks ever made, yet when he 'retired' to Abingdon, where he died in 1727, he made simple country thirty-hour clocks such as this one.

The dial (Plate 196) uses up-to-date London features such as male maskhead spandrels, matted centre without decoration, 'floating' half-hour markers of the fleur-de-lis style, i.e. detached from the quarters band. The hand is a modern replacement. The movement is conventional, but well made (Plate 197). His dial fastens by true dial feet, a much better system than the lug-and-pin method so often used and derived from lantern clock times. The stripped

151

PLATE 196. *Ten-inch dial of a longcase clock of about 1710-20 by Joseph Norris of Abingdon, a former London maker now making country clocks. Note the graining of the random matting.*

pine case (Plate 198) is simple yet elegant and was probably originally ebonised or perhaps painted or lacquered.

The case of the clock by John Peck of Bolnhurst, Bedfordshire, is not dissimilar from the Norris case, being in stripped pine and of the same general style (Plate 201). Both stand about 6ft.5in. The ten-inch (solid) dial of the Peck clock (Plate 199), dating about 1720-25, is of the same nature as that by

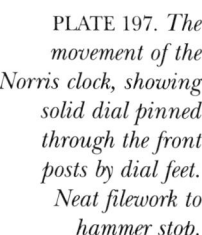

PLATE 197. *The movement of the Norris clock, showing solid dial pinned through the front posts by dial feet. Neat filework to hammer stop.*

PLATE 198. *Simple pine case of the Norris clock standing about 6ft.3in with original sound fret.*

PLATE 199. *Very fine ten-inch dial from a longcase by John Peck of Bolnhurst dating from about 1720-25 but looking deceptively earlier. Engraving between spandrels is unusual on thirty-hour clocks. Absence of calendar is a typical southern feature.*

Norris, both showing the undecorated matted centre influence from the London eight-day clock of the day. Again Peck uses the new 'floating' fleur-de-lis half-hour marker, though he still uses the now slightly old-fashioned twin cherub spandrel of the small crown (earlier) pattern. Floral engraving between the spandrels on all four edges of the dial is a pleasant and unexpected extra touch more normally found on eight-day work of a slightly earlier period. The simple blued steel hand is believed original.

Interestingly, the solid-sheet dial bears an engraved legend on its back (Plate 200) – 'Saml Bridger Clock Engraver Oldbily London', 'Oldbily' being his attempt at spelling Old Bailey, a district of London, pronounced in a Cockney (London) accent as Byley. Very rarely is an engraver's name recorded and this helps to date the clock as we know Samuel Bridger went bankrupt in 1729. It is interesting, too, that Peck had his engraving done as far away as London and that Bridger engraved the whole dial, not just the chapter ring, and also presumably carried out the fine matting work in the dial centre. The quality and evenness of such matting is apparent when contrasted with more feeble attempts such as Plates 204 and 249.

PLATE 200. *Back of the solid sheet dial of the Peck clock showing the London engraver's name. The engraving is expertly done and indicates that he was using this blank space for practice work. Note square-headed spandrel screws.*

PLATE 201 *Pine case of the Peck clock standing 6ft.5in., very similar in style to the Norris case in Plate 198.*

PLATE 202. *Ten-inch dial of a longcase clock of about 1700-1710 by Joseph Pickering of Lutterworth, No. 92. The original steel hand is exceptionally sturdy and finely sculpted. Random matting grain.*

PLATE 203. *Movement of the Pickering clock showing solid dial sheet attached by the sensible method of screwing into the top plate. Decorated hammer stop with a 'bird-beak' lip.*

The ten-inch solid dial of the thirty-hour longcase clock by Joseph Pickering of Lutterworth in Leicestershire dates from about 1700-1710 (Plate 202). The original single hand is of iron (we usually refer to it as steel) and is exceptionally sturdy and of an interesting pattern. The half-hour markers are of the meeting arrowhead style and both this and the twin cherub spandrel pattern suggest a date at the beginning of the new century. The spandrels on this clock still bear their original (mercurial) gilding and it is surprising that the maker took the trouble to gild spandrels on a thirty-hour clock in the first place and that this has not been worn away with polishing over such a long period. No positive dates are known for this maker, but other clocks are recorded dating from a similar period.

The decorationless matted centre with no calendar suggests a styling affinity with the South, though Leicestershire is a Midlands county. This dial bears the serial number 'No 92' below the chapter ring on the main dial sheet. Number 151 is also known. Clock numbering was a practice in which a few provincial makers indulged, but seldom as early as this and this maker must be one of the first to have done so.

The posted movement (Plate 203) is sturdily built with some attractively decorative filing to the hammer spring. Note that the dial attaches to the movement by flanges at the top and a lug at the bottom, lantern clock fashion. However, four small holes can be seen above the top flange where the maker had originally planned to position them and then, realising his error, he fitted them in the present position. These sort of alterations are quite often seen in clocks and show us that the makers were human and quite capable of making mistakes.

The dial by Peter Hathornthwaite (sometimes spelled Haythornthwaite or Hawthornthwaite and abbreviated for convenience to Hawthorn by a later member of the family) dates from about 1710-20. The maker worked at Kirkby Lonsdale in Westmorland. This clock (Plate 204) was made as a single-hander, as is apparent from the single-handed calibration on the chapter ring, and was

PLATE 204. *Single-handed longcase dial of about 1710-20 by Peter Hathornthwaite of Kirkby Lonsdale, the movement later modified to carry two hands. Original steel (hour) hand, much later minute hand.*

PLATE 205. *Ten-inch longcase dial of about 1700-1710 by John Richardson of Malton with original but unfaceted steel hand. Very obvious circular matting grain. The three roundels are simply to break up the otherwise plain surface.*

later modified to carry two hands. The hour hand appears to be the original single hand, but the minute hand is a mid-nineteenth century hand from a painted dial longcase. The signature still includes the word 'Fecit', which by now is a little antiquated but which did continue in fashion with some makers for another generation or more. Meeting arrowhead half-hour markers are small and unexciting and very conventional. The matting is random and not improved by the fact that the photograph shows the clock in dirty condition. Rivets hold the chapter ring at XII and VI.

The calendar box above VI is shaped and its border engraved just as was the practice with eight-day clocks of the time. Many thirty-hour clocks hitherto were made without calendars, especially, it seems, those from central and southern England. Eight-day clocks almost always did have them. Of course, the clockmaker would make a calendar or not at the wish of the customer, but up to this point northern clocks often have them and southern clocks do not. From about 1730 calendars become increasingly common on thirty-hour clocks from all regions.

The spandrel pattern on this clock is very up to date as this new style, being one version of several showing a female head, came in just at the beginning of the new century and remained popular in varying forms for about forty years. It is therefore a difficult pattern from which to pinpoint dating.

The ten-inch dial by I. Richardson of Malton, North Yorkshire is from a plated thirty-hour movement, with the top two pillars of brass and the lower two of iron – a system sometimes used for cheapness. The top two pillars were more visible, the lower two less so. The letter I with a crossbar is a fashion of the day for J, the maker being John Richardson, about whom no biographical facts seem to be known. This clock is a little difficult to date precisely but probably dates about 1700-1710.

The circular 'grain' of the matting shows clearly in Plate 205. The three ringed holes in the dial centre are found on some dials which would otherwise have an undecorated matt centre and are thought to be to break up the plainness, often positioned where calendar and winding holes might be on an eight-day clock, though in this instance they don't quite fit that configuration. The steel hand could well be original, though it lacks the usual faceting and bevelling to the face surface.

The chapter ring has rather strange half-hour markers, a sort of cross between a trident and a fleur-de-lis. The naïvety of this dial might tempt us into dating it too early and it could well have been made in the 1720s or even later.

The two-handed dial shown in Plate 206 is from a hook-and-spike wall clock of the zigzag type as made by a known group of Oxfordshire Quakers – see page 134 for a fuller account of this type.

The spandrels are of lead, which was used occasionally for cheapness, being much less expensive than the conventional brass. Spandrels of pewter and occasionally of tin were also sometimes used for the same reason, but are uncommon.

Here the dial centre has six concentric rings of zigzags, done with an engraving tool in a guider by someone incapable of true freehand engraving. The half-hour markers are primitive asterisks. Unusually the quarter-hours are *not* marked on the inner chapter ring edge, only halves. Minute numerals are very small in size at this early period and the engraving here suggests a hesitant hand, less confident than that of a professional engraver (compare the Sadler dial in Plate 212). The date is probably about 1710-20, but again the primitive style may

PLATE 206. *Anonymous two-handed hook-and-spike wall clock of the Oxfordshire Quaker zigzag centre type. This dial sheet is copper rather than brass. Original steel hands.*

PLATE 207. *Longcase dial of about 1710 by Edmund Bullock of Ellesmere, No. 208, nine and a half inches square. Very coarse matting. Original steel hand.*

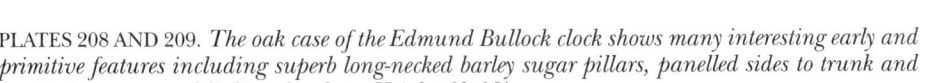

PLATES 208 AND 209. *The oak case of the Edmund Bullock clock shows many interesting early and primitive features including superb long-necked barley sugar pillars, panelled sides to trunk and hood, strange carved linking brackets. Height 6ft.10in.*

mislead us into too early a date and it might have been made some years later.

The nine-and-a-half-inch dial by Edmund Bullock of Ellesmere, Shropshire in Plate 207 dates from about 1710 or a little later. The maker was working from about 1703 and died in 1734. Naïve fleur-de-lis half-hour markers are almost of the old trident form. Spandrels are a form of early cherub head. The matting is random and the multi-looped iron hand is original. This maker numbered his clocks on the dials, this being No. 208. I know of examples numbered between 133 and 417.

The original oak case of this clock (Plates 208 and 209) is charmingly primitive with long-necked barley sugar pillars of unusual free-standing nature and strange carved ear-pieces attached to the top sides of the trunk, making the hood join reminiscent of a hooded wall clock. It stands 6ft.10in. There is no hood door so the hood must be removed to set the time. Beading on the trunk sides is an odd feature, made to match the left-handed door in the same way that the hood side beadings match the beaded glass surround.

PLATE 210. *Ten-inch longcase dial by Richard Bullock of Ellesmere, No. 27, c.1735. Unusual dial centre with fan pattern engraved lines.*

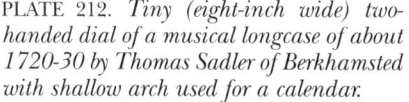

PLATE 212. *Tiny (eight-inch wide) two-handed dial of a musical longcase of about 1720-30 by Thomas Sadler of Berkhamsted with shallow arch used for a calendar.*

PLATE 211. *Primitive oak case of the Richard Bullock clock. No pillars. Ogee top-of-trunk mould.*

The ten-inch dial signed 'Richd. Bullock, Ellesmere, 27' (Plate 210) is from a longcase clock by the son of Edmund Bullock. The meeting arrowhead half-hour markers might suggest an early period, but the female head spandrels indicate a little later. This maker, like his father, numbered his clocks and I know of numbers between 27 and 424. This clock is clearly one of his earliest but, as he was not born till 1719 and is believed to have taken over from his late father about 1735, this would date the clock about 1735-6. It is positioned here as an indication of how stylistics alone can be misleading. The unmatted centre is decorated by straight ruled lines fanning out from the centre, suggesting this was done by the maker himself, whereas the crisply-engraved chapter ring looks like professional work.

The original oak case (Plate 211) is primitive and has several very early features, including the half-round door beading, under which the wrought-iron hinges are applied externally. The hood and trunk sides overlap the hood door and trunk framework, producing an effect like a thin pillar.

The dial of the thirty-hour clock by Thomas Sadler of Berkhamsted, Hertfordshire (Plate 212) is eight inches wide and is an example of an early arch; the arch itself is shallow, considerably less than the semi-circle which later became standard. The earliest examples of arches were often either shallower or higher than standard or had a span which was either narrower or wider than standard, or both.

The arched dial came into fashion sooner on eight-day clocks than on thirty-hour examples, probably because they cost more and a thirty-hour clock would not bear that extra cost. Throughout their entire history the great majority of thirty-hour clocks had square dials, though some later ones were round. When early examples of arched dials *were* used on thirty-hour clocks it was often because they were special clocks, as here.

Clockmakers were still experimenting with this newly available arch space, both as to its size, proportion and function. Here it is used for a calendar, something seldom seen in that position on a thirty-hour clock. But then this is

PLATE 213. *Movement of the Sadler clock from behind showing musical train set parallel to the dial (to avoid massive movement and case depth). Six bells with eleven hammers.*

no ordinary thirty-hour clock, but a musical one, playing every four hours three verses of either of two tunes on six bells, the tune being selected by means of a lever below the dial edge at seven o'clock.

The early two-handed dial has a London influenced undecorated matted centre, floating diamond half-hour markers, small minute numerals and sphinx head (female bust) spandrels, all of which suggest a date of about 1720-30. The enormous movement (Plate 213) is of the posted type measuring seven and a half inches deep. Going and strike trains are positioned normally, but the musical train is placed at the back with the pin-barrel parallel to the dial and its wheelwork therefore swung round at right-angles to the dial. It is very sturdily made, which is essential to prevent such a deep movement from flexing and distorting with its own weight, a fault to which posted movements are more prone than plated ones.

The tiny case is of red (Virginia) walnut and stands only 6ft.9in. This wood looks a little like mahogany but tended to be more widely available earlier than

PLATE 214. *The case of the Sadler clock stands 6ft. 9in and is of red walnut with fine original sound fret and original caddy top.*

PLATE 215. *The side windows of the Sadler hood allow a full height view of the interesting mechanism.*

mahogany was. It has its original caddy top – many clocks have had their caddies removed to reduce height (Plate 214). The case is simple and plain in style and deep relative to its narrow width (ten inches wide, nine inches deep), a deep case being essential for any posted movement but especially so with this extraordinarily complex one.

The hood has large side windows and with a movement of this highly complicated nature we can more easily understand how such a view into the 'works' could have been of interest to an owner. Note how the window is positioned exactly to offer a view of the movement (Plate 215).

The eleven-inch single-handed longcase clock by Thomas Clark of Warrington, Lancashire, was made in 1716 (being dated on the case). Little is known about the maker except that he was working there till at least 1736. Early period indicators on the dial (Plate 216) are: the *circular* calendar box (circular boxes, when they do appear, tend to be early); twin cherub spandrels; anchored half-hour marker of the fleur-de-lis style, the marker here being of the bolder type we associate with northern work; small area of engraving around the calendar box; fine early pattern steel hand.

The dial is cast in the cartwheel fashion, typical of northern dials from this time onwards.

PLATE 217. *Movement of the Thomas Clark clock showing cartwheel cast dial, offset Westmorland style calendar wheel, strange external single spoke bellstrap and dial-to-top-plate fixing by double pins through lugs.*

The movement (Plate 217) is of the posted type with rectangular brass posts, an unusual and almost always early feature on a northern clock. This movement has most unusual brass top 'finials' (in the lantern clock tradition) holding the top-plate in place. Also very unusual, perhaps unique even, is the iron bellstrap which fits *outside* the bell, rather like the single spoke of a four-legged lantern clock bell cage. This strange bellstrap is a good idea as it leaves more hammer space inside the bell and it is surprising that other clockmakers did not think of this system or use it. Two lugs on the upper dial pin into the top-plate by means of *two* pins in each rather than the normal one, a sensible safety measure. Clearly this is the work of a clockmaker who thought for himself.

PLATE 216. *Eleven-inch dial of a longcase clock of 1716 by Thomas Clark of Warrington, seen here in rather dirty condition. Bold half-hour markers. Primitive engraving to datebox.*

The oak case (Plate 218) is a rustic affair with several primitive and early features – wrought iron 'butterfly' hinges fitting externally, D-mould to the door, interesting attached three-quarter pillars to the hood door which too is iron-hinged. The hood slides into place, but certain aspects suggest this was made by a carpenter still new to clock casemaking, an interesting indication that we are still in the infancy of provincial clock casemaking.

First the lower hood mould has no protruding lip to set it firmly into the 'home' position when fully assembled, so that the hood sits as far on as you happen to push it – push it too far home and the glass fouls the hand (see Plate 219). Secondly there is no means of holding the hood in place (no hood restraint slats), so that when the door is opened the hood is in danger of falling forward. The casemaker has not left himself any space to fit such slats as the inner hood side timbers fit flush up to the upper case side extensions which form the seatboard 'cheeks'. This is the same omission we saw on the Richard Midgley case and is not unusual with carpenters still unfamiliar with the new skill of casemaking. The cabinetmaker, or some later owner, has tried to rectify this problem by fitting a small turnbuckle behind each side of the backboard, locking into the hood sides, but this solution is barely satisfactory.

Most interesting of all is that fact that the hood fret is in the form of initials and the date of making – RSE 1716 (Plate 219). Very seldom are clocks dated and, where they are, it is hardly ever on the case. In fact dates carved into cases are notoriously unreliable, but here this is original work. The form of initials is that done for husband and wife when the clock was made for such an

PLATES 218 AND 219. *The oak case of the Thomas Clark clock dated 1716 in the fret with the initials of the first owners. Height about 6ft.5in. Slender turned integral hood pillars.*

PLATE 220. *Eleven-inch single-handed longcase dial by Andrew Knowles of Bolton dating in the period 1720-30 and showing three extra divisions per quarter hour, therefore giving five-minute indication, set on the outer edge of the chapter ring for clarity. Original brass hand, unusually without a tail.*

PLATE 221. *Ten-inch longcase dial dated 1723 by Robert Davis of Burnley with fine tulip-based centre.*

PLATE 222. *Hood of the oak case of the Robert Davis clock showing crude sound fret, carved rather than fret-sawn.*

occasion as a wedding (sometimes for an anniversary) and this is almost certainly a 'wedding clock', perhaps bought as a wedding present. The central initial is that of the surname, the first being the forename of the man and the third that of his wife. We do not know for whom this clock was made but the initials would be a contraction for a couple such as Robert and Ellen Smith, followed by the date of the marriage. When the clock was made by a village maker for a local couple it is sometimes possible to trace the actual marriage entry in the parish registers, as with luck only one marriage would appear in that year with the right configuration of initials. In a town the size of Warrington such a search is less practical.

A similar practice undertaken by a few clockmakers was to engrave the full names of husband and wife on the clock dial as first owners together with the year. The one clockmaker who excelled at this was William Porthouse of Penrith and on investigating some twenty odd instances of such names on Porthouse clocks the great majority proved to be in commemoration of wedding years. Occasionally a single name on the dial is that of the first purchaser.

The eleven-inch dial of a single-handed longcase clock by Andrew Knowles, who worked at Bolton in Lancashire, shows a feature not seen on previous examples (Plate 220). The quarter-hour units on the chapter ring are sub-divided into threes, thus giving a five-minute reading facility. This was a calibration used by some makers of single-handed clocks on an occasional basis, perhaps when a customer requested it. The dial is cast in cartwheel fashion which by this date we expect from a northern maker.

This clock probably dates between about 1720 and 1730, though it might just be earlier. The maker is known to have been there in the 1724-38 period. The matted centre is so worn from polishing that the matting is now very thin and barely shows in the photograph. The spandrel is one of the many forms of female head. The fleur-de-lis half-hour marker is here a floating one,

PLATE 223. *Eleven-inch two-handed longcase dial dated 1728 by John Belling of Bodmin. Most unusual centre with quadrants in a form of zigzag work surrounded by cup-and-ring turnings.*

PLATE 224. *Ten-inch longcase dial by John Lucas of Newbury c.1735. Typical calendarless southern dial with floating half-hour markers.*

developing into the larger, bolder type associated with northern styling.

Ringed 'winding holes' give a superficial eight-day appearance, probably done initially to break up the otherwise rather plain centre, but as time went by these developed into a deliberate attempt at simulating eight-day appearance. The engraving around the date box adds to the decorative effect of the centre, so that even though this dial is of the matted centre type, the dial centre has far more going on than with a similar period southern clock such as the Lucas example in Plate 224.

The original hand has no tail (like Richard Midgley's in Plate 188), which is an unusual quirk of some makers, and this hand is made of cast brass. Brass hands were used on brass dial clocks by a few makers principally in Lancashire and West Yorkshire in the period roughly between 1720 and 1770. They tend to be in the same style as the more usual steel hands of the same period. The engraving is bold and confident suggesting professional work. The movement is of the plated type and is unremarkable.

The single-handed plated-movement thirty-hour longcase clock by Robert Davis of Burnley in Lancashire is dated 1723 on its ten-inch dial (Plate 221). Dated dials are unusual but ideal evidence for the study of dial style development. The finely engraved centre is based on the tulip theme of lantern clock origin and this and the chapter ring are most competently done by an experienced engraver – though note the slip of the tool at III. The fleur-de-lis half-hour marker is still anchored to the quarters ring by a stalk, in contrast to the more modern style of 'floating' half-hour marker by now fashionable in southern clocks of this time and is bolder in execution than a southern one. The spandrels are of the later twin cherub type with the large crown. The original hand is of steel and is blued.

The case is of oak and has a heaviness about the hood (especially the crudely-carved fret) characteristic of many early north-country cases (Plate 222). The door top is shaped into an arch, marking the beginnings of a stylistic feature soon to become a regular characteristic of northern work – the shaped door top.

The eleven-inch dial of a posted-movement two-handed thirty-hour longcase

PLATE 225. *Posted movement of the Lucas clock showing the solid sheet dial and plain springwork.*

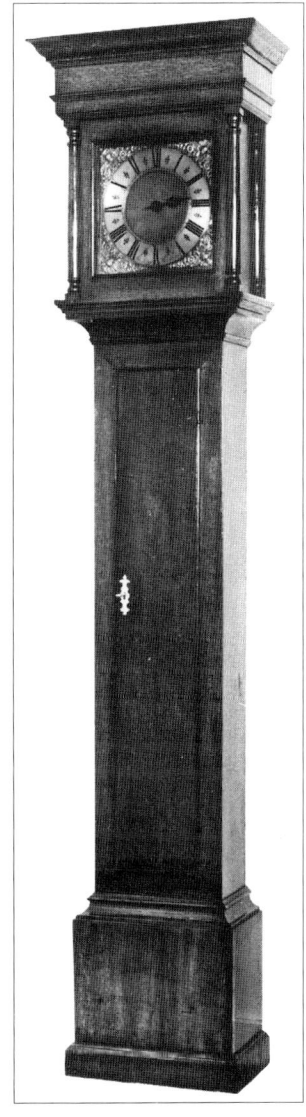

PLATE 226. *The Lucas oak case is plain and simple with integral pillars and southern mouldings. About 6ft.6in.*

clock by John Belling of Bodmin in Cornwall (Plate 223) is also dated – 1728. Interestingly, the 172 numerals seem strongly engraved whilst the 8 is weak and less confident. The dial sheet is decorated with cup-and-ring turns and with a strange form of zigzag engraved pattern formed into four quadrants, all work which was capable of being done by someone who could not engrave. On the other hand, the chapter ring is clearly engraved by the competent and probably different hand of a professional engraver. This makes me wonder whether the maker bought his chapter ring work in 'dated' 172 and added the final year number himself. This impression is strengthened by the fact that the only freehand engraving in the centre is a rather feeble little decoration above the calendar box.

The half-hour markers are a variation of the meeting arrowhead theme, perhaps a little old-fashioned by now. The spandrels are a form of female head pattern popular at the time but are cast in *tin*, which was readily available locally and cheaper than brass. The calendar is of the twenty-four hour wheel drive type, as on an eight-day clock, and it is interesting that Belling took this trouble, when a northern maker would have used the wheel-less twelve-hour knock-on system which was cheaper to make. John Belling was born in 1685 and worked till the 1750s, when his son, John II, succeeded him.

The single-handed longcase clock by John Lucas of Newbury, Berkshire, has a solid sheet dial and dates from about 1735 (Plates 224 to 226). This maker finished his apprenticeship in 1732 and died in 1750. The dial size is still small at ten inches; by now a northern dial would be larger. The dial centre is matted

PLATES 227 AND 228. *Oak-cased clock of the 1730s by John Fielder of Alton showing many similarities with the Lucas case in Plate 226. The lenticle glass is here retained a little later than usual.*

with no calendar or decoration of any type, another regional indicator, though not an absolutely firm one. The spandrel is a cherub head with two flowerheads, a pattern favoured by the famous London maker Thomas Tompion and used by a few other southern makers in the early eighteenth century, but a pattern not usually seen on northern clocks. This might be because this pattern would be too small for the eleven-inch (or larger) dial then in thirty-hour fashion in the North.

Fleur-de-lis half-hour markers are of the 'floating' type and are smaller and more restrained than a northern equivalent. The chapter ring is finely engraved, indicating professional work. The original blued steel hand is decorative but also restrained and based on a London style. The movement is of the birdcage type with flat bar posts and is well made but without the ornateness of earlier clocks.

The difference in style and construction between southern and northern clocks is by this time becoming increasingly apparent, though not unfailingly evident. Dial size, type of dial decoration and movement construction are three areas where this difference is usually to be seen. The oak case of the Fielder clock (Plates 227 and 228) is very similar to that of the Lucas clock (Plate 226), both being very simple and slender and standing about 6ft.6in. They display indications of southern styling in the slimness (produced by the small dial?), simplicity of line, simple mouldings, exceptionally long trunk and door.

Both these cases have an elegance of proportion based on the basic London case shape of the day, with which carpenters in these southern counties would be familiar. By comparison, the north-country thirty-hour cases of this time are often much more rustic, with less well-balanced proportions, in which lies

PLATE 229. *Single-handed longcase of about 1740 by Francis Pile of Honiton, the matted dial centre having more engraved work than usual in this region – a sun face, butterflies and a bird.*

much of their charm and interest. This is probably because familiarity with London case styling decreased in proportion with increasing distance from London – in these early times, that is.

The clock by John Fielder of Alton, Hampshire, dates from the 1730s. The dial has many similarities of style with the Lucas clock, involving an overall simplicity of style, although here having a circular calendar box with a little engraving around it which incorporates a basket of fruit and/or flowers.

This engraved feature formed the basis of a whole series of dial centre designs, popular for half a century or so from about 1710 to about 1770, incorporating a basket of fruit and often one or more birds, these usually flying and each frequently carrying a branch in its beak. This 'birds and basket' theme presumably portrayed the dove as a bird of peace and the basket as a symbol of plenty, the combination representing 'peace and plenty'. It was a very popular theme on matted dial centres over a wide area of central and southern England, though less often seen in the North.

The eleven-inch single-handed thirty-hour longcase clock by Francis Pile of Honiton in Devon (Plate 229) dates from about 1740. The maker is recorded there by 1731 and died in 1763. The movement is of the posted type with decorative brass corner posts, lantern-clock fashion. The dial has small and restrained floating half-hour markers, based on the London style. However, the calendarless dial centre has a particularly large and splendid sun motif below XII, which seems to be a whim of the particular maker or engraver. It is further engraved with two butterflies and two birds carrying branches, an extension of the birds-and-baskets concept, making it overall a very much busier dial centre than a London one. The blued steel hand is original and of a pleasing pattern not remote from London style.

167

PLATE 230. *Eleven-inch longcase dial of about 1735 by Hampson of Wrexham with original* brass *hand. Note the random grain of the irregular matting.*

PLATE 231. *Very primitive pillarless pine case of the Hampson clock with ogee mouldings and late use of a lenticle glass. Height 6ft 7in*

The oak case stands about 6ft.6in. The integral pillars have a pleasing swell to them and suggest southern styling as by now northern pillars are usually free-standing. The flat door top is also a southern feature. The cloth-backed sound-fret is not indicative of any particular region, nor is the broad topmould which is more an indicator of period than region.

The eleven-inch dial of the Hampson of Wrexham (then in Denbighshire, North Wales) thirty-hour clock appears to date from about 1735 (Plate 230). Here the matting tool marks can be seen clearly in the irregularly matted dial centre. Two of the dial feet ends can be seen at XII and VII when with more careful planning the maker could have concealed these behind the chapter ring, as he has with the third foot which must be somewhere near the V numeral. The circular calendar box has engraving spreading some considerable way from it as far as the hand aperture, a style which later developed more fully in this region to spread throughout the dial centre over the matted ground. The hand is of cast brass – compare that on the Andrew Knowles clock in Plate 220.

The female head spandrel is one of several versions used around this period. The chapter ring is well engraved using a fleur-de-lis half-hour marker with a hint of the earlier trident form to it. The clock is numbered 907, numbering being a regular trait of this maker. Clocks by him with the following numbers are also known: 16, 254, 504, 700, 910 (dated 1736), 936, 962, 1027 (believed dated 1744), 1146, 1152, 1284, 1360, 1439, 1538 and 1563. This maker is known to have been working in Wrexham by 1718 and to have died in 1754 and these are believed to be the actual serial numbers of the clocks he made, though some researchers doubt whether he kept to a strict chronology. If his working life spanned a period of fifty years the production rate works out at just over thirty clocks a year, not such an impossible task as these very high numbers may first suggest.

The case (Plate 231) is in stripped pine, stands 6ft.7in. and has an unusually shallow base, which may imply that the clock has been cut down. However, a

PLATES 232 AND 233. *Two single-handed longcase dials by Samuel Whit(t)aker of Middleton dating from about 1710 and 1735 respectively. Several retained features and several changed. Identical original iron hands.*

number of early primitive cases do seem to have had a much shallower base than we now regard as normal and a base of the more usual proportions would have raised the height to something like 7ft.6in., unusually tall for a thirty-hour clock.

The hood door has no pillars, a feature which gives the clock a very early look. However some primitive cases were made without pillars well into mid-century, so this is not a reliable dating feature. The ogee top-of-trunk mould may look misleadingly early, resembling as it does the convex mould of the late seventeenth century. The hood topmould is strangely bland and is more a feature of the primitiveness than age. The lenticle glass to the door and the D-mould around it are more reliable dating features.

Two single-handed thirty-hour longcase dials by Samuel Whittaker of Middleton, near Manchester, are pictured together in Plates 232 and 233 for comparison. Samuel Whittaker's birth year is unknown, but his elder brother, James, also a clockmaker, was born in 1666 and died in 1720. Samuel lived on until 1746, a bachelor. The first clock dates from about 1710, the second perhaps 1735-40. What is interesting is that in that period Samuel has changed his style in some respects and not in others.

The matted dial centre with a calendar box surrounded by engraving is very similar in both clocks. So too is the female head spandrel pattern and, surprisingly, the single iron hand, which is of the same unusual styling as used much earlier by his brother James. The chapter ring, however, is very different, the earlier (Plate 232) having meeting arrowhead half-hour markers and the later (Plate 233) a form of bold anchored fleur-de-lis. The signatures vary too from the heavy block lettering single T spelling of the earlier to the cursive flowing script of the double T spelling of the second. This suggests the two chapter rings were engraved out of house, probably by two different engravers each familiar with the up-to-date styling of the day, whilst the dialwork was done by Samuel himself and remained unchanged as if stuck in a timewarp.

I have no details of the movements but it is known that brother James used plated movements with latched pillars (cf. the John Hough clock in Plate 177),

PLATE 235. *Two-handed longcase of the 1740s by Thomas Moore of Ipswich. The pitched pediment is one form of hood finish, more often seen later in the century. The diamond-shaped inlays are very much a regional feature.*

PLATE 234. *The oak case of the second Whittaker clock stands 6ft.11in. and has walnut crossbanding. The carving to the door may be later.*

a surprising technique for an early provincial maker.

The oak case of the later clock (Plate 234) is crossbanded in walnut, which was often used in this way until mahogany took over as the normal crossbanding wood just after mid-century. The original purpose of crossbanding was for strength, to present an end grain to the edges of all panels, doors, etc., as this stood up to wear and tear better than the long grain, which was prone to splitting. However, the crossbanded feature very soon became a decorative feature rather than a functional one.

Many features of this case style now display characteristics which are north-western rather than general. These include the ogee top-of-trunk mould; the free-standing hood pillars, which in the South were still attached to the door; the half-round rear hood pillars, in this instance attached to flat splats; the shaped door top, whereas southern door tops usually remained flat for longer; the complex and deeply recessed trunk-to-base mould; the caddy top, which was far more common here than in the South.

The pierced cloth-backed fret to the hood appears original and of respectable design. The upper frontal areas of the case, however, are carved. Carved

PLATE 236. *Carved date stone dated 1760 from Birchenlee Farm, Great Marsden, near Colne, Lancashire, home of John Spencer and his wife, Sarah. The typical clock dial of the day is still shown as a single-hander.*

cases are a difficult subject, as very many were carved years after construction. In fact some authorities take the view that *all* carved cases were carved later. At this juncture it is perhaps simplest to say that the carving *may* be later.

The thirty-hour two-handed longcase clock by Thomas Moore of Ipswich (Plate 235) dates from the 1740s. The plain (matted), calendarless dial centre is typically south-eastern. The oak case retains from earlier periods the integral pillars and flat-top door. The pitched pediment, however, is a relatively 'modern' feature, unusual as early as this (and more commonly seen in the north-west in the 1780s), but a style which could crop up at any time from now on. Pitched pediments often have dentil moulds, sometimes known as dog-tooth moulds, running beneath the upper lip, as here. The centre finial is missing.

The face of the uppermost horizontal mould is inlaid with diamonds of alternate woods, a time-consuming and attractive feature which seems to be limited to casework of the later eighteenth and early nineteenth century in East Anglia (Norfolk, Suffolk and Essex). This can also appear on the curved mouldings on arched dial casework.

The thirty-hour clocks we have so far seen have been mostly single-handers, the most numerous type made before the middle of the century in all provincial regions. The great majority were made in what were then rural areas; before mid-century town and city clockmakers were the ones more likely to make two-handed thirty-hour clocks (or, of course, eight-day ones). There were occasional clockmakers whose work ran contrary to this general trend, but the above generalisation summarises the overall picture.

Illustrated in Plate 236 is the date stone from the house of clockmaker John Spencer of Birchenlee Farm, Great Marsden, near Colne in Lancashire. His father, Henry Spencer, a farmer, had bought the farm in 1715, but John decided to commemorate his presence there by having this stone carved in 1760 with initials ISS, the central S standing for Spencer, the I for John, the S for his wife, Sarah. Incorporated into the carving is a single-handed clock dial. A two-hander would have been more difficult to carve, of course, but the reason a single-handed dial was chosen was probably because that was how the rural public thought of a clock. Most church clocks and most domestic clocks in rural Britain had a single hand. That situation was rapidly to change in the coming years as two-handed clocks began to outnumber single-handers in a population increasingly familiar with domestic clocks, even those which registered minutes by means of two hands.

PLATE 237. *Ten-inch longcase dial of about 1735 by Jonathan Savage of Clifton, Cumberland, showing initials of first owner.*

PLATE 238. *Movement of the Savage clock showing offset 'Westmorland' calendar wheel still having casting rag, cheese-headed collets, and cartwheel dial casting.*

The thirty-hour single-handed clock by Jonathan Savage of Clifton, near Penrith, Cumberland, is very difficult to date accurately. The ten-inch dial (Plate 237) is of the cartwheel type with offset 'Westmorland' calendar wheel. Most details of the dial suggest a date of perhaps 1720-30: chapter ring with indescribable half-hour markers of a style popular in the area; early style of blued steel hand; circular ringed calendar aperture. However, the seashell spandrel pattern was more popular in the 1750s and, though original to this clock, I feel the date here must be earlier. My guess is for the 1735-45 period. The maker was working in the 1730s and 1740s, but little is known about him.

The circular name-plate is clearly an ownership plaque for husband and wife, probably for a marriage clock. Three drill holes in the chapter ring at XI, III and VII are rivet holes used when the chapter ring was fastened to a bed during engraving. The matted dial centre is evenly worked with an apparently random graining but shows traces of a circular tool direction.

The plated movement (Plate 238) is well made with crisp wheelwork using a shouldered collet of a shape sometimes called a cheese-head, being of a squat cylindrical outline with rounded shoulders and looking like a whole cheese in shape. The lower left dial foot is also used as a calendar wheel post, an economy of manufacture often employed with a 'Westmorland' calendar wheel. Note the casting rag clearly visible on the calendar wheel spokes, and to a lesser degree on the dial spokes, showing that normal practice was to cast these rather than cut them out from solid sheet brass.

The clock by William Reynolds of Wigton has some unusual features about it making it less easy to date than some. The eleven-inch cartwheel-cast dial

PLATE 239. *Two-handed thirty-hour longcase dial with engraved penny moon made about 1740 by William Reynolds of Wigton.*

PLATE 240. *Reynolds movement showing moon and calendar wheels knocked on twelve-hourly by the same hour-pipe wedge.*

(Plate 239) has blank corners, a feature used by some Quaker clockmakers who eschewed frivolous decoration. However, Reynolds appears *not* to have been a Quaker! The dial sheet has a circular, ringed datebox and ringed dummy winding holes, giving mock eight-day appearance. All these features and the 'penny' moon tend to suggest an early date.

Certain other clocks by Reynolds also have some features associated with Quaker work and there may have been some influence from one-time Quaker John Sanderson, founder of the Wigton 'school' of clockmaking (see page 392) and still very much alive at the time of this clock's making. Reynolds' clocks, however, do not show any obvious Sanderson influence in their movement construction and his engraving is more professional than much of Sanderson's.

This dial centre is matted and of such a nature that it could be done by someone incapable of engraving. The engraved work of the name-plate and chapter ring could well have been done by an outside professional. The engraving is skilled and sharp, and also incorporates some early features (flat-topped 8 and half-quarter markers, both a little outmoded by this time). The floating fleur-de-lis half-hour markers are up to date and are approaching the boldness of style which later so often became a northern characteristic.

We can deduce the period of the clock as around 1740. The maker married at Wigton in 1739 and was there till about 1743, when he moved to nearby Kirkbride and signed his clocks as from that place. He returned to Wigton again in 1761 and remained there till his death in 1792. The clock must therefore date from his first Wigton period.

The movement (Plate 240) shows that the calendar is of the offset

PLATE 242. *Longcase dial of the 1740s by Mos(e)ley of Penistone with ringed 'dummy winders' type roundels. What at first sight appears to be a penny moon is in fact a fixed face with rocking eyes.*

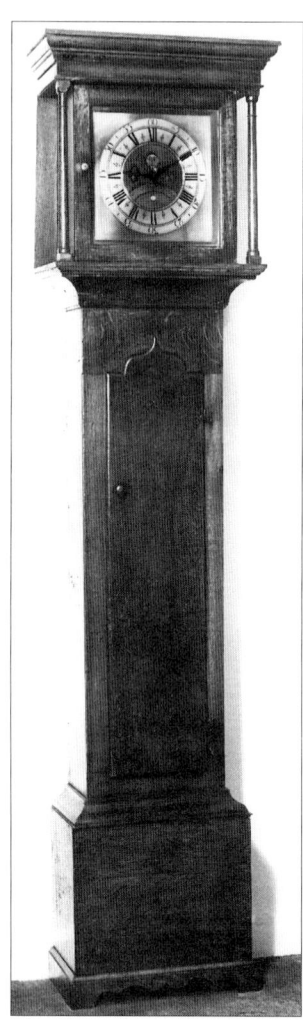

PLATE 241. *Oak case of the Reynolds clock with free-standing pillars and cupid-bow door top, standing about 7ft. high.*

'Westmorland' type, knocked on twelve-hourly by a wedge on the hour pipe. The penny moon disc drives from the same wedge, this being the usual method for penny moon drive.

The oak case is sturdy and simple in style (Plate 241). Hood pillars have by now become free-standing on many northern clocks, though still attached to the door in the South. The single 'Gothic' point door top (sometimes called a cupid's bow shape) is one of many door top shape variations which by now have become a regular feature in the North-west. The trunk-to-base moulding is of an ogee shape, which is unusual anywhere but is suggestive of north-western styling. Otherwise the case is so simple in style that it has no strong regional features.

The dial of a thirty-hour clock signed 'Mosley, Penistone' (in Yorkshire) dates from the 1740s and is believed to be by Joshua Mos(e)ley (there were several clockmakers in the family). This dial (Plate 242) still marks half- and quarter-hours, the half-hour marker being a rather strange combination of meeting arrowheads and spidery flowerheads, well engraved but artistically lacking. This dial shows an early use of the serpentine 'Dutch' minute band. The steel hands are original and well styled. Mouth calendar work is the usual form after mid-century on those thirty-hour clocks which have calendars. Two ringed decorative circles are set in dummy winder positions, not so much to simulate winding holes as to break up the otherwise plain (matted) centre, and this is an extension of the cup-and-ring decoration principle in so far as it could be done by someone not capable of engraving. All the engraved work is skilled and probably done by a specialist engraver.

The interesting 'penny moon' face below XII is not in fact a moon at all, though the aperture and moon face are exactly as when used for that purpose. The fixed face has rolling eyes, whereby the eyes are waxed on to a silvered backdisc, which sways from side to side on an extension of the anchor arbor. The rocking eye motion is uncommon – it is difficult to see it ever having been popular.

CHAPTER EIGHT
THIRTY-HOUR CLOCKS AFTER MID-CENTURY

The thirty-hour longcase clock by William Porthouse of Penrith dates from about 1750 and has an eleven-inch dial (Colour Plate 13, page 217). Porthouse's dials usually have a rather plain centre, conventionally matted and with the usual square calendar box. Here he has signed the clock on a boss in the arch, with an unusual Gothic style of lettering so fanciful that the capital letters are difficult to make out. The engraving is professional throughout, his chapter ring using the bolder northern half-hour marker in a sort of elaborate fleur-de-lis style. Corner spandrels are of the pattern with two eagles supporting an urn. Arch spandrels are dolphins, the commonest pattern for arches over a considerable period of time, hence not much help towards dating.

The plate below XII bears the names of the first owners, Thomas and Mary Arnison, and is the kind of ownership plate often used on what we usually think of as a wedding clock. Porthouse was the principal clockmaker in the land who specialised in personalising his clocks and more of his clocks carry 'marriage plates' than of any other clockmaker I know. Many also have the year alongside the names. I know of thirty-five clocks by him with this sort of name-plate, the majority, though not all, commemorating marriages. I once tried to trace the actual marriage entries of many of these Porthouse first owners and succeeded with some. The result was that, although many clocks proved to have been made at the time of the marriage (perhaps wedding presents from parents?), others were clearly bought some years later, perhaps when the couple had saved enough for the purchase of what many saw as the ultimate object to possess to complete their homes. For most households the longcase clock was *the* most costly single item they would ever buy and often therefore its purchase was last on the affordable shopping list.

I discovered that Thomas Arnison was married in 1729 to Mary Pattinson at Alston, some few miles from Penrith. Such a confirmation makes one examine the possibility that the clock may have been made at that time (1729), which was conceivable as William Porthouse was working in Penrith from about 1726 till his death in 1790. However, too many aspects of the clock's styling are just not that early, and it seems as if the couple must have bought the clock twenty years or more after their marriage

The clock is one of those uncommon 'luxury' thirty-hour clocks which chimes the quarter-hours, in this instance on six bells. The bell set-up, known as a 'nest' of bells, can be seen in the movement picture in Plate 243, though the pin barrel is hidden behind the hammer support piece. The movement frontplate shows clearly the rack striking system for the hourly striking, used to allow a repeating facility. The repeater bladespring trigger is missing; it once was screwed to the lower left corner of the frontplate in what is now an empty screwhole just above the movement pillar. The bladespring passed up vertically between two pins in the curved ends of the lifting piece and would have had a hole at its top to take a repeater cord, which passed either through the hood side or hung down inside the case. The purpose of the *two* pins is to allow the spring to pull against one of them when the cord is pulled to trip the repeat, and to enable the spring to push back against the other on release to ensure that the lifting piece is returned to its original position.

PLATE 243. *The Porthouse movement from the side showing the nest of bells and the hammer layout. Two pins on the lifting piece are to accommodate a repeater trigger spring missing in the photograph. See Colour Plate 13, page 217.*

The marking out circles show very clearly scored into the frontplate (Plate 244) – most makers made no attempt to polish these out and they can often be a very obvious indication of a handcrafted movement. On the other hand, the absence of such marking out lines is not necessarily an indication of the contrary, since some clockmakers would have made up a template for drilling their wheel arbor holes. Like many clockmakers of this period Porthouse scored his wheels with decorative rings just to give them a little finish, a nice touch which was completely unnecessary mechanically but showed a pride in his work. From our point of view such ringing can help us recognise an undecorated replaced wheel.

The case is of oak and stands about 7ft.3in. (Plate 245) It shows certain northern features of a general nature – the shaped door top, canted corners to the base and basemould, the latter being a complex mould. It also has some features local to this immediate area – the heavy topmould to the hood, a sort of mix of a dometop style and a swan-neck, a kind of swan-necked dome, having very heavy roundels which are undecorated on their fronts; the top-of-trunk moulding being of a convex outline, by this time an antiquated shape

PLATE 244 . *Frontplate of the movement of the Porthouse clock, which chimes the quarters on six bells. Note scribing out lines for the wheels.*

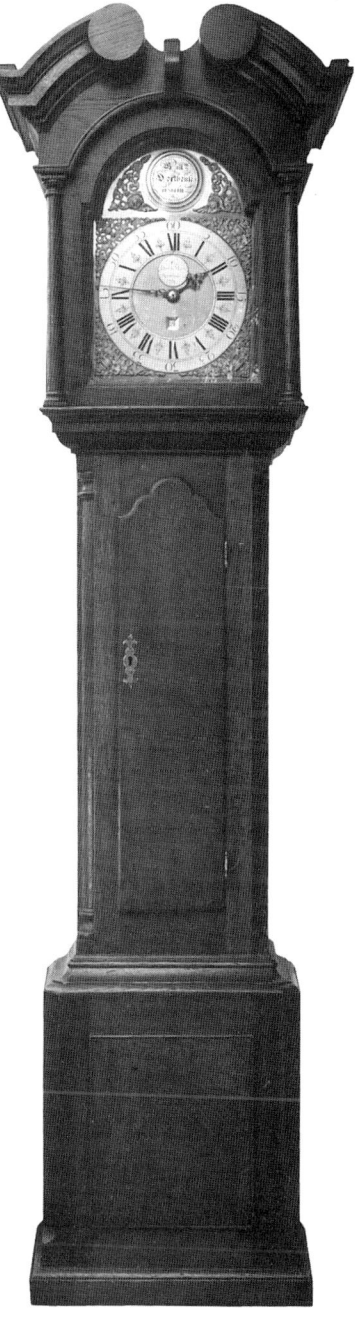

PLATE 245. *The Porthouse case is of oak with some of the heavier mouldings associated with this maker's cases at this time.*

long discontinued by most casemakers.

The hood pillars are integral with the door, again a slightly old-fashioned feature by now in the North and probably just a whim of the carpenter. Their capitals and bases are of gilded (gold-painted) wood rather than brass, this being a feature of many country cases from all regions.

The fact that provincial clockmakers were capable of producing both fine work and crude work from the conventional to the downright quirky is often a major factor attracting collectors to their work. Few clockmakers were more eccentric in their work than the little-known Henry Mason who worked at Egremont in Cumberland in the middle eighteenth century. He is believed to have been married at Moresby in 1735 and to have been buried at Whitehaven in 1789. Not more than five examples of his work have so far come to light, all of them eccentric in that the movements were made using 'skeleton' plates, as is the one pictured in Plate 247.

Only half a dozen clockmakers in the land are known to have used skeleton plates for longcase clocks, though the well-known Victorian skeleton bracket clocks embody the same principle of having brass only in those plate areas

PLATE 246. *Eleven-inch single-handed thirty-hour longcase dial by Henry Mason of Egremont dating from about 1750, showing eccentric registration of five units per hour of twelve-minutes each.*

where they would be carrying some pillar or pivot and having gaps elsewhere. Henry Mason seems to have hit upon the idea independently of those few other contemporary clockmakers who used skeleton plates. The point was that this method used less brass and was therefore cheaper. What is exceptionally unusual about Henry Mason however is that he used *bronze* rather than brass for his plates, the reason being (I am told by those who know these things better than I do) that bronze would be more hard wearing. The movement illustrated is typical of Henry Mason's unique bronze skeleton plates.

The Henry Mason clock shown in Plates 246 to 248 dates from about 1750 and has an eleven-inch single-handed (cartwheel-cast) dial. The spandrels are a form of crowned cherub head (quite different from the earlier cherub heads), a pattern sometimes used at this period principally on northern clocks and of which several variations exist. The dial centre is coarsely matted, but a strange feature of this clock is that the calibrations are on the *outer* edge of the chapter ring and the original blued iron hand reaches out this far instead of stopping short, like most single hands do, at the limit of the dial centre.

This eccentric style may have been determined by the strangest factor of all, namely that each hour division is split not into four quarters but into fifths. This means that the clock registers twelve minute units instead of the usual fifteen minute ones. My guess at the reason for this is that by reading a split division the time can be seen at a glance in units of six minutes rather than the seven-and-a-half minute split reading from a normal quarter-hour registering single-hander. It is perhaps a different approach to the problem of getting a single-handed dial to show smaller units than quarter hours (which some clockmakers solved by subdividing their quarter-hour units into three to give five minute reading – see Plate 220). This is certainly an unusual method which I have never seen on any other clock, even on other clocks by Henry Mason! Also unusual is Henry Mason's half-hour marker in the form of a small

PLATE 247. *Movement of the Henry Mason clock having bronze plates which are skeletonised, a combination unique to this maker.*

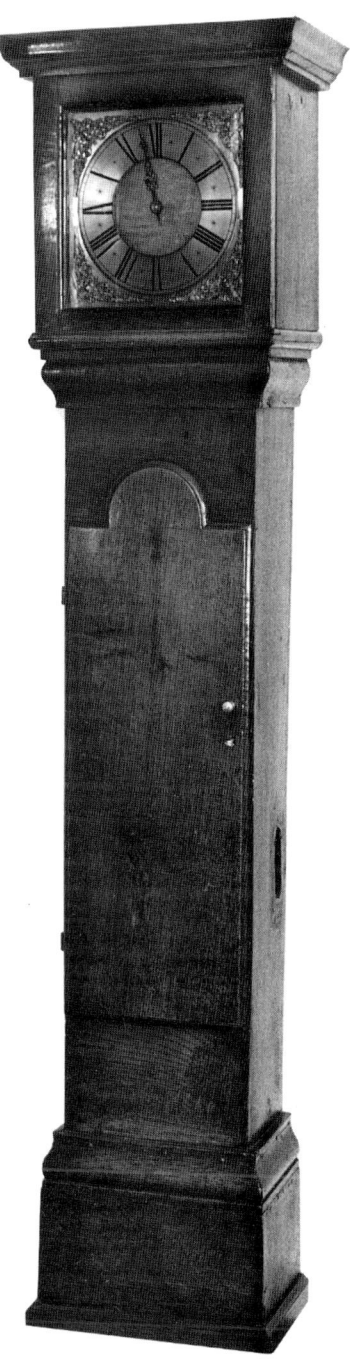

PLATE 248. *The primitive oak case of the Henry Mason clock suggesting at first a greater age than it has. Height 6ft.5in.*

asterisk. Of course the half-hour marker could be in any shape or form as long as it is eye-catching. The amazing thing is that clockmakers throughout the land mostly adhered to the general style of the day – except for free spirits such as Henry Mason, who did their own thing.

The oak case (Plate 248) stands 6ft.5in. and is of a charmingly primitive nature. Primitive cases are not necessarily the earliest; this example's crude styling might suggest a greater age than mid-century. Such features include: absence of hood pillars, large unused trunk space above and below the door, crude ogee moulding below the hood with that same mould used again as a trunk-to-base moulding (to save the trouble of making two different mouldings), crude arched top to the trunk door. The hinges to the main door and the hood door are of the wrought-iron butterfly type fitted internally. Both doors open left-handedly, a feature found on some cases presumably at the request of the customer.

179

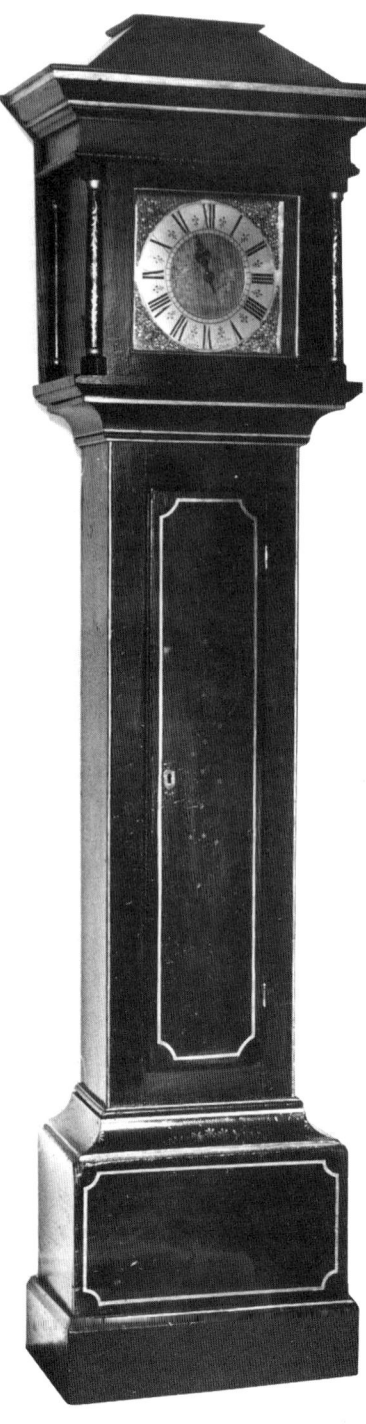

PLATE 249. *Ten-inch single-handed dial from a clock of about 1750 by John Gilkes of Shipston. Note the crude matting.*

PLATE 250. *The pine case of the Gilkes clock is painted black outlined in gold panels. Height 6ft.10in.*

The holes in the case side have been cut by some past owner to prevent the pendulum bob knocking, after incorrect restoration had given the clock too wide a swing for such a narrow case.

The ten-inch single-handed longcase clock by John Gilkes of Shipston on Stour, Warwickshire, dates from about 1750. Gilkes is believed to have been a Quaker clockmaker, though his dial shows none of those specialised characteristics we associate with some Quaker work. He was working between about 1740 and 1770.

The dial centre is crudely matted, the vertical grain being very obvious in this photograph of the unrestored dial (Plate 249). Feeble attempts at 'engraving' a bird on a bracket at each side suggest that Gilkes was no engraver and the stylish and crisp engraved work on the chapter ring with the very up-to-date floating fleur-de-lis is clearly the work of a professional outside engraver. The spandrels are one of the myriad versions of the female head. The birdcage movement is typical of the work of many rural clockmakers in central and southern England. Notice the absence of a calendar feature, a common southern trend, whereas northern clocks usually have them.

The case is of pine, standing about 6ft.10in. and still in its painted form, the paintwork being black with gold highlights to pick out the edges of panels and mouldings (Plate 250). Painted on the door is a most interesting flower pot from which grows a tall upright plant (Plate 251). Any door decoration, including that on carved cases, is obliged to feature an upright design by virtue of the door shape, and very often this will be a growing flower or tree or grapevine. However, the Quaker ethic, and that of the Shakers in America, included a 'tree of life' theme, whereby the harder you worked, and the more devoutly, the closer you grew towards God and the more your flowers blossomed. This is sometimes expressed visually in the form of a flowering tree or plant and that may well be what this door design symbolises.

The hood caddy is here a very strangely sharp-angled affair like a truncated pyramid. All mouldings are simple, even severe – simpler moulds, of course, were easier to make! The trunk door is flat-topped, though the gold highlighting suggests clipped corners. Hood pillars are still integral with the door as a normal feature of country work in this area.

An oddity of construction in the case is that the hood door glass area measures ten inches high by eleven inches wide. The case is original to the clock and it looks as if the casemaker was using a design planned for an eleven-inch dial, adapted it for this order for a ten-incher, and realised too late that to have shrunk the width measurement further would have left him with an even more unbalanced hood door. This is yet another and surprisingly late example of the beginnings of casemaking, showing that the casemaker is still learning his craft in what is still a relatively new product.

Thomas Pinfold worked in Banbury from about 1760, or perhaps a few years earlier, until 1789, when he died there. The ten-inch (cartwheel) dial longcase clock pictured in Plates 252 to 254 dates from about 1760. The style of single-handed dials changes less rapidly than that of two-handers and those of a simple nature, such as this one, can at first sight appear considerably older than they are. The crudely-matted dial centre, with random matting and matting roller marks clearly visible here and there, is then 'engraved' by chasing a windmill into the centre between two flowers. This is clearly the inexpert work of the maker himself, against which the crisp chapter ring engraving is far more professional and obviously engraved out of house. The diamond half-hour markers used here are exceptionally slight. The blued iron hand is original and typical of later 'single' hands.

The absence of a calendar is more typically southern than northern. The cockle-shell spandrel pattern is the 'latest' feature on which a visual dating of this dial would depend, as all other features could just as well date from ten or twenty years earlier. Single-handed clocks lingered longer in the South than in the North, as too did the posted movement, which by now is somewhat simplified stylistically. The dial is cast in the cartwheel fashion, a little unusual in the South, and this may have been a deliberate choice of the maker because the dial was less likely to warp during his rather heavy-handed matting process.

The case is of very simple style made in fruitwood. The absence of hood pillars (for ease of construction and cheapness?) and the still-heavy hood topmould with its wide overhang are both old-fashioned features which at first sight suggest greater age. This is an example of the cheapest type of casework of the period and is therefore made without any fancy frills – such as pillars. The height is about 6ft.6in. Some clocks by this maker have black and gold painted cases in the manner of the Gilkes example in Plate 250.

The eleven-inch dial of a plate-movement longcase clock by Samuel Parrat of Grassrigg, Westmorland, dates from about 1750-60. The dial is seen in Plate 255 before being cleaned. Half-hour markers are of large foliate form, the successor to the fleur-de-lis in many northern clocks after mid-century. The quarter hour units have now ceased to be marked on the inner chapter ring edge, as increasing familiarity with minute-marked dials made them obsolete. The maker is known to have been born about 1729 and to have died in 1783.

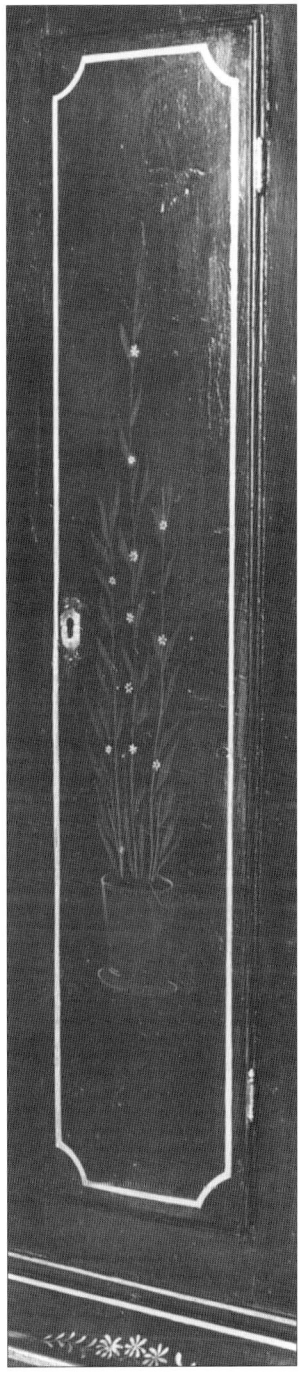

PLATE 251. *Detail of the door of the Gilkes case, painted with a vase and flower, perhaps to symbolise the Quaker 'tree of life' theme.*

PLATE 252. *Ten-inch dial from a single-handed longcase clock of about 1760 by Thomas Pinfold of Banbury. Note the grain of the matting tool.*

PLATE 253. *Posted movement of the Pinfold clock by now somewhat simplified in form.*

PLATE 254. *Fruitwood case of the Pinfold clock standing about 6ft.6in. Absence of hood pillars was a feature occasionally continued late in this area.*

PLATE 255. *Eleven-inch dial from a longcase clock of about 1750-60 by Samuel Parrat of Grassrigg, Westmorland. Circular grain obvious in the matting.*

He is believed to have been a Quaker though the clock styling is conventional and shows none of the sometimes eccentric styling we associate with some Quaker makers.

The matting of the centre can be seen to have a circular grain, with a little decorative engraving beginning at the calendar box but now spreading further into the dial centre than with earlier periods. The dial foot end shows near the IIII numeral, probably because it has been loose at some time and re-riveted carelessly. That the other two dial feet are not obvious is an indication of the care taken with their fixing.

The original hands are of blued iron and a bit 'thinner' than at earlier periods. The minute hand is fitted by a screw through its centre, a method used occasionally by some northern clockmakers, though seldom in southern clocks. Pin fitting was the normal method in all areas at all periods. A version of the female head pattern is still being used for spandrels, though now decidedly old fashioned.

The case is a simple cottage style of case in elm (Plate 256). Those familiar with elm will recognise the distinctive figuring to the door. The only wood easily confused with elm is chestnut, which was even more prone to woodworm and in my experience was not used for casemaking other than occasionally for backboards. Elm seems to have been rated as a cheap wood, along with pine, and many elm cases have perished because of worm. Surviving elm cases often show traces of having been painted, like pine ones, but this case does not seem to have ever been painted.

No hood pillars are used – probably for cheapness. The pillarless hood is unusual in northern casework, less so in southern work. The shaped door top is typically northern. Otherwise such a simple case has little in the way of styling to enable any regional trait to be deduced. The height is about 6ft.3in.

The ten-inch (solid sheet) dial longcase clock by John Ford of Arundel in

PLATE 256. *The Samuel Parrat case is in elm, with highly distinctive grain showing on the door. Height 6ft.3in.*

PLATE 257. *Ten-inch dial from a birdcage thirty-hour longcase of the 1760s by John Ford of Arundel. Stylised scenes were sometimes engraved in the polished centre of southern dials. Original blued steel hands.*

PLATE 258. *The neat oak case of the John Ford clock stands 6ft.3in.*

Sussex dates from the 1760s. The scroll-engraved centre (Plate 257) shows a feature sometimes used at this period in the South, less often in the North, of including a miniature landscape scene, here of a cottage which, judging by its hanging signboard, would seem to be a tavern. Spandrels are now of the foliate scroll style, popular in the 1760s and 1770s. The movement is of the birdcage type.

The plain oak case (Plate 258) is of traditional style and stands about 6ft.3in. Features retained from earlier times are principally the flat-top door and the integral hood pillars. Mouldings are simple throughout and the case lacks any decoration in the form of crossbanding, which by now would be normal in the North.

The ten-inch single-handed plate-framed longcase clock signed 'Donisthorp' (Plates 259 to 261) is believed to have been made by Joseph Donisthorpe of Normanton-le-Heath in Leicestershire, and is dated 1758 on the movement. The numbers are stamped into the brass frontplate by means of punches, implying that he could not engrave and that he probably bought his engraved name-plate and chapter ring work from a professional engraver. It is also marked behind the dial as being clock number 10. Donisthorpe is believed to have been born in 1703 and to have died in 1774, being a blacksmith in his earlier years. If we can judge by his numbering, it would seem he did not take up clockmaking until the 1750s.

The dial style is quite plainly matted and without calendar (often a southern trait). The chapter ring has only vestigial traces of half-hour markers in the form of a tiny inverted V. Single-handed clocks were by this time becoming old fashioned but there was still some calling for them, especially in rural areas. The iron hand is original. Spandrels are the cockle-shell pattern popular for a decade or more hereafter.

The movement frontplate is left virtually raw from the casting (or at least is not well finished) and clearly shows his marking-out lines and the punch-lettered year (Plate 260). The striking countwheel is of a form used by a few makers, especially in the area, where the normal situation is reversed. Whereas on the traditional countwheel locking takes place when the lifting piece drops into the predetermined slot, here it occurs when the lifting piece is raised by one of the appropriate pins (Plate 261).

PLATE 259. *Ten-inch dial of a thirty-hour longcase by Joseph Donisthorpe of Normanton-le-Heath in Leicestershire dated 1758. Original blued steel hand.*

PLATE 260. *Frontplate of the Donisthorpe clock with wheelwork removed to show the date 1758 and the number 8 (suggesting his eighth clock?). However, the hour wheel is scratched 'for the 10th clock'!*

PLATE 261. *Rear view of the Donisthorpe clock showing his form of pinned countwheel.*

The single-handed plated-movement longcase dial by John Agar of York dates from about 1760 and is unusual in several respects (Plate 262). The dial has a plain matted centre with one of several patterns of foliate spandrel popular in the third quarter of the century. The chapter ring shows a very late use of meeting arrowhead half-hour markers, old-fashioned by this time, as indeed is the single-handed clock in this area, where by now most clocks have two hands. However, the inner band of the chapter ring has each quarter sub-divided into three, thus giving a five-minute reading facility. This was one method of

PLATE 262. *Single-handed longcase clock with alarmwork made about 1760 by John Agar of York showing subdivisions of each quarter hour into three to give five-minute registration. Original hand of blued steel. The brass hand is for setting the alarm time.*

PLATE 263. *Dial from a single-handed longcase clock of about 1760 by John Glover of Bungay in Suffolk. Polished dial centre decorated only by the engraved name.*

showing closer time registration with a single hand, which one can meet occasionally in any part of England, though it never became very popular.

The original hand is of blued iron. The secondary hand, of brass, is not a minute hand, but a hand to set the alarm time, as this is a longcase with an alarm function, something which is very unusual in British longcase clocks. The alarm mechanism is the same as used in lantern clocks, hook-and-spike and hooded clocks, though a setting hand is used instead of a disc.

The single-handed longcase dial by John Glover of Bungay in Suffolk (Plate 263) also dates from about 1760, perhaps even a little later. It might at first sight appear earlier, but the maker was not born till 1740 and lived on till 1810. This dial has a chapter ring very like the one on the Thomas Pinfold clock in Plate 252, with a very restrained diamond half-hour marker, perhaps even done by the same engraver. The dial centre is polished with no decoration except the engraved name and, of course, no calendar. Spandrels are the small urn with foliage surround. Single-handed clocks remained popular for longer in south-east England than in most other areas.

The ten-inch thirty-hour clock by William Emerton of Wootton, Bedfordshire, dates between about 1760 and 1770. The maker was born in 1738 and died in 1789. The dial centre and chapter ring are both well engraved (Plate 264), but whether the work of the maker or an outside engraver is impossible to say. The dial centre is engraved on to a polished ground, a style generally more popular in the North than in the South but exceptions are always possible, as here.

The chapter ring by 1760 has usually lost its half-hour markers and quarter-hour divisions from the inner edge, making a much plainer style overall. The larger area of engraved dial centre design often found from this time on (with this polished-centre style of dial, of course), may perhaps be a compensation for the plainer chapter ring. In this particular example the minute band is arcaded in a style sometimes called a 'serpentine' minute band, but usually known as a 'Dutch' minute band because it was at one time particularly

PLATE 264. *Ten-inch dial of a thirty-hour longcase of about 1765 by William Emerton of Wootton, Bedfordshire. The arcaded ('Dutch') minute band was a variation on a theme. Original blued steel hands.*

popular on Dutch longcase clocks. The 'Dutch' minute band was predominantly a style popular just after mid-century in northern England, but there are obviously southern exceptions.

The hands are of blued iron and are original and typical of the now slightly more delicate style than earlier. The spandrels are one form of the numerous scrollwork patterns of this period, this one based on C-shapes.

The case is of pine (now stripped) and of simple cottage style (Plate 265). Several features are old fashioned and little different from a cottage case of a generation or two earlier – simple lines, flat-topped door, surface-fitted H-hinges, pillars attached to the hood door, simple mouldings. The dentil mould is the only permitted 'extravagance' and gives the case a little individual personality. The height is about 6ft.5in.

The twelve-inch thirty-hour longcase clock by William Parkinson of Lancaster, Lancashire, also dates from the 1760s. The maker succeeded his father in 1759 and worked there till his death in 1800. The dial centre (Plate 266) is matted with a vertical grain visible; the matted centre style was now becoming less popular in this north-western region but some makers retained it. Better planning would have meant that the two lower dial feet at IIII and VII would have been hidden by the chapter ring. Northern dials are often larger than equivalent southern dials of the day.

This may at first sight appear to be an eight-day clock, as is the intention of the maker, but the ringed winding holes and the winding squares visible in them are in fact dummies. The clock is a thirty-hour masquerading as an eight-day, made that way presumably to impress the neighbours. This 'dummy winding hole' style is a feature very largely confined to the north-west of England, though occasional southern examples occur. In fact experience enables us to guess at this as there is no seconds dial (present in virtually all eight-day clocks without moonwork) and the calendar is of the thirty-hour knock-on type used predominantly in thirty-hour work.

The chapter ring is now quite plain, lacking half-hour and quarter-hour

PLATE 265. *Simple pine case of the Emerton clock, standing 6ft.5in. Classic country style.*

PLATE 266. *Twelve-inch dial from a longcase clock of the 1760s by William Parkinson of Lancaster. Vertical grain matting. Ringed 'dummy' winding holes and winding squares to simulate an eight-day clock.*

PLATE 267. *Oak case of the Parkinson clock, standing 6ft.8in., with much mahogany crossbanding. A considerably costlier case than a simpler southern counterpart of the day.*

indicators, being the new style of the period. The engraving of this and the name-plate is crisp and professional. Note the curly tail to the five numerals, done to widen the only single numeral on the dial to match up better with the other double numbers. Several engravers did this, some, such as Jonas Barber of Winster, notoriously so and almost to the point of being a trade mark. The spandrels are a form of floral scrollwork which I call the C-scroll pattern.

The case (Plate 267) is of oak with mahogany crossbanding and trim and stands 6ft.8in. Crossbanding surrounds most panels such as the door and base and often the hood door. By 'trim' I mean the additional mahogany work, typically found, as here, in the quarter trunk pillars and their support blocks, hood pillars, frieze across the hood top (where some clocks have a fret) and in the dentilwork on the hood topmould. This degree of mahogany crossbanding on an oak case is more typically northern than southern at this period, though there are many exceptions.

Typically northern stylistic features are: shaped door top, base higher than square, ogee top-of-trunk moulding, ogee and complex trunk-to-base mould. Many north-western cases have a noticeable absence of brasswork as, for example, in the hood pillar caps, even in cases of above cottage quality, as here. Although this is a cottage style of case, the quality of cabinetwork, including the use of oak rather than the more usual pine for backboards, thickness of timbers, and quantity of mahogany used, makes this a much more costly case than many a southern equivalent of the day.

The twelve-inch (cartwheel-cast) dial thirty-hour longcase clock by John Stancliffe of Barkisland, near Halifax, West Yorkshire, dates from the 1760s. Stancliffe was born in 1706 and died in 1780. Here the maker uses the slightly old-fashioned dial centre of matted ground into which an engraved design is cut after matting (Plate 268). The random matting can be seen to be uneven and the centre engraving crude, suggesting Stancliffe did this himself. The chapter ring, on the other hand, is expertly done, still keeping the half-hour marker but omitting quarters.

The moon dial below XII was popular amongst certain northern and north-western makers (only occasionally elsewhere) and was driven by the same

PLATE 268. *Twelve-inch dial from a thirty-hour longcase of the 1760s by John Stancliffe of Barkisland, Yorkshire. Note penny moon dial and crude chasing on to coarsely-matted centre.*

twelve-hourly knock-on system as the mouth calendar, in fact usually by the same wedge. The numbers show the lunar date and the circle shows the shape or phase of the moon, here a thin crescent. This kind of moon dial is often called a penny moon, sometimes a Halifax moon or a twelve-o'clock moon. This system was used as a means of incorporating a moon dial into a square dial. Early forms of penny moon discs were engraved, waxed and silvered; later ones were painted, even on brass dials.

The carved oak swan-necked case in Plate 269 houses a different thirty-hour clock by John Stancliffe dating from the same period, the 1760s. The clock in this case is of very similar style with the same type of coarse matted ground on to which is chased a scroll design. The case has many features typical of northern central England of the Lancashire/Yorkshire borders. These include: deep moulds to the swan-neck, complex ogee top-of-trunk moulding, square-to-high base and generally rather stocky proportion derived from the fact that many such clocks have dials of eleven or twelve, even thirteen, inches with a relatively restrained height of around 7ft.3in. The free-standing pillars, normal throughout the North by this period, make for a broader style than the southern integral pillar form. These cases are often of very thick timbers, frequently using oak for the backboards, and are consequently very heavy to move.

Carved cases are regarded suspiciously by many collectors who believe them to have been carved much later, principally about a century ago when there was a taste for black carved 'Gothic' furnishings. Undoubtedly many early pieces *were* carved at that time and then stained black, perhaps to enhance the carving. Many of these later-carved cases are very obvious with pathetically poor carving performed on timber which was too thin for the purpose or through crossbanding, which was clearly unintended. However it is difficult to write off all such examples as having later carving and to do so would presuppose that there was no taste for carving at any period in the history of longcase clocks with the sole exception of the few years at the turn of the last century.

The question of the age of the carving can be a very difficult one to resolve, especially when there was a strong taste for it in certain regions, such as the Lancashire/Yorkshire Pennines, where carved cases were often of unusually thick

PLATE 269. *The oak case of the Stancliffe clock is lavishly carved and stained to near black. Height is about 7ft. Such cases are very strongly built.*

PLATE 270. *Eleven-inch dial of a thirty-hour longcase of the 1760s by Tobias Fletcher of Barnsley. Blued steel hands probably original.*

PLATE 271. *Eleven-inch dial of the 1760s from a longcase by John Steel of Killamarsh in South Yorkshire. Note the engraver's error at 25 past. Dummy winding holes to simulate eight-day appearance. Blued steel hands probably original.*

timbers. The carving on some of these cases is highly skilled, as in this example.

The eleven-inch thirty-hour cartwheel-dial longcase clock by Tobias Fletcher of Barnsley in South Yorkshire dates from the 1760s (Plate 270). The maker was working from about 1760 till his death in 1813. The style is that of the polished and engraved centre, here professionally done. The chapter ring omits half- and quarter-hour markers. The blued iron hands are original and of the style we term 'non-matching' in that hour and minute hands are distinctly unalike to avoid confusion. Hands of 'matching' pattern begin towards the end of the century. The scrollwork spandrels are of a style I call castle gateway – if you can't see why don't worry about it as we have to call them something!

The eleven-inch cartwheel-cast dial by John Steel (Plate 271) is from a thirty-hour longcase clock and dates from the 1760s. John Steel worked at Killamarsh near Rotherham in South Yorkshire. He is believed to have been born in 1711 and died in 1792. Here the maker uses the matted form of dial centre with mouth calendar and dummy winding decorations to break up the plainness which northern makers seldom liked in their dial centres. Two of the three dial feet ends show at XI and IIII, probably because they have at some time worked loose and been re-riveted tight again. The small mark above VI is where the pivot post of the calendar disc is fixed.

The chapter ring is very similar to that on the Fletcher dial in Plate 270, and could even be by the same engraver. There is an error on the engraved work in that the twenty number has been engraved twice in error for twenty-five. Then, seeing his slip, the engraver has inserted a five inside the nought, an amusing way of remedying the error. It is surprising how often such mistakes seem to have been accepted by the clockmaker and allowed to pass. This chapter ring is riveted into place (at III and IX), something less often found as time progresses, as chapter ring feet were undoubtedly more professional. Non-matching hands are of blued iron and are original. Castle gate pattern spandrels are typical of the period.

PLATE 272. *Eleven-inch thirty-hour longcase of the 1760s by Archibald Lawrie of Carlisle. The oak case has a pitched pediment, a form occasionally used in this area at this period. Note half-round forward-facing rear pillars.*

The oak-cased eleven-inch thirty-hour clock by Archibald Lawrie of Carlisle (Plate 272) dates from the 1760s. The floral engraved centre incorporates a square calendar box with the maker's name. The old W looks rather like a modern N and such names are often misread as, for example, Lanrie. Eagle-and-urns spandrels here continue a little later than normal.

The hood has what is known as an architectural pediment, a style occasionally found anywhere but perhaps more popular in the North-west than in general. Some such pediments run directly to the hood top corners, but more often there is a shoulder, as here, with blocks to take side finials as complements to the central one, normal with such hoods. Here, of course, all three finials are missing. The ogee top-of-trunk mould and shaped door top are distinctively north-western.

The fourteen-inch dial longcase clock by James Gandy of Cockermouth in Cumberland (Plates 273 to 275) is actually dated 1761 on the dial centre, a very unusual feature. This clock, however, is unusual in very many respects, in

PLATE 273. *Musical longcase clock by James Gandy of Cockermouth dated 1761, the engraving in Gandy's typically bold and eccentric style. Centre calendar, rocking Father Time, and rocking ship and moon dial! The clock is key wound and runs for two and a half days.*

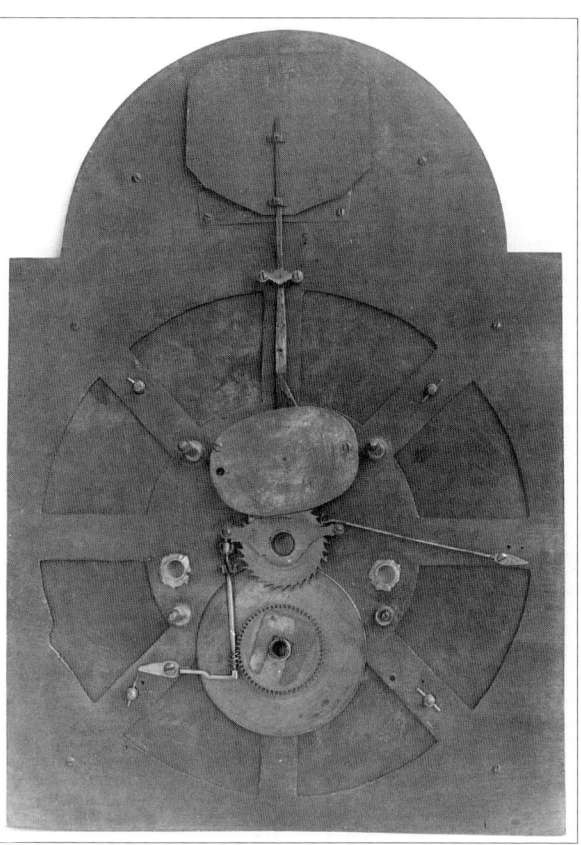

PLATE 275. *The back of the Gandy (cartwheel-cast) dial showing fitting method of some of the numerous features.*

that it is a musical clock playing a tune on six bells every fourth hour. Musical clocks usually play every third hour, some every fourth hour, especially those of thirty-hour duration, principally because they cannot contain enough power to perform hourly which in any case might have been rather too much of a good thing for some owners. This in fact is not a thirty-hour but has a duration of two and a half days, an unusual option chosen just occasionally by provincial makers.

The large dial contains a number of features not found on most clocks – centre calendar work, whereby a third concentric brass hand laps the dial once a month behind the two conventional iron time hands; a rocking ship in the arch driven on an extension of the anchor arbor; a rocking Father Time below XII driven from the same arm; a moon dial above VI. The engraving is in an eccentrically Gothic style which seems to be Gandy's own. His chapter ring has the arcaded 'Dutch' minute band and a large and bold half-hour marker deriving vaguely from a fleur-de-lis shape. Quarter-hour units are retained a little later than on most dials of the period.

In the dial centre is engraved the monogram of the first owners, so 'Gothic' as to be undecipherable. A border round the dial is of the herringbone engraved type, here used long after that feature had passed from general fashion. The spandrels are one version of the two-eagles-with-urn pattern, with the ubiquitous dolphins in the arch. This was clearly a special clock and Gandy has included all kind of extra features and decoration to spice it up. This maker moved to Cockermouth from parts unknown about 1726 and worked there till

PLATE 274. *Movement of the Gandy clock with the dial removed showing the nest of six bells and hammers and seventh bell for the strike.*

his death in 1779. He seems to have aimed at the upper end of the clock market and even his thirty-hour clocks tend to be of unusually high quality.

The eleven-inch thirty-hour clock by Daniel Dickerson of Framlingham in Suffolk dates from about 1775. The maker is known to have been working there from about 1755 until he moved to the town of Eye about 1778. The dial centre has fine engraving of the rococo style with typical pillars, urns, flowers, swags, all very finely engraved on a polished ground (Plate 276). The chapter ring too is professionally engraved in up-to-date style without any half or

PLATE 276. *Eleven-inch dial of a longcase clock of about 1775 by Daniel Dickerson of Framlingham in Suffolk. Fine engraved centre. Four-wheel train to show seconds. Original blued steel hands. Birdcage movement.*

quarter markers. The presence of a seconds dial is unusual on thirty-hour clocks and the movement has a four-wheel train instead of the normal three, the extra wheel being to give correct direction drive for the seconds feature. The movement is of the birdcage type, still in use in some of these south-eastern regions. Spandrels are the string-of-pearls pattern, one form of late eighteenth century foliate style. The original hands are of iron in non-matching pattern.

The oak case stands about 6ft.6in. and is of simple cottage style (Plate 277). Features indicative of its region are the simple moulds, pillars still attached to the hood door, arched door top, side windows to the hood (long obsolete in the North) and the strange cresting piece which is shaped into a simple form of

PLATE 277. *The Dickerson case is of plain oak with traditional features such as side windows. Height 6ft.6in. Rudimentary form of swan-neck pediment.*

PLATE 278. *Twelve-inch thirty-hour longcase dial of the 1770s by Thomas Whipp of Rochdale. Typical Lancashire scroll-engraved centre with birds.*

PLATE 279. *Arched thirty-hour longcase dial by Christopher Caygill of Askrigg in North Yorkshire. Simple engraving set around dummy winding holes. Steel hands probably original.*

PLATE 280. *Oak case with mahogany banding of the Caygill clock. Height about 7ft.3in. The break-arch pediment was popular with some makers in this area.*

swan-neck between three blocks intended for finials (the finials now missing).

The twelve-inch thirty-hour dial from a longcase clock by Thomas Whipp of Rochdale in Lancashire is all that survives of a clock now lost (Plate 278). At some time the dial has been drilled through with two winding holes to take a married eight-day movement, itself since lost. The finely-engraved centre on to a polished ground incorporates floral sprays with two jay-like birds, which are in fact imaginative, and similar birds feature on a number of northern dials, especially in Lancashire.

The chapter ring has developed to a broader style, a trend which happened earlier in the North-west than elsewhere. Half-hour markers are still present, though fast becoming an old-fashioned feature. The newest stylistic feature of the day is the marking of the minutes by dots instead of the traditional double track. This seems to have been copied from the first painted (i.e. japanned) dials, which began in the early 1770s, and to have very soon become the generally accepted style throughout the land, with the exception of London and those makers strongly influenced by London style. The spandrels are a late form of cherub head within tracery. The clock appears to date from the 1770s. The maker was born in 1738 and died in 1780.

The strangest feature of the dial is the scene above VI of two game cocks spurred on to fighting by their owners, together with the legend: 'Weather for Sixpence'. Weather here is a survival of the use of the Old English word *waether,* meaning 'which one of the two' (would you back for sixpence?).

The great majority of thirty-hour clocks at all periods were of the square dial type, but arched dials are met with occasionally after about 1720. This example in Plates 279 and 280 by Christopher Caygill of Askrigg in North Yorkshire dates from the 1770s. Caygill was born in 1747 and died in 1803. The chapter

PLATE 281. *Twelve-inch single sheet brass dial from a longcase of about 1770 by George Miles of Chipping Sodbury. The Four Seasons corners are most unusual on this type of dial.*

PLATE 282. *Twelve-inch longcase dial of the late 1770s by Grundy of Whalley in Lancashire, having dummy winding squares to give eight-day appearance. Some late spandrel patterns have borders, as here.*

ring, by this date lacking half or quarter markers, is professionally engraved, as also is the name-plate over the arch. The dial sheet itself may have been engraved by Caygill, as the centre design (on to a polished ground) is relatively simple and signs can be seen of dotted lines alongside some of the scrolls such as would be made if using marking-out dots from a template.

The painted scene in the arch is a little unusual, as an arched dial with this sort of cut-out arch would more often have moonwork, but the painted scene is one occasional option chosen by clockmakers (or their customers, of course). The two winding holes in the dial are dummies to give the impression of eight-day work. The original iron hands have a screw fitting, a method preferred by several makers in this area. The spandrels are of the string-of-pearls pattern.

The case is of oak with mahogany crossbanding and trim of a style sometimes called a break-arch or broken arch. The centre upstand is for a finial, now missing. The case is of considerably higher quality than the usual cottage thirty-hour clock and approaches that of an eight day, which strengthens the dummy winding holes effect to add to the eight-day appearance. It is certainly well above the usual village carpentry and of a sophisticated style suggesting familiarity with case design. Certain features of its style are general northern ones – fret below the hood, separate pillars, complex trunk-to-base mould, small ogee bracket feet. The half-round door top without a shoulder was much used by certain Lancashire cabinetmakers such as Gillows of Lancaster, but was occasionally used elsewhere too, though never widely popular. This type of break-arch hood style, usually with three finials, was used by some northern casemakers (it was popular in Leeds, for example), less often in the South. The combination of both half-round door top and break-arch hood is associated in my mind with Darlington casemaking, not far distant from Askrigg, and it may well be that Caygill got his case there. The height is about 7ft.3in.

A different type of brass dial appeared in the 1760s which consisted of a single sheet of brass having all details engraved on it instead of a separate

PLATE 283. *Late 18th century thirty-hour longcase regulator dial of the single sheet type by Emanuel Burton of Kendal. Steel hour and minute hands original; seconds hand may not be. Very fine engraving.*

chapter ring. This type of dial is known as a single sheet dial or sometimes a one-piece dial, and it is believed such dials were silvered over the entire surface. The single sheet dial is believed to have been introduced as a cheaper alternative to the conventional multiple or composite brass dial, which of course continued to be used for the great majority of clocks until the end of brass dial clockmaking. The single sheet dial was more popular in London and the South in general than in the North, although it was popular in Scotland.

The twelve-inch single sheet thirty-hour clock by George Miles of Chipping Sodbury in Gloucestershire (Plate 281) dates from about 1770. This is unusual in two respects, in that it has the 'Dutch' minute band, not often found on single sheet dials, and also in that it has the Four Seasons theme engraved in the corners, also extremely unusual on a single sheet dial. By the time single sheet dials appeared the fashion for half-hour and quarter-hour markers had passed. The dial centre features stylised houses and a church and such small scenes were a feature of some polished centre composite dials too, principally in the South-west of England but on a smaller scale nationally.

The presence of a seconds dial means that this clock has a four wheel going train to give clockwise rotation of the seconds hand. Seconds are numbered every tenth unit, a period indicator, as earlier examples numbered them every fifth. The original hands are of non-matching pattern in iron. The engraving is naïve, but very professional and sharply done.

The twelve-inch longcase clock by Grundy of Whalley in Lancashire (Plate 282) dates from about 1775-80. It has dotted minutes, always a late sign for a brass dial, and an unusual fan pattern of spandrel with a border – bordered spandrels seem to be always late. This clock has dummy winding holes and dummy winding squares (the latter usually screw into the frontplate to add to the eight-day appearance). Here the winding holes are incorporated into the floral design quite neatly. The engraving is naïve but competently done. Half- and quarter-hour markers are outmoded by this time. The (probably) original hands are of iron in non-matching pattern but by now the minute hand has developed into the serpentine shape much used in the latter stage of brass dial clockmaking.

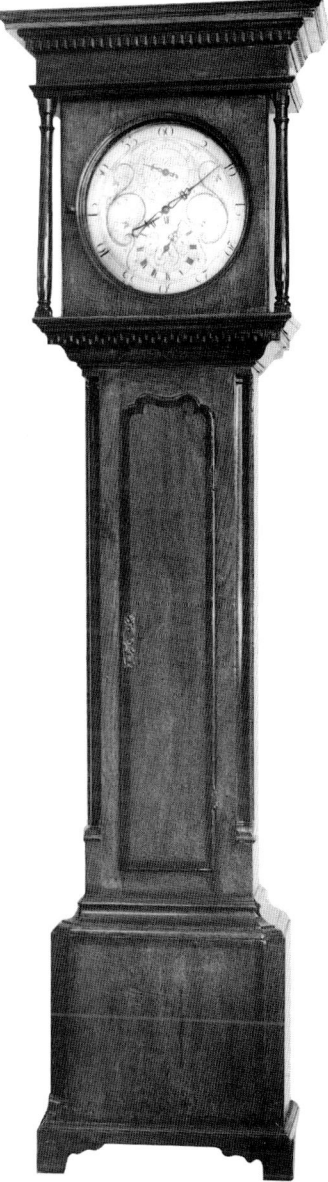

PLATE 284. *Oak case with mahogany trim of the Burton regulator. Height about 7ft.*

PLATE 285. *Thirteen-inch dial of a thirty-hour musical longcase by James Monkhouse of Carlisle made in the 1780s for Joseph and Mary Watson. Four tune selector below XII. Original blued steel hands and brass centre calendar hand.*

The thirty-hour clock by Emanuel Burton of Kendal (Plates 283 and 284) dates from the 1770s and is unusual in being a regulator, probably made for the clockmaker's own use as a master clock by which to test the timekeeping of those he made for sale. Regulators are not often made in thirty-hour form, although such a clock has built-in maintaining power, whereas on eight-day examples maintaining power had to be specially constructed as an additional feature if required. The central hand is the minute hand, here counter-balanced to avoid drag. The lower dial shows the hours, the upper one seconds. All hands are of iron and, although they have lost their bluing from over-polishing, they are of excellent design. The seconds hand appears to be a replacement.

The dial is of the single sheet type, one piece of brass without a separate chapter ring. This kind of dial came into use in the 1760s coincidentally with the first round dials. If the dial was to be circular, then there was little point in having a separate chapter ring for just its outer section. This type of single sheet dial then spread into square and arched dial use too. This example has floral engraving which is superbly executed.

The case of the Burton regulator is of oak with mahogany banding and is of considerably higher quality than the average cottage case. Proportions are very good with an overall elegance derived from the slender lines of a case which stands only 7ft. high. The pillars have a classic swell known as entasis. The dentil (or dog-tooth) moulding on the hood and the Greek key moulding on the top-of-trunk mould are well done.

Regional features include ogee top-of-trunk moulding, complex and canted trunk-to-base mouldings, canted base (which may once have been a little taller), shaped door top, plus the fact that the round dial itself at this time was

PLATE 286. *Thirty-hour longcase of the 1780s by James Monkhouse of Carlisle. Original blued steel hands.*

PLATE 287. *Simple oak case of the Monkhouse thirty-hour clock standing about 7ft.*

found more in the North than in the South.

James Monkhouse of Carlisle was born in 1713 and died in 1793. The thirteen-inch arched dial clock pictured in Plate 285 dates from the 1780s and is a thirty-hour pull-wind musical clock, the three dummy winding 'holes' being engraved on to the dial in a most convincing way to give the impression of a key-wind three-train eight-day clock. The winding holes are neatly designed into the overall dial pattern as if they are flowerheads, always a sign of thoughtful workmanship. This is a very complicated and high quality clock, but the dummy winding holes help it masquerade as something even more impressive.

The engraving is professional, though a little naïve. The recessed disc below XII gives a choice of four tunes by moving the lever – here set on tune H. The clock has a centre calendar shown by a brass arrowheaded hand rotating once a month. The time hands are in blued iron to give a deliberate contrast to avoid confusion between time hands and calendar hand. The minute hand is of the serpentine shape popular towards the end of the eighteenth century.

The spandrels are of the large question mark style suitable only for large dials of thirteen or fourteen inches in width; they were especially popular in the North-west. Dotted minutes and relatively plain and wide chapter ring are late indicators. The names of the first owners, Joseph and Mary Watson, are engraved across the arch, and this was probably a marriage clock, made for their wedding. Rolling moon dials are uncommon on thirty-hour clocks, but then this is a special clock.

The square dial clock by this same maker (Plates 286 and 287) dates from the 1780s and is more typical of his thirty-hour work. Dotted minutes are late indicators for a brass dial, as are the string-of-pearl spandrels. Here Monkhouse makes comparatively late use of the half-hour marker of a 'floating' flowerhead nature. The blued iron hands are original and of the non-matching type. The centre has a matted ground with the mouth type of calendar often used on thirty-hour clocks for cheapness. The fact that two of the three dial foot ends show at VI and I (but not the third near XI) suggests this is not the maker's poor workmanship but that these may have come loose at some time through rough handling and may have been re-riveted. The

PLATE 288. *Twelve-inch dial of a longcase clock by George Moyle of Chester made in the 1780s. Blued steel hands probably original. Centre engraving not as skilled as the chapter ring work.*

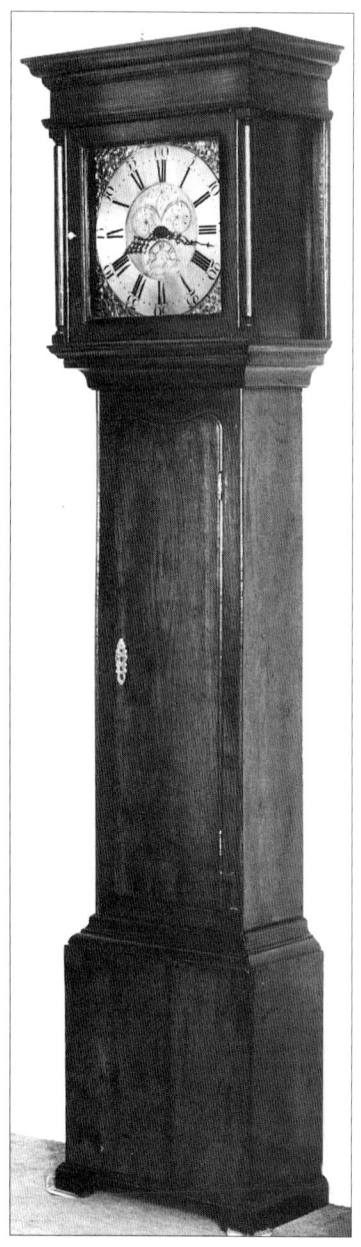

PLATE 289. *Oak case of the George Moyle clock standing 6ft.4in.*

engraving is professional and crisply done. The movement is of the plated type, as were all northern thirty-hours by this time.

The simple oak cottage case is original and highly typical of many from the region in the later eighteenth century. Shaped door top, busy moulds and slightly higher than wide base are the only distinctive features of the region. The turnbuckle door catch was often used on thirty-hour cases as it was easier than daily locking. The case is otherwise much like any simple cottage style of case of this period from almost anywhere.

The twelve-inch dial of the thirty-hour plated-movement longcase by George Moyle of Chester is very typical of many later eighteenth century examples, dating from the 1780s (Plate 288). Dotted minutes are virtually standard by now. The spandrels are a slightly unusual version of general C-scroll type. The mouth calendar is the normal system by this date. Chapter ring engraving is crisp and well done, but the dial centre is slightly off balance and shows less skill, which suggests Moyle may have done his own centre engraving. The blued iron hands appear original, the minute hand being of the serpentine style, the last style to be used in the non-matching form.

The case (Plate 289) is of oak, stands 6ft.4in. and is of the general north-western simple style of the day, the shaped door top, high base and complex moulds being the only regional features. Such simple cottage cases, however, seldom show the strength of regional styling to be seen in a more sophisticated case.

The clock by Emanuel Burton of Kendal (Plate 290) is a thirty-hour with dummy winding holes to give an eight-day appearance. A blued brass backing piece behind the holes adds to the effect of shadow. The subdial below XII, where a seconds dial might normally be placed on an eight-day clock, is in fact a calendar dial, but positioned here deliberately to add to the overall eight-day look. Dotted minutes are typical of many clocks of this period (1780s), especially in the North. All the engraving is finely done, especially the floral centre which is set on to a polished ground. Non-matching iron time hands are original, held by a screw fitting. The calendar hand is original but is of brass for deliberate contrast. The spandrels are of the late wreath pattern.

The dial by Lawrence of Lancaster (Plate 291) also dates from the 1780s and bears quite a resemblance to that by Emanuel Burton of Kendal (Plate 290). This too has dummy winding holes (with a blued backing sheet) and the

PLATE 290. *Thirty-hour dial of the 1780s by Emanuel Burton of Kendal with dummy winding holes to give eight-day appearance. Calendar positioned below XII to imitate a seconds dial. Superb engraving.*

twelve o'clock calendar positioned to resemble an eight-day seconds dial which also has dotted minutes and a fine floral engraved centre. The spandrels are the bordered wreath pattern. The original blued iron hands are held by a screw fitting. Again the original calendar hand is of brass for contrast.

The case of the Lawrence clock (Plate 292) is of oak with mahogany trim, including unusually narrow crossbanding. It has certain features of northern styling such as the ogee top-of-trunk moulding, complex trunk-to-base mouldings, tiny original ogee bracket feet. This case has some features favoured by the cabinetmaking firm of Gillow of Lancaster, such as the half-round (unshouldered) door top, and may well be by them, though their clock cases are not signed.

PLATE 291. *Similar dial of the 1780s by Lawrence of Lancaster, also with dummy winders and calendar looking like a seconds dial.*

PLATE 292. *Oak case with mahogany banding made for the clock by Lawrence of Lancaster. The case may be Gillows work. Height about 7ft.*

CHAPTER NINE

EIGHT-DAY SQUARE DIAL WORK

At the start of the eighteenth century London clocks were still being made with square dials, an example of which is shown in Plates 293 and 294 – a twelve-inch dial by John Miller of London dating from about 1700. Miller trained under Samuel Knibb, then under Joseph Knibb, and so was schooled in the best disciplines of the day. He worked from 1676 till 1702, after which date he is not heard of again. All the characteristics of the seventeenth century remained, though these changed slightly as time passed. Notably the minute numbers became larger, the half-quarter markers soon fell from use, the herringbone border passed from fashion (though not all makers used this feature anyway), and the half-hour marker soon changed in pattern. Within a very few years the square dial was replaced almost totally amongst London makers by the arched dial, and square dials only appeared occasionally in London work after about 1720 and then usually for clocks of a special nature such as regulators.

This clock by John Miller in its original walnut-veneered case is typical of the London clock of the day, an archetypal example. We can take this as a model not only for London square dial work (by those few London makers who continued square dial work) but for provincial eight-day work too by those numerous makers who carried on the square dial form.

At the start of the eighteenth century the provincial eight-day dial was based on that of London, rigidly by some makers, less so by others. Those makers closest to London were naturally more strongly influenced than those far distant, with town and city makers probably feeling more inclined to offer

PLATE 294. *The walnut-veneered case of the John Miller clock stands about 6ft.10in. and includes many typical early features of London cases of this period.*

PLATE 293 (ABOVE RIGHT). *Twelve-inch eight-day dial from a longcase by John Miller of London made about 1700. Seconds marked in 5s. Ringed winding holes with plain matted centre.*

PLATE 296. *Twelve-inch long-case dial of about 1710 by John Smallwood of Chelford. Many features of the London dial style. Fine original blued steel hands. Calendar box ringed to match winding holes.*

PLATE 295. *Oak-cased eight-day clock by John Ettry of Bishop's Canning, the base somewhat reduced. Very much a country case but clearly based on the London principles.*

work similar to that of the capital than would rural makers. Some clockmakers would even have bought their dials (or engraved work) direct from London, but we must assume that all provincial clockmakers knew what was happening in London and that they opted to vary their styles by choice and not by ignorance.

Clock casemaking in the provinces was also based on the London model but that too would follow or depart from it to whatever degree was decided by either the clockmaker himself or his customer in instances where the clock was made to advance order. The provincial casemaker was almost always using more humble woods than the burr-walnut veneered (oak carcase) cases being made in London – ebonised casework had largely fallen from fashion there by now, lacquer cases were yet to blossom and mahogany would seldom be available before mid-century. Oak was rarely used in London clock casework (except where covered by veneer or lacquer or ebonising) other than in a few special examples such as regulators. It is almost safe to say that oak was *never* used in conventional London domestic clocks. Oak examples seen today tend to fall into two categories: clocks re-housed in non-original cases and clocks in cases that were once lacquered and have been stripped down to bare wood on account of poor surface condition.

The oak case of this early eighteenth century eight-day by John Ettry of Bishop's Canning in Wiltshire, though reduced somewhat in the base, can be seen to follow the London concept, but in oak (Plate 295). This clock could be as late as the 1730s in fact. It stands about 6ft. Wrought iron hinges fitted on the surface (some were fitted internally) are a country touch, as London clocks by now always had high quality cast brass hinges, often of the cranked type, and of course neatly fitted internally. In general, however, this case

PLATE 297. *Twelve-inch dial from a longcase of about 1710 by Joseph Kirk of Skegby. Fine original blued steel hands. Matting tool grain shows here and there.*

PLATE 298. *The oak case of the Kirk clock stands 7ft.1in. and has numerous early features including exceptionally long trunk.*

follows the shape, proportion and principles of the London walnut case of the day, but in simplified form.

The twelve-inch eight-day clock by John Smallwood of Chelford in Cheshire dates from about 1710, and follows London styling in almost every respect other than its circular calendar box (Plate 296). London calendar boxes were almost always square, but some provincial makers seem to have felt that a ringed circular one better matched the ringed winding holes.

The twelve-inch dial by Joseph Kirk of Skegby in Nottinghamshire also follows London principles but varies in, for instance, the arrow markers for the seconds dial and the rather strange half-hour markers which are a cross between a fleur-de-lis and a flowerhead (Plate 297). It dates from about 1710. The original blued iron hands are particularly fine.

The oak case of the Kirk clock (Plate 298) stands 7ft.1in. and has the customary early features of: convex top-of-trunk moulding, cloth-backed fret to the hood, small hood side windows, integral hood pillars, shallow base, half-round beading to the door (and hood door in this instance). As a very general guide it may be helpful to think of the early (seventeenth-century) convex top-of-trunk moulding as changing to concave in the eighteenth century, but it seems to me to be not nearly so firm a guide as some books would suggest and there are many exceptions.

The twelve-inch dial of the clock by William Farrer of Pontefract in Yorkshire (Plate 299) dates from about 1710-20. The clockmaker was working there by 1707 and is believed to have died in 1726. Some of his work is of highly

PLATE 299. *Twelve-inch dial from a longcase of about 1710-20 by William Farrer of Pontefract. Original blued steel hands. Birds-and-basket theme to the calendar box. Early use of floating half-hour markers.*

sophisticated nature and this present clock is no mean example. The dial is very up to date and has several stylistic features of London work – floating fleur-de-lis markers, winding holes now *un*ringed (as became the new fashion there), spandrels of the latest type (female head).

Certain features, however, are provincial – the retention of half-quarter markers currently dropping from London use, the original blued iron hands of non-London pattern and the birds-and-basket engraved theme on the matted centre. This birds-and-basket feature appeared on many provincial dials of the matted centre style from about the beginning of the century to mid-century, principally in southern regions, and in Pontefract we are getting towards its northern limit. The birds (doves) signify peace and the basket of fruit signifies plenty; this theme was most popular in central England, though little used in London. Its use was so widespread over such a long period that it cannot have been simply the work of one engraver. The matting of the centre is more coarsely done than with a London dial. What might appear to be rivets at XII and VI are in fact drill holes where the chapter ring was pinned to a bed during engraving. This dial then is based on the London concept, though clearly not a London dial.

The oak case (Plate 300) stands about 6ft.6in. and is of a most interesting and handsome style, again based loosely on the London principle but very provincial, though as yet showing no features distinctive of any particular region. It is slightly old-fashioned when compared with contemporary London tastes. The topmould is still convex and the trunk-to-base mould is almost a

PLATE 300. *Oak case of the Farrer clock standing about 6ft.6in. Handsome style with many early features including full height hood side windows and heavy topmould overhang.*

PLATE 301. *Twelve-inch longcase dial of about 1710-20 by George Mills of Ripon. Original blued steel hands. Four leaf clover winding squares. The half-quarter markers have now been dropped. The four pillar movement has inside countwheel striking.*

PLATE 302. *Longcase dial of about 1710-20 by Daniel Tantum of Nottingham. The unusually thin dial sheet has been repaired here and there. Blued steel hands probably original. Irregular matting.*

quarter-circle mould, both early London features but used later here. The D-mould door lip and lenticle window are also retained from the earlier London fashion. The hood still has side windows, as some country cases still did in many areas, but this example has very large windows taking up almost all the available hood side space, as happened only with very early London cases such as those on some Fromanteel clocks.

The integral hood pillars, with quarter rear pillars, have narrow, long necks and a well-balanced and graceful entasis (a tapered swell), quite different from those on London cases. The hood topmould is large and heavy with a wide overhang, a typical early provincial feature, and this is exaggerated by the fact that the hood width narrows just above the glass door, the latter an odd feature of some early country cases, where the narrowing can be much more pronounced than here. Altogether this is a handsome and quite dainty case – compare with the much heavier example on the clock by Robert Davis of Burnley in Plate 306.

The twelve-inch dial, four-pillar eight-day clock by George Mills of Ripon (Plate 301) dates from about 1710-20 and has inside countwheel striking, a system retained by many provincial eight-day clocks until mid century and in some cases long after. This is a mixture of old and new styles. Old-fashioned by now are the *circular* calendar aperture, the meeting arrowhead half-hour markers and also the use of 'of' in the signature, 'Geo. Mills of Ripon'. Modern features of style are the female head spandrels and the absence of half-quarter markers. The matting of the dial centre is coarse. The winding squares have their ends filed four leaf clover fashion, a nice touch and

PLATE 304. *Hood detail of the Tantum clock showing integral pillars and much use of D-mouldings.*

PLATE 303. *Simple oak case of the Tantum clock standing about 6ft.6in.*

something done only by occasional makers here and there though over quite a wide period. The original blued iron hands are of good style. Note the chapter ring is riveted at XII and VI, a sign of rustic work.

The clock by Daniel Tantum of Nottingham (Plates 302 to 304) is very difficult to date but could be about 1710-20. The very thin dial sheet can be seen to have been buckled here and there, even in the dial centre, where the matting can be seen to be uneven and where the datebox engraving and perimeter chasing can be seen to be less than perfect – all indicative of the man doing his own work. The chapter ring engraving is more professional, suggesting outside work. Twin cherub spandrels are here of the (second) later type with the crossed maces and large crown. Fleur-de-lis half-hour markers show up-to-the minute styling.

The Tantum oak case (Plates 303 and 304) stands about 6ft.6in. and retains the early features of lenticle window (here with bull's eye glass), D-mould door lip, and hood side windows. The trunk moulds are now of the concave type. Integral hood pillars are conventional. The hood topmould is unusual in having a D-mould shape worked into its outer edge.

The Robert Davis of Burnley, Lancashire, twelve-inch dial shows some very countrified stylistics (Plate 305). It could date between 1720 and 1730 or a decade later. Nothing is known about the maker except that he was working there in 1723, the date on the thirty-hour clock by him in Plate 221, and that another longcase by him is said to be dated 1695 on the dial. The female head spandrels, sophisticated calendar box engraving, and ten-digit numbered seconds dial (instead of the usual five) are all 'modern' features. However, his

PLATE 305. *Twelve-inch longcase dial of about 1720-30 by Robert Davis of Burnley. Engraver's slip between XII and I. Blued steel hands probably original. Seconds numbering now in units of ten.*

half-hour markers are a strange fleur-de-lis with a crossed top (note the engraving slip between XII and I), and his half-quarter markers are of the old meeting arrowhead type. The somewhat quirky engraving is probably his own – he runs the hour line right through the minute numerals as, for example, between 1 and 0 in 10 and between 1 and 5 in 15, and this was not usual practice. The chapter ring is riveted at XII and VI. The original blued iron hands are of his own decidedly eccentric pattern, which adds to the interest.

The oak case is unusually tall and heavy (Plate 306). It retains the lenticle window and glass side windows to the hood, yet has an arched top to the D-mould edged door, a feature more usual with the newly-fashionable arched dial clock. His integral hood pillars have heavy caps and bases, which most unusually are gilded. The hood fret runs along the sides as well as the front, an unusual treatment. The whole is reminiscent of the Kirk case in Plate 298, though treated more heavily.

James Speight was working in Skipton, Yorkshire, from at least 1713 till his early death in 1721. His working life probably spanned no more than these

PLATE 307. *Twelve-inch longcase dial by James Speight of Skipton about 1715. The 'de' in the signature is an early sign. Clover leaf winding squares. Birds and basket calendar.*

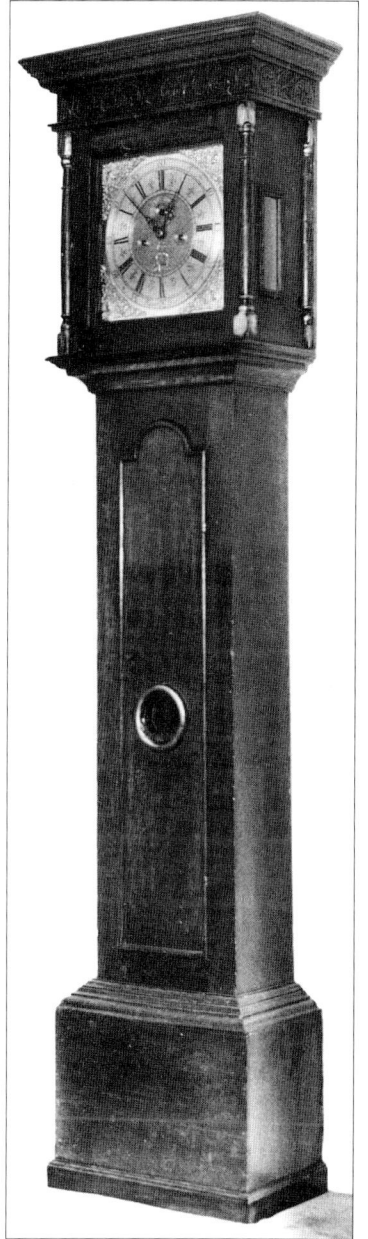

PLATE 306. *The oak case of the Davis clock is quite heavy in proportion and stands about 7ft.3in. Fretting on the hood sides is unusual.*

eight years and his clocks are seldom seen, the one pictured in Plate 307 being the only one I have ever seen by him. The twelve-inch dial eight-day clock dates from about 1715, give or take a few years, and is signed 'Jam. Speight De Skipton', the use of 'de' as an alternative to 'of' being almost always an early sign. We know the clock cannot date later than 1721 and some features of it are surprisingly modern for so early a date.

The birds-and-basket theme is unusual this far north. The floating half-hour fleur-de-lis and the female head spandrels are both very up to date stylistic features. The four leaf clover winding squares (sometimes found in early London work but by this date still in occasional use as a provincial feature) are here present with ringed winding holes on what is a very stylish dial with highly competent engraving. The blued iron hands, though the right size, look too late in period to be original.

It will be obvious from those examples examined so far in this chapter that we are looking at clocks strongly based on London influence yet beginning to show a variance and variety according to the taste of the individual provincial clockmaker. That variance as yet has barely any regional pattern, the departure from London fashion being more a matter of the individual whim of the clockmaker than of the area where he lived.

PLATE 308. *Twelve-inch longcase dial of about 1710 by Leyson Williams but the chapter ring re-engraved some years later in the newer style of perhaps 1750-60 to 'modernise' the clock. Original blued steel hands. Very fine matting.*

PLATE 309. *The Leyson Williams dial with the chapter ring reversed to show its original forward side, including earlier stylistic features such as half-quarter and half-hour markers and quarter-hour internal chapter ring divisions.*

The clock illustrated in Plates 308 to 311 involves a little detective work, being a twelve-inch eight-day clock signed 'Leyson Williams', a clockmaker hitherto unrecorded and therefore offering no clues as to period. Several aspects of this dial (Plate 308) suggest a clock based strongly on London styling of about 1710 – twin cherub spandrels, ringed winding holes, square box calendar with a little scroll engraving around it, fine original blued iron hands. The seconds ring is marked every tenth unit rather than every fifth, a feature possible for this date, but more commonly found later.

However, we must date a clock from its *newest* stylistic feature and in this case that is the chapter ring itself, which shows floating fleur-de-lis (possible over a fairly wide period) but without any half- or quarter-hour features on the inner edge of the chapter ring, a style not usually appearing until the 1760s. The movement (Plate 310) is of fine (London) quality with five finned pillars and inside countwheel strikework (countwheel on the main barrel), and that too shows every sign of high quality work of about 1710, a style which had moved on considerably by the 1760s. The case is of solid walnut and stands about 6ft.6in. (Plate 311). This is clearly a provincial case and it too seems by its style to date from the 1760s. Was it perhaps a provincial clockmaker working in the 1760s in an old-fashioned style and using spandrels and hands that were long outmoded – a possible but unlikely explanation?

This was a very puzzling clock. On dismantling for cleaning it was found that the chapter ring was engraved on what is now its reverse side, but had originally been its face side, and this did have those earlier features we would expect about 1710 – meeting arrowhead half-hour markers, half- and quarter-hour chapter ring divisions and even half-quarter asterisk style markers again using arrowheads (Plate 309). Clearly the clock dates from 1710 but had been 'modernised' in the 1760s to the style of the day, the present case being built then perhaps to update its style or to replace a damaged original. Examination

PLATE 310. *Fine London-quality movement of the Leyson Williams clock with five pillars, solid dial sheet, half-round wheel collets.*

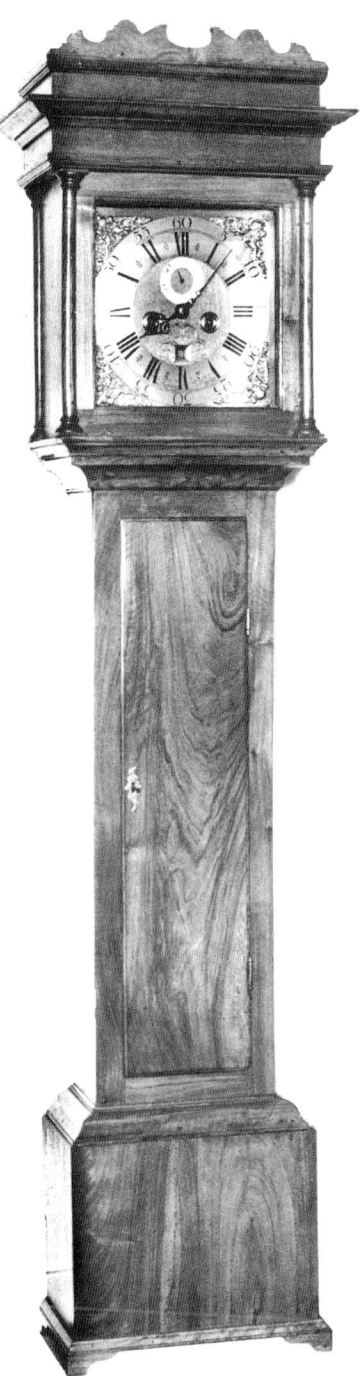

PLATE 311. *Solid walnut case of the Leyson Williams clock dating from perhaps 1760. The double pediment is typical of cases from the Bristol area. Height about 6ft.6in.*

showed that it had the original (1710) seatboard, but a sidepiece had been added to widen it a little to fit the new case.

The case has a feature which is traceable to a certain region, namely the Bristol Channel area, in that its flat top has a second top, an extra pediment, itself topped with a cresting of an embryonic swan-neck shape. Research too complex to explain here showed that Leyson (sometimes Leyshon) was a surname used as a first name, and that by 1710 that surname was known only in South Wales. The clock, then, was made in South Wales by the previously unrecorded clockmaker Leyson Williams (a good Welsh surname too) about 1710 and modernised in a most unusual way in the 1760s.

PLATE 312. *Twelve-inch dial of fine quality made about 1710-20 by John Stevenson of Stafford. Birds-and-baskets theme again. Floating half-hour markers with herringboning represent modern and traditional features respectively.*

PLATE 313. *The oak case of the Stevenson clock has many early features such as barley twist pillars and lenticle glass. The carving on this example looks ancient.*

PLATE 314. *Twelve-inch (cartwheel-cast) dial from a longcase by Roger Parkinson of Richmond, Yorkshire, dating from about 1710-20. Crude matting. Skilled yet naïve engraving. Spandrels lost.*

PLATE 315. *Twelve-inch dial from a longcase by Edmund Bullock of Ellesmere, Shropshire, dating about 1720 and numbered 417. Rough matting. Bizarre half-quarter markers.*

The twelve-inch eight-day clock by John Stevenson of Stafford (Plate 312) dates from about 1710-20. The work is of London quality and to some degree London styling too, but incorporates the birds-and-basket theme, which is almost always provincial. Herringbone bordering was by now beginning to fall from use in London. This is a fine dial with excellent engraving. The hands are modern replacements in correct style.

The original case is of oak, carved and stained black, and stands 7ft.1in. (Plate 313). The heavy overhang to the hood, slightly unusual barley twist integral hood pillars, D-mould beading to the door and to the hood side windows, and convex top-of-trunk moulding are all interesting provincial features. The age of the carving is uncertain.

The twelve-inch dial by Parkinson of Richmond, North Yorkshire, is all that remains of this particular clock (Plate 314). It dates from about 1710-20. Roger Parkinson is recorded at Richmond from at least 1711 till the 1740s. The spandrels are now lost but their absence reveals that the dial sheet is of the cartwheel-casting type and the gaps can be seen behind the slightly undersized chapter ring in the corners, where once spandrels would have concealed these.

Half-hour markers are large and ornate, usually a northern indicator, and so too are the unusually large half-quarter markers. The matting is uneven. The herringbone border is far less expertly done than in the dial by Stevenson of Stafford (Plate 312). Ringing around the seconds aperture was to match the other holes, but here the maker has got his seconds hole slightly off-centre with his seconds chapter ring. Such features, together with the unusual figure style of, for example, the 1 and 2 numerals, suggest the work of Roger Parkinson himself and not an outside engraver.

Edmund Bullock of Ellesmere, Shropshire, was working from at least 1708, maybe earlier, till his death in 1734. He numbered some of his clocks, this twelve-inch eight-day clock (Plate 315) being number 417, the highest number I know of his. Bullock's style is a bit idiosyncratic and his clocks are difficult to date. This one would seem to date about 1720, though his numbering system might suggest a little later.

PLATE 316. *Twelve-inch dial of about 1720-30 by Francis Tantum of Loscoe in Derbyshire. Bold engraving, the centre pattern wound round the winding holes. Unusual wide-mouth calendar. Original blued steel hands.*

PLATE 317. *Movement of the Tantum clock showing cartwheel dial sheet, finned pillars now less ornamental than earlier.*

Stylistic features on the dial are conventional for the day with the exception that his half-quarter markers are a little bizarre, a sort of sprouting stalk (based on an arrowhead) with the same motif repeated on the seconds ring. Oddly enough he uses that same motif on another clock (the arched dial in Colour Plate 2) and I am inclined to think that Bullock did his own engraving. His calendar box border is barely even and not quite square. The general boldness and heaviness of the dial style is almost always a feature suggesting northern work.

Francis Tantum of Loscoe, a small village in Derbyshire, is a well-known and respected clockmaker of the early eighteenth century, though nothing seems to be known about his life. He is presumably related to Daniel Tantum of Nottingham (see Plate 302) but we do not know in what way. The twelve-inch dial eight-day example shown in Plates 316 to 318 dates from about 1720-30, perhaps a little later even – it is very difficult to pin this one down on account of his unusual style. The half-hour markers and half-quarter markers are unusually large and bold, though based on a fleur-de-lis. His minute numbers are large, perhaps suggesting a date towards the 1740s. Large, bold markers and large minute numerals this early tend to suggest northern work. Indian head spandrels and original blued iron hands were patterns used over a considerable period. Seconds marked every tenth unit form the newer method gradually replacing five unit marking, but both calibrations are possible at this period.

The newest feature of the dial is the calendar which is now of the 'mouth' form, but probably a prototype version as the actual mouth itself is used just for decorative scrollwork with the number of the day still shown through a square box, here three-sided. This is the twelve-hourly knocked-on calendar system used widely in thirty-hour work and less often with eight-day clocks. The matted centre (showing tool marks) is engraved with encroaching scrollwork twisting round the winding holes, a style more often associated with the North-west. The flower head above VI is there to balance that in the seconds circle, but also to conceal the riveted end of his calendar wheel post.

The movement (Plate 317) shows the dial to be a cartwheel type of casting which we can begin to expect as normal now in northern work. Pillars are still decorated with some finning, but less boldly than with earlier ones.

The oak case (Plate 318) stands 7ft.4in. and has a shaped pediment not much used on square dial work, known usually as a break-arch top and more often found on arched dial clocks. Some early features still persist – for example, side windows to the hood, pillars attached to the hood door. We have now reached a period where walnut crossbanding was sometimes used in quite a wide band about an inch wide and divided, as here, from the oak it surrounds by a double yellow string line within which is set a yellow and black chequered string. The banding can be seen clearly on the hood door, main door and base, and the stringing alone as a decorative motif within the hood pediment.

The most striking thing about the case is its starburst inlay to the door and base area. These are sometimes called sunbursts, though that term is more appropriately used to describe a half circle, as a sun's rays might appear above the horizon. Starbursts are usually inlaid in several fancy woods of varying colours (yellow contrasting with black was popular), often purpose stained and therefore unidentifiable. Some have straight 'rays' compass-rose fashion and others may have wavy 'rays'; some have a mixture of both, as here. Some may have a cartwheel style centre, as this one does. Starbursts were popular, principally to enliven oak cases, in the 1730-50 period, but examples are found outside that period.

The twelve-inch eight-day clock by Richard France of Warrington (Plate 319) dates from perhaps 1725-35. The maker died in 1740, leaving in his will 'to the clockmakers that come to my burial, each a pair of gloves' (black gloves were worn for mourning). The fine original blued iron hands are based on a London pattern, but the dial is very provincial, even very northern. The two tiny spandrels alongside the seconds dial are actually arch (maskhead) spandrels for a bracket clock but were occasionally used in this manner by a handful of clockmakers early in the eighteenth century, mostly in the North. The ringing of the winding holes is now in a rather different, plainer manner than early ringing, a form which lingered with some northern makers, especially in Lancashire. Half-quarter markers are still in use, though gradually falling from fashion. The calendar border engraving is spreading.

The case is in walnut, bookmatched on the door (unusually top-to-bottom) and crossbanded on door and base (Plate 320). Ogee top-of-trunk mould and complex trunk-to-base mould are pure Lancashire styling. The box top to the hood is not so common with a flat-top (i.e. square dial) case as on some southern clocks. Hood side windows are a traditional feature, seldom used this late in the North. Hood pillars are free-standing with forward facing half-pillars against the backsplats, the latter mostly a northern feature. The height is about 7ft.8in.

PLATE 318. *The oak case of the Tantum clock is crossbanded in walnut and stands about 7ft.4in., having a break-arch pediment, rather uncommon on a square dial. Starburst inlays are a feature of some clocks of the second quarter of the century.*

PLATE 319. *Twelve-inch dial by Richard France of Warrington dating from about 1730. The two small spandrels beside the seconds dial were used by just a few makers at this period. Fine original hands in blued steel.*

PLATE 320. *The France case is in book-matched walnut veneer and stands about 7ft.8in. It has certain regional features including ogee mouldings and forward facing half-round pillars to the backsplats.*

PLATE 321. *Longcase dial of about 1730 by William Nicholson of Whitehaven. Good engraving. Four leaf clover winding squares. Original blued steel hands.*

PLATE 322. *The case of the Nicholson clock is in walnut, solid in construction but bookmatch veneered in quarters (on pine) for the door. About 6ft.6in.*

COLOUR PLATE 13. *Eleven-inch dial of a thirty-hour longcase of about 1750 by William Porthouse of Penrith, made for Thomas and Mary Arnison. The bold Gothic lettering was used by some Lake District makers at this period. See page 175.*

William Nicholson of Whitehaven in Cumberland is a maker about whom no facts seem to be known, but his eight-day clock pictured in Plates 321 and 322 dates between about 1720 and 1740. The engraved work is well done in a relatively restrained style, though still bold enough in the half-hour and half-quarter markers to be recognisably northern. The ringing of the winding holes has also simplified somewhat, a northern characteristic with those makers who continued to use ringing. Four leaf clover filing of the winding square ends is now only a whim of certain makers. The well-shaped calendar box retains the wheat-ear border from earlier times but the engraving now

PLATE 323. *Oak-cased longcase of about 1730 by Joseph Shepherd of Sheffield. Exceptionally long in the trunk. The door has a tapering double-D mould, higher on the outer edge.*

spreads further into the matted centre zone. The good original iron hands are relatively simple.

The case of the Nicholson clock is in walnut, used as solid for most of the construction but in bookmatched veneer on the trunk door, being the main eye-catching area, here bookmatched in quartered form. The hood pillars are still attached to (integral with) the hood door, an old-fashioned style now for this area. This is an unusually simple case, perhaps because it is slightly old fashioned for its day and retains the simple style more often seen in earlier clocks. The height is about 6ft.6in.

The oak-cased eight-day clock by Joseph Shepherd of Sheffield, Yorkshire, dates from about 1720-30 (Plate 323). The maker is recorded as working there by 1711 and was still there in 1730, perhaps later. The case shows no particular regional styling and retains the early features of hood side windows and integral hood pillars, as well as the long slender trunk. The arched top to the trunk door, even on a square dial clock, is usually indicative of the arrival of the arched dial period, i.e. post 1710 or even post 1720, though there are exceptions. The door lip moulding, which appeared as a half-round D-mould on earlier cases, has now developed into a more complex mould, still projecting slightly beyond the door, its purpose of course being to conceal any gap between door and frame. Brass H-hinges in this instance are fitted half inside and half outside, the earlier type of wrought-iron hinge being now outmoded except on primitive, rustic cases. The height is about 7ft.4in.

The twelve-inch dial by Thomas Sillito of Uttoxeter in Staffordshire (Plates 324 and 325) dates from the 1730s. Here the maker deviates from the London style by his large half-hour markers, his continued and late use of half-quarter diamond markers (sometimes called lozenge shape) and the greater spread of his date-box surround engraved work into the matted centre. The matted centre itself no longer has winding holes ringing and his name is on an attached plaque, both the latter being London influenced features.

The oak case stands 6ft.3in. and is of simple country style yet has a restrained proportion based on the London (walnut, of course) concept of a few years earlier. The pillars are integral with the hood door and the arched door top has a D-mould edge bead, both being London derived features. The base looks a little shallow and may once have been slightly higher. It has nothing strongly regional about it (such simple cases seldom have), but these latter features indicate a more southern than northern influence.

It is interesting to compare the twelve-inch dial by John Vise of Wisbech, Cambridgeshire (Plate 326), with that by Sillito (Plate 324), dating as it does from a similar period or perhaps slightly later (1730s or early 1740s). Cambridgeshire is slightly further south than Staffordshire and considerably closer to London, so that it is not surprising to find a stronger London influence here. The dial centre is very plain, simply matted with no winding hole ringing and no engraving at all to the plain calendar box. Name-plate signing is more common in London-influenced clocks than those further afield, especially those further north. John Vise's half-hour marker is much smaller than Sillito's and 'floats' higher. This too derives from London influence, as in some London clocks the half-hour marker has disappeared completely by this time.

The simple oak case (Plate 327) which stands 6ft.7in. is not unlike Sillito's (Plate 325), the main difference being the door lip mould which on John Vise's clock is a thumbnail mould worked into its edge rather than an applied D-mould. This thumbnail mould became the most widespread manner of edging a door from now on.

PLATE 324. *Twelve-inch dial of the 1730s by Thomas Sillito of Uttoxeter. Modern in the ten-digit seconds marking, traditional in retaining half-quarters. Original blued steel hands.*

PLATE 326. *Twelve-inch longcase dial of the 1730s by John Vise of Wisbech, making interesting comparison with the Sillito dial (Plate 324) – plainer centre, smaller 'floating' half-hour markers, no half-quarters, five unit seconds. Original blued steel hands.*

PLATE 325. *Oak case of the Sillito clock standing about 6ft.3in. Traditional and simple in style, but 'modern' in the arched door top shape. Hood side windows are by now old-fashioned in the North.*

219

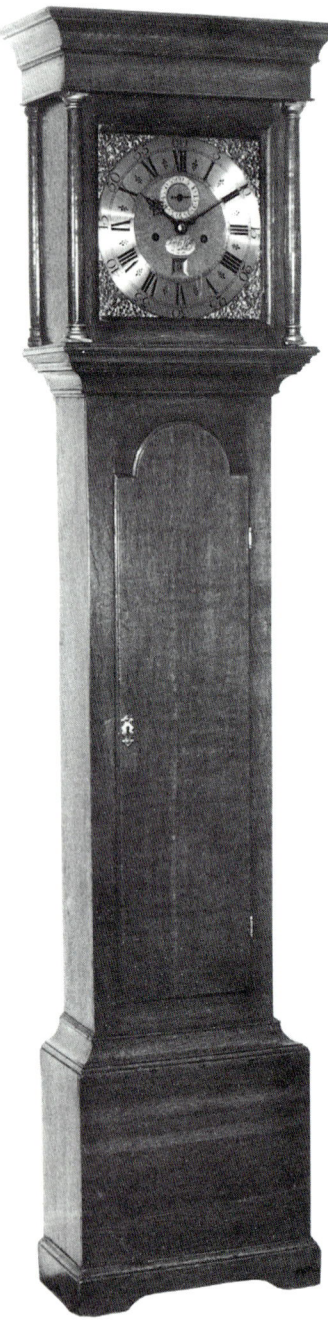

PLATE 327. *Simple oak case of the Vise clock standing 6ft. 7in. Similar to the Sillito case (Plate 325) but without D-mould to the door and with concave rather than ogee mould to the base.*

COLOUR PLATE 14. *Twelve-inch dial of the 1750s by James Sandiford of Manchester in a style highly typical of the area. This and the dial in Plate 338 were probably engraved by the same man, yet the numbering shapes differ. The overlong hour hand may be a replacement. See page 226.*

The twelve-inch dial eight-day clock signed 'John Taylor' (Plates 328 and 329) is extraordinary in several respects. This John Taylor is believed to be the one who worked at Ashton under Lyne, near Manchester, Lancashire, who normally signed without a place name. The clock probably dates from the 1730-40 period, though just possibly a little earlier – the maker died in 1744.

First it has the calendar below XII, a position seldom employed as it was normally used for a seconds dial. The fact that this is a single-handed eight-day clock, an exceptionally unusual combination, may have a bearing on that, as a customer who didn't want minutes would hardly have wanted seconds registration. This high-positioned calendar may account for the low positioning of his winding squares, to try to make a more balanced dial. His winding holes are still ringed, a feature retained longer in the North. His calendar hole is also ringed and is circular, both traditional features lingering longer in the North.

His half-hour markers are based on the fleur-de-lis but are large and, although of the floating type, in fact almost meet up with the inner quarter-hour band in the manner of attached ones. On the outer chapter ring edge is a very unusual decorated border of herringbone type adapted to a circular pattern. This was probably done to fill the vacant space usually taken up by minute numbering on a two-handed dial (single-handed dials are not usually this large with such a broad chapter ring). The fine blued steel hand is original, based on a London hour-hand pattern, and has an unusual curved tail for leverage when setting the time. The engraving is strong and sharply executed, but note the small slips on the upstrokes of XII and IIII.

COLOUR PLATE 15. *Thirteen-inch longcase dial of the 1760s by Thomas Richardson of Weaverham. Blued steel hands probably original. See page 228.*

PLATE 328. *This twelve-inch single-handed dial of the 1730s by John Taylor of Ashton under Lyne is exceptionally rare being an eight-day single-hander. Fine original blued steel hand. Still datable by style but this is a mixture between eight-day and thirty-hour styling. The mark on the half-hour marker between XII and I is an imperfection in the metal.*

PLATE 329. *Oak case of the John Taylor clock standing 6ft.4in, with original caddy top. Several features locate this as a distinctly north-western case.*

PLATE 330. *Unusual thirteen-inch longcase dial of about 1740 by William Davenport of Stockport showing an early example of the matted background with overall engraving. The place names between hours are exceptionally uncommon and constitute a simple form of world time dial. Original blued steel hands.*

PLATE 331. *The oak case of the Davenport clock stands about 7ft.3in and is very reminiscent of the John Taylor case (Plate 329). Typically north-western.*

The oak case stands only 6ft.4in. and is distinctly north-western in several respects: caddy top, free-standing hood pillars (though of early slender tapered form), multi-stepped arch to the door top, slightly heavy and stocky proportion.

The oak case of the eight-day clock by William Davenport of Stockport (Plate 331) makes an interesting comparison with that of the John Taylor clock (Plate 329), having many features in common. These are: multi-stepped arch door top, separate hood pillars, distinctive caddy top. Here the caddy front has round holes to let out the sound of the bell. Stockport is very close to Ashton under Lyne and these two cases show distinct similarities of style, though the Davenport clock dates from perhaps ten years later than the Taylor one. The raised base panel of the Davenport clock here has clipped corners, a feature of many later north-western cases. Such base panels usually also have the same thumbnail mould as found on the door of both of these cases.

The thirteen-inch dial by William Davenport (Plate 330) has progressed in several ways from that by John Taylor and dates between 1735 and 1745. The maker was apprenticed to John Shepley of Stockport in 1728 and worked there till he died in 1760. This dial shows its later period and its northern styling in: larger size, extensive engraved work on to the matted background of the dial centre, use of the large cherub head spandrel of a type especially popular in Lancashire and sometimes known as a Lancashire cherub head.

The most interesting feature of this clock, however, is that it is a 'world time dial'. A dot on the inner chapter ring marks the position when the hour hand indicates noon in each country specified. For example, when the clock shows 12.30 British time, the hour hand reaches the appropriate dot to indicate noon at Cadiz. Places round the world have been selected to fall conveniently within each hour band. It is probably for this reason that the lozenge half-hour

PLATE 332. *Fine twelve-inch longcase dial of the 1740s by Jonathan Lees of Bury with finely-matted engraved centre. The engraving is interrupted by the winding holes. Typical north-western half-hour markers of the period. Original blued steel hands.*

markers are exceptionally small – to leave space for the lettering of the country. It may have been that the clock was made for some merchant with foreign interests, or perhaps just to impress the neighbours. This is only one quite simple form of 'world time dial' and others are known, some very complicated indeed, but any world time dial is an item seldom met with.

The eight-day clock by Jonathan Lees of Bury, Lancashire (Plates 332 and 333), dates from the 1740s. The maker was working there by the 1730s but later moved to Middleton where he died in 1785. The case is very much of the same style as the preceding caddy top cases of the Taylor and Davenport clocks (Plates 329 and 331) and has many stylistic features in common with them, all very north-western, including the ogee top-of-trunk moulding. This case, however, is carved very finely all over – whether at the time of making or later is undeterminable. The case stands about 7ft.6in.

The dial is superbly engraved in the north-western style of lavish floral engraving on to a matted centre. It will be noticed that there is no seconds dial (which was a standard feature on most square dial eight-day clocks) and that the winding holes cut through the engraved pattern. The latter can be a sign of a clock with a non-original movement, but that is not always the case and is not the case here. The implication is that when the clockmaker engraved (or bought) his dial he had intended this for a thirty-hour clock and had a last minute change of mind so that he has made do with a thirty-hour dial for this eight-day example.

The chapter ring has fine and bold half-hour markers based on a fleur-de-lis but developing more into a twisted leaf effect, which became very popular in the North-west. The half-quarter markers, here retained later than with some makers, take up the same theme. The original blued steel hands are well made and based roughly on a London design. Such bold and busy engraving is almost exclusively north-western.

PLATE 333. *The caddy-topped oak case of the Lees clock stands about 7ft.6in. and is finely carved and blackened to the front. The age of the carving is undeterminable.*

COLOUR PLATE 16. *Twelve-inch longcase dial with rocking Father Time made about 1755 by Francis Moore of Ferrybridge. The original London influence is now so varied as to be hardly recognisable. Original blued steel hands. See page 292.*

PLATE 334. *Longcase dial of the 1740s by John Holroyd of Wakefield. The calendar is an engraved dummy. Ringed winding holes continue with some northern makers.*

The eight-day clock by John Holroyd of Wakefield, Yorkshire, dates from the 1740s. He spells his name Holroide on this dial (Plate 334) but he was not consistent in that. In this example he too did not bother with a seconds dial, nor even with a calendar – his calendar 'box' is not pierced through but is simply a blank zone with an engraved border to give the superficial appearance of a calendar. A dummy calendar is very unusual and suggests he was penny-pinching. His winding holes are still ringed, a fashion now only common in localised areas. The bold half-hour markers are typically northern. The large area of undecorated dial centre shows southern influence.

The oak case of the Holroyd clock (Plate 335) is crossbanded in walnut in a strange way in that the main door has an unusual fielded panel and it is the edge of the raised inner section which is trimmed in this way. The case has a form of caddy top, a little plainer and simpler than the Lancashire version we have just seen in Plate 333. Pillars are still integral with the hood door, by now becoming an old-fashioned feature in the North, as too is the lip mould to the door. Some casemakers moved much more slowly with fashion than others. The height is about 7ft.3in.

The twelve-inch solid-dial eight-day longcase clock by Thomas Furnival of Taunton in Somerset dates from the 1740s. This is a little known maker and one of this same name recorded in the 1760s as working in Sheffield must surely be the same man. This dial is very much in the London styling, and could almost be a London dial (Plate 336). The typical features which indicate this are: undecorated matted centre with unringed winding holes, engraving limited to the square calendar box rim, signature on an applied plaque, tiny restrained floating half-hour markers of the fleur-de-lis type, solid dial sheet (northern ones by now are almost always of the cartwheel casting type), eagles-with-urn spandrels (used in and out of London). Somerset is a long way from London, but this styling is very much London based. The blued steel hands

PLATE 335. *Oak case of the Holroyd clock standing about 7ft.3in. The door has a fielded panel, unusual in longcase work. Walnut crossbanding.*

225

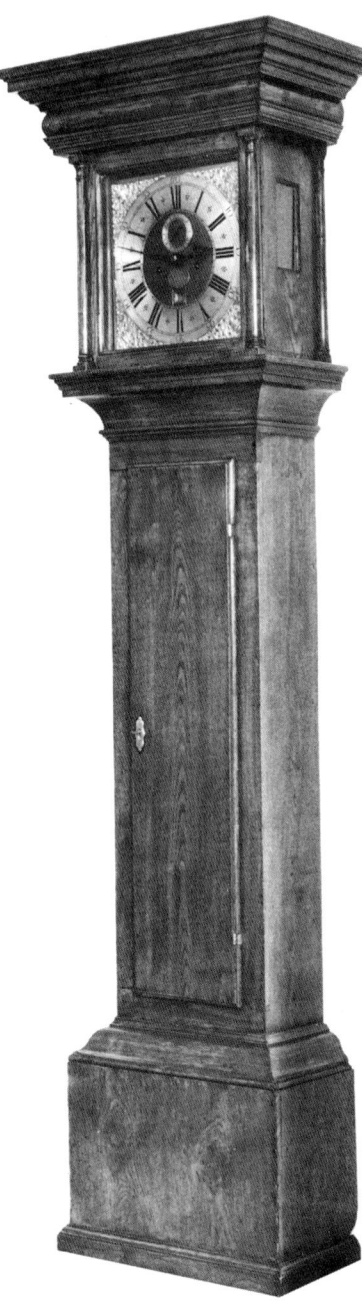

PLATE 337. *The case of the Furnival clock is in elm and stands 6ft.10in. Very simple and somewhat old-fashioned styling in retaining side windows and a very late use (for the South) of cushion moulding.*

PLATE 336. *Twelve-inch (solid sheet) dial of the 1740s by Thomas Furnival of Taunton with original blued steel hands. Southern indicators include tiny, floating half-hour markers and unengraved matted dial centre.*

are original and based only vaguely on London patterns. The seconds are marked every tenth unit which is now gradually becoming the new fashion in all areas (replacing the former five unit marking), starting earliest in London.

The case (Plate 337) is of elm, a wood not often seen today in early longcase work, though probably much more commonplace originally – woodworm must have destroyed many such cases. It is of simple country style, as we might expect. Some early features are retained in this example a little later than usual, as we might expect in a rural, traditional county – heavy overhang to the hood topmoulds, cushion mould in the upper hood. Still in fashion in the South are integral hood pillars, hood side windows, flat-top door.

Two twelve-inch eight-day clocks by James Sandiford of Manchester are illustrated in Colour Plate 14 (page 220) and Plate 338. They date from the 1750s and are of the cartwheel dial type which was now virtually standard in northern clocks. The maker was born in 1725 and died in 1775. His dials are similar, with matted dial centres enlivened by a considerable spread of floral engraving. In neither instance is his engraved design broken by the winding squares, implying planning in his layout. One dial incorporates the strange exotic bird which was popular on many north-western dials and is not a representation of any particular species. His half-hour markers are the twisted leaf form which became a feature of very many north-western dials. Ten unit seconds marking is by now universal standard practice. The spandrels are the string-of-pearls pattern, very popular on eleven and twelve inch dials especially in the North. The blued steel hands are of the correct style but the hour hand in Colour Plate 14 is overlength and may well be a replacement.

The oak case of the James Sandiford clock in Plate 339 is crossbanded

PLATE 338. *Twelve-inch dial of the 1750s by James Sandiford of Manchester very similar to the one in Colour Plate 14 (page 220). Original blued steel hands.*

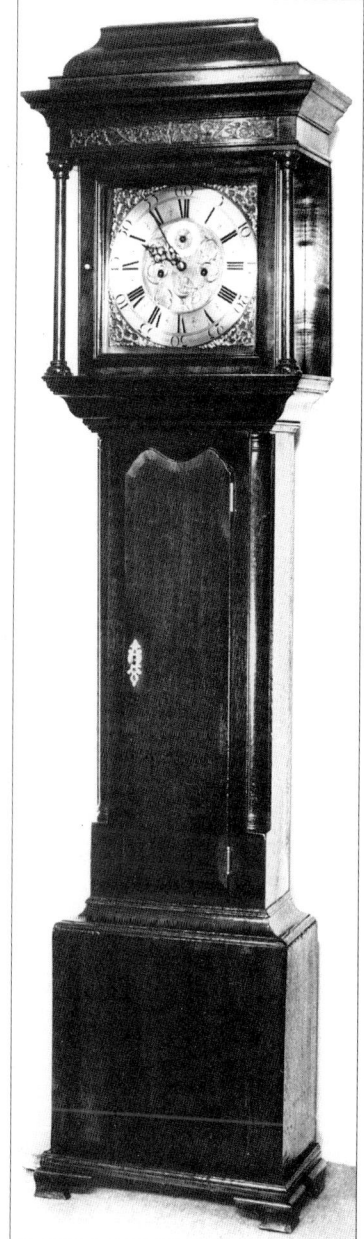

PLATE 339. *Oak case with mahogany crossbanding of the second Sandiford clock (Plate 338), standing 7ft.2in. Highly typical of the region.*

liberally with mahogany (door, base, hood door) and has mahogany pillars and trunk quarter-pillars. The caddy top is one form of hood pediment popular in the North from now on, though many cases now flat topped have had the original caddy removed later to reduce height. The hood fret is of the 'blind' type, i.e. has a solid backing, hence is purely decorative and no longer fulfils its original purpose in letting out the bell sound.

This is the grander form of square dial clock which can hardly be termed a 'cottage' clock any longer. This one stands 7ft.2in. These better quality north-western cases almost always have oak backboards, were made of sturdily thick timbers and are quite heavy to move, as you will find if you try lifting one single-handed. The whole concept of such cases is north-western, but individual details indicative of this are: ogee top-of-trunk moulding, shaped door top, tiny ogee bracket feet, caddy top, heavy proportion, liberal mahogany trim.

The thirteen-inch cartwheel dial eight-day longcase clock by Thomas Richardson of Weaverham in Cheshire (Colour Plate 15, page 221 and Plate 340) dates from the 1760s. This shows another form of the scroll engraved matted centre, though here the winding holes pierce the engraved pattern. This factor, together with the applied plaque signature, might suggest that the maker bought his engraved work from an outside engraver. The bold half-hour marker (here still attached to a stalk) persists, but by this date he is no longer marking quarter-hours on the inner chapter ring edge, this latter feature falling from use with many clockmakers in the 1760s. The spandrel is now a later form of cherub head. Seconds marking was now normally of the ten unit division type. The blued steel hands are believed original and of typical pattern for the period.

The oak case of the Richardson clock has considerable mahogany trim and stands just 7ft. high. This has a traditional north-western caddy top and, apart from the now larger dial, has progressed a little in style. Top-of-trunk moulds are of the normal concave form, increasingly used in almost all regions by now. The trunk-to-base moulding is still made in the highly complex north-western manner and is canted, leading to canted base corners. The base has a raised panel, popular in many areas and especially so in the North-west. The shaped door top completes what is still easily recognisable as a distinctly regionalised style. Pillar capitals are in wood, as was often preferred in the North-west, though not unfailingly – see the next example.

A second eight-day longcase clock by Thomas Richardson of Weaverham, dating from the 1770s-1780s, is seen in Plates 341 and 342 slightly out of sequence for comparison. This has a fourteen-inch cartwheel-cast dial, now using what I call the question mark spandrel pattern, only used on large dials of thirteen or fourteen inches, particularly in the North-west and especially popular in Lancashire. The chapter ring now shows minutes by means of dots, a feature I always think was copied from the first painted (japanned) dials which began about 1770 and therefore an instantly recognisable dating sign. A particularly interesting feature is that the engraving of the chapter ring is by the very same hand as that on the previous dial, recognisable by odd flourishes – such as the tail of the 5, the upstroke on the tail of the 2, the tapering crossbar on the 4. Whether this means that Richardson did his own engraving or was simply using the same engraver fifteen to twenty years on I cannot say.

In this clock the dial centre is of the polished ground type with floral engraving, the design in this instance planned round the winding squares which are incorporated into the pattern. His signature is on a reserve, engraved on to the centre itself. This clock has a third hand (here indicating V) which is a centre calendar hand, the numbers 1 to 31 being engraved on to the dial centre. The date shown here is the 13th. The calendar hand is in brass to avoid confusion with the main time hands which are original and are of blued steel. The centre calendar could occur anywhere in the country but was more often used in the North-west than elsewhere and was mostly used in the last quarter of the century.

The oak case of this clock stands about 7ft.2in. and has considerable mahogany trim in the usual places but additionally has a mahogany strip across the hood front, where the fret might otherwise be. A mahogany cushion mould runs round the case above the door – at this period a feature found only in Lancashire and Cheshire and in some ways continuing the principle of the now discontinued ogee top-of-trunk moulding. The top-of-trunk moulding is of the nationally conventional concave type, but has an unusual feature seldom found outside the North-west – crossbanding on its edge.

Brass capitals and bases are used here, the caps being of the Corinthian type, unusual for the area. More unusual still is that this case has brass caps and bases to the trunk quarter pillars too. The hood door has an ogee-moulded surface, a feature seldom found

PLATE 341. *Fourteen-inch longcase dial by Thomas Richardson of Weaverham dating from the 1770s. This one has centre calendarwork and a different centre treatment by engraving on to a polished ground. Original blued steel hands but original calendar hand of brass for contrast.*

PLATE 340 (opposite). *Oak case with mahogany crossbanding and trim of the Richardson clock in Colour Plate 15 standing about 7ft. Many typical features of the area.*

PLATE 342. *Oak case with mahogany crossbanding of the second Richardson clock (Plate 341) standing about 7ft.2in. The Corinthian capitals are an unusual feature on a style which is otherwise very typical of the area.*

PLATE 343. *Thirteen-inch dial from a longcase of the 1760s by William Greenall of Parr, Lancashire. By now the quarter-hours have ceased to be marked on the inner chapter ring edge. Blued steel hands probably original.*

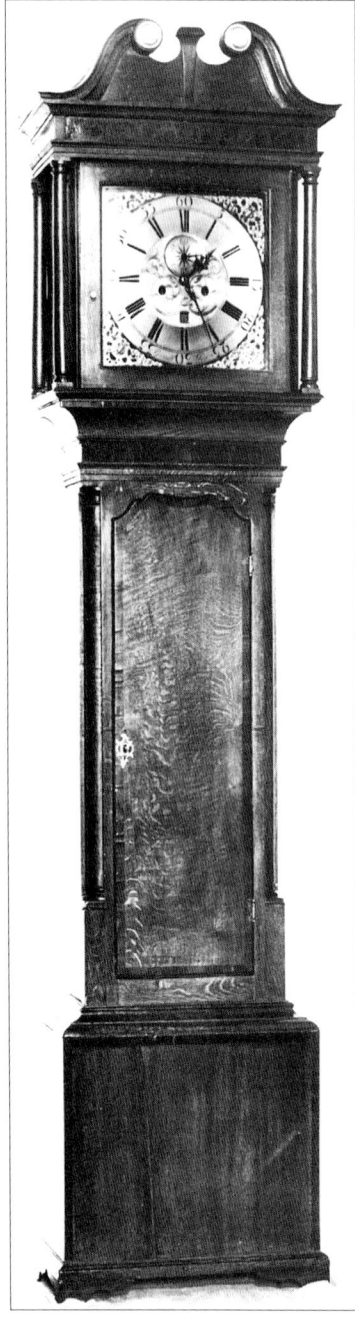

PLATE 344. *The case of the Greenall clock is in oak with mahogany crossbanding. Many features of local styling, including the cut of the swan-neck and absence of brass fittings.*

outside the North-west. The raised base panel here has clipped corners, a theme repeated in the clipped-corner door top. Overall, then, although this case has one or two features which are unusual for the area, its region of origin is still readily recognisable.

The thirteen-inch cartwheel dial eight-day clock by William Greenall of Parr, near Wigan in Lancashire, dates from the 1760s. The overall dial style (Plate 343) is not unlike that of the first Richardson clock in Colour Plate 15 (page 221), having a matted centre with overall engraving, here again pierced by the winding squares. The biggest difference is that Greenall has ceased to use either half-hour or quarter-hour markers, making for a plainer style which is the more obvious as chapter rings broadened later in the century. The recessed seconds dial was a method used by some makers and avoided the chance contact of the seconds hand with the passing hour hand. The blued steel hands are original, based only vaguely on London hands of the day. The spandrel is here yet another version of the later cherub head style.

The oak case (Plate 344) has mahogany trim and is of the swan-neck pediment type, a version of which was sometimes used on square dials as well as arched ones. Shaped door top, complex trunk-to-base mouldings, wooden-capped hood pillars and bold swan-neck moulding are all north-western features. The case stands about 7ft.6in.

The 'fret' is a gold-painted tracery on to a solid painted background and not an actual pierced or blind fret. This type of painted 'fret' on to a solid painted wooden background was often used in Lancashire and Cheshire and occasionally in other north-western areas in the late eighteenth century and resembled *verre églomisé* which was painted on to a glass panel. The ground was usually blue (occasionally green or black) and the tracery gold, which when

PLATE 345. *Thirteen inch longcase dial by Edward Clementson of Melton Mowbray made in the 1765-75 period, the dial centre of the polished ground type with floral engraving. Blued steel hands may be original.*

worn on the surface can appear as a yellow gesso colour. The purpose was to achieve the *verre églomisé* effect but without the fragility of the glass panel which often cracked through shrinkage or rough handling.

The thirteen-inch cartwheel-cast dial of an eight-day longcase clock by Edward Clementson of Melton Mowbray in Leicestershire dates from the 1760s or perhaps the 1770s and shows a quite different type of dial centre treatment. This dial centre (Plate 345) has lavish floral engraving on to a polished ground, an alternative style to that just examined with the matted ground and one which could in theory occur anywhere in the country with the normal exception of those two strongly opposite poles of style, London and Lancashire. By this period London had set the initial trend with a matted and undecorated dial centre; Lancashire filled that matted centre to bursting point with engraving on to the matting. Most other regions followed the lead of London but a few areas closest to Lancashire followed the Lancashire style.

Those were the two camps which used dial centres based on matting, but the alternative style seen in the Clementson dial, of engraving on to a *polished* (i.e. unmatted) ground might occur anywhere in the country at this period, though was least likely to occur in areas most strongly influenced by London or Lancashire, i.e. the South-east and North-west. This engraved-on-to-polished-ground style derived ultimately from the old tulip-engraved lantern clock dial centre, though by now the engraved designs were usually of the leafy scroll type, as here.

On this particular dial the engraving is beautifully executed both in the dial centre and on the chapter ring, which by now has ceased to show half-hour or quarter-hour markers. However, the centre engraving is pierced by the winding holes, perhaps suggesting the work of an outside engraver.

PLATE 346. *Twelve-inch longcase dial by Litton of Ashburn dating from the 1760s. Plain matted centre based on the London influence, but the mouth type of calendar is very provincial. Blued steel hands could be original.*

PLATE 347. *The oak case of the Litton clock is crossbanded in walnut and stands 6ft.4in. Unusual form of caddy top is original.*

The calendar is of the mouth type used mostly on thirty-hour work, an inferior system to the normal eight-day box calendar, but requiring one wheel less in the making. The mouth calendar was therefore chosen for cheapness and can be a sign of lesser quality when used in eight-day work. It is less effective because it shows the correct date for twelve hours followed by a half-way point through the date during the second twelve hours, whereas a box calendar shows the correct date for the full period of the day in question.

This dial uses the string-of-pearl spandrel pattern, which was designed for, and used mainly on, eleven-inch and twelve-inch dials. This is a thirteen-inch dial and these spandrels are not an ideal fit for it, leaving a larger than usual dial sheet space showing around the corners. The blued steel hands could well be original, as the serpentine pattern minute hand is just coming into use about now.

The twelve-inch eight-day longcase clock by a little known maker, Litton of Ashburn, Derbyshire (Plates 346 and 347), dates from the 1760s and shows a clock which is similar in many respects to the Clementson one (Plate 345) but uses the London-influenced undecorated matted centre. Again this clock uses the mouth calendar system. The blued steel hands are probably original.

The London style influenced some (though relatively few) northern makers as well as some, relatively numerous, southern ones. Certain clockmakers made some of their dials showing the influence of London styling and others of their dials showing the influence of Lancashire styling. It might even be that such makers offered their customers a choice of either style, so that the fact that a certain regional styling appears on one dial by a particular clockmaker does not exclude the possibility that a quite different regional styling might

PLATE 348. *Twelve-inch longcase dial of the 1760s by William Wallen of Henley on Thames. Very much the London look to it, as applied to a square dial. The two marks beside the datebox are bearer studs for the calendar disc.*

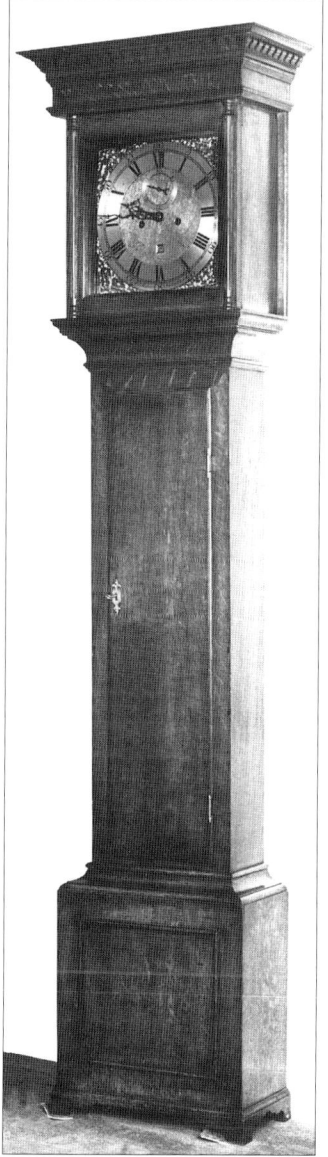

PLATE 349. *The plain oak case of the Wallen clock is 6ft.11in. high. Southern stylistics include the flat-topped door and to some degree the beaded panel-effect base.*

appear on other dials by the same maker – that styling perhaps being a deliberate choice rather than an exclusive trait. This means that it is not always safe to attribute a clock to a certain region by its dial styling alone, although this will generally be possible.

The oak case of the Litton clock stands 6ft.4in. and has broad crossbanding in walnut to the door and similar treatment to the hood door, which is in fact entirely walnut faced, with a little chequered stringing to surround the dial. There is no banding to the base, though this is original, and it does not always follow that crossbanding (or other decoration) will be repeated in each section of the case, as we would normally expect.

The overall style is relatively neutral, though the shaped door top and free-standing pillars suggest northern styling. The hood has an unusual form of caddy top, somewhat wider than most and actually overlapping what would normally be the flat of the topmould. This is not a typical form of caddy, more a whim of the individual cabinetmaker.

The twelve-inch eight-day longcase clock by William Wallen of Henley in Oxfordshire dates from the 1760s. The working dates for this maker are confused, as he is listed at Henley from 1725 but at Reading in 1756, which may simply indicate that he had a second shop for a while at Reading. I don't think this clock dates before 1756, as it is typical of the 1760s. The dial (Plate 348) is very much based on London styling, being very plain and formal, the matted centre completely without decoration, most noticeable round the calendar box and winding holes. The engraving is crisp and clear and expertly done. The spandrels are the string-of-pearl pattern widely popular in the third quarter of the century.

233

PLATE 350. *Twelve-inch dial by Thomas Clare of Warrington dating from the 1760s. The pointer form of calendar this early is uncommon. The two tiny spandrels were used by occasional makers but seldom this late.*

PLATES 351 AND 352. *Two oak cases of broadly similar type made in the 1760s and 1780s respectively. That on the left.is by John Lawson of Bradford; the right one by John Barnish of Rochdale. Both have original caddy tops. Height of each about 7ft.3in.*

The oak case stands 6ft.11in. and is similarly plain (Plate 349), still retaining the integral pillars as part of the hood door, though here fluted, which is a little unusual on a cottage clock. The dentil moulding gives just a little touch of style to the hood. The flat door top, simple moulds and absence of any crossbanding or trim are southern features.

The eight-day longcase clock by Thomas Clare of Warrington, Lancashire (Plate 350), dates from the 1760s. This has spandrels of a later variation of the female head type, which I call the head-and-flowers pattern, a form more often used in the South-west of England, but occasionally found elsewhere. The two tiny 'spandrels' beside the seconds dial of elves riding eagles are unusual and are the sort of tiny spandrel used in the arch of a bracket clock dial. The use of tiny spandrels in this position was uncommon at any time but especially as late as this, as the taste for this positioning of them was more or less confined to the early years of the century (see Plates 319 and 706).

The chapter ring, without half or quarter markers, is typical of this period. The seconds ring is marked in units of ten, again typical of the period. The use of a pointer for the calendar with its own separate chapter ring was a method chosen by only a few clockmakers at this time, but became more common later. The ringing of the seconds aperture and winding holes is unusually late.

The oak-cased eight-day longcase by Lawson of Bradford, Yorkshire (Plate 351), dates from the 1760s and makes an interesting comparison with a vaguely similar one by John Barnish of Rochdale, Lancashire (Plate 352), of the 1780s. John Barnish of Rochdale is recorded as being born in 1760. Northern features are: complex mouldings, especially the ogee top-of-trunk moulds, shaped door tops, raised clipped-corner base panel on the Lawson case, small ogee bracket feet, caddy tops, free-standing hood pillars. The pierced fret on the Barnish caddy is a little unusual, though many such crests

PLATE 353. *Thirteen-inch dial of about 1780 by John Barnish of Rochdale, the minutes by now marked by dots, a style which probably came in soonest in the North and is believed to have been copied from the first japanned dials.*

PLATE 352.

may have been removed over the years. A *verre églomisé* panel, seen on the Barnish hood, occurs on some of these north-western cases, principally in Lancashire and Cheshire. There is an overall chunkiness on such cases, especially when compared with a southern one such as that by Wallen of Henley (Plate 349). Both cases stand about 7ft.3in.

The eight-day dial by John Barnish (Plate 353) measures thirteen inches and shows dotted minutes (almost always a post 1775 sign) yet still retains the half-hour marker, by now an old-fashioned feature. The dial centre has scrollwork engraved on to a polished ground.

Eight-day clocks were made with round dials as well as square ones. Round dials are seldom seen before the 1770s and are mostly of the single sheet type (see Plate 356). The eight-day clock pictured in Plates 354 and 355 by John Agar of Malton dates from the late 1760s or early 1770s and is not a single sheet dial but a 'composite' dial with separate chapter ring, having very fine engraving to the floral centre. Stylistic features are identical with those on a square dial of the same period, including the non-matching original hands in blued steel. Note that the fine original steel seconds and calendar hands are deliberately different from each other.

The case of the Agar clock (Plate 355) is of oak with mahogany trim, standing about 7ft.6in., and has a short pagoda top, a form of pediment most often used on arched dials but occasionally found on a square or round one. This sort of pagoda is usually confined to East Yorkshire and Lincolnshire, though could conceivably occur elsewhere. Apart from the pagoda itself, northern features include free-standing pillars, wavy topped door, higher than square base, and the round dial itself is more often found in the North than elsewhere, southern examples being quite uncommon.

The eight-day longcase clock by Robert Holborn of South Cave in East

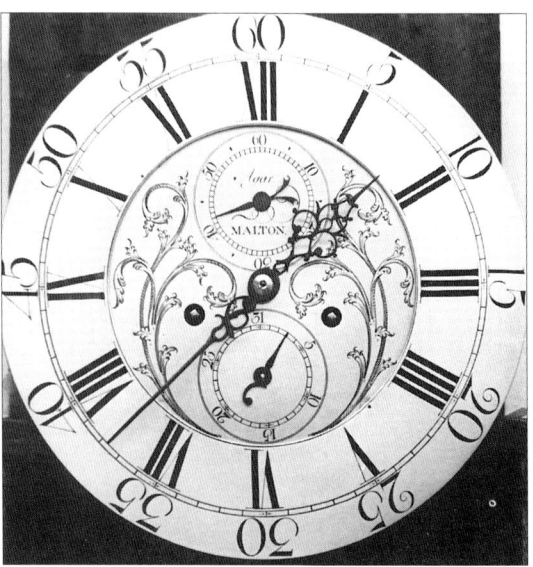

PLATE 355. *The oak case of the Agar clock has mahogany crossbanding and trim and stands about 7ft.6in. A pagoda top is uncommon on a round (or square) dial clock.*

PLATE 354. *Round dial longcase of the 1770s by John Agar of Malton, with very fine engraved centrework. Original hands in blued steel with seconds and calendar hands deliberately different from each other.*

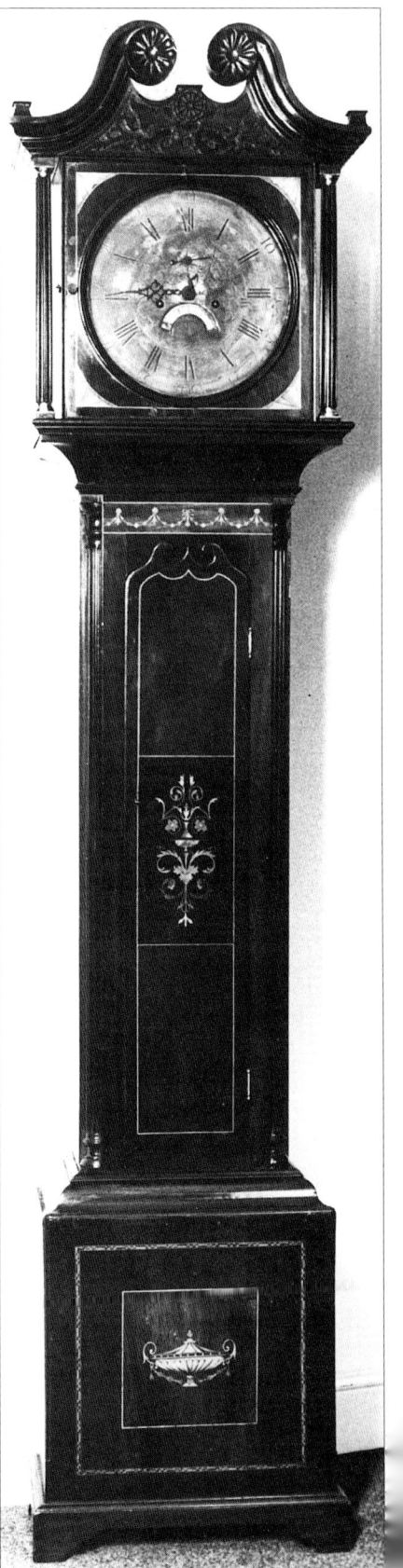

PLATE 356. *Mahogany round dial clock of the 1770s by Robert Holborn of South Cave and having a single sheet dial. Height about 7ft.6in.*

PLATE 357. *Boldly-engraved though naïve twelve-inch longcase dial of the 1770s by Adam Costen of Kirkham. Dotted minutes are by now the norm. Blued steel hands probably original.*

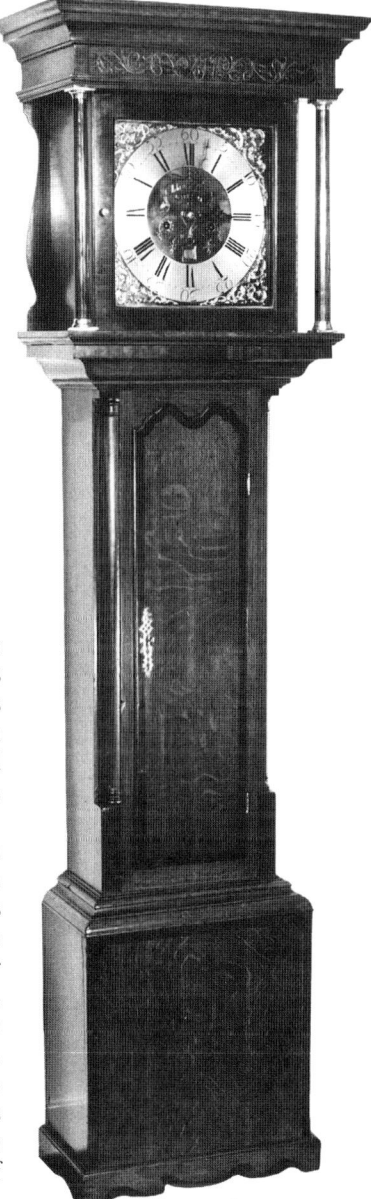

PLATE 358. *The oak case of the Costen clock has mahogany crossbanding and trim and stands about 6ft.7in. Many features indicative of regional styling.*

Yorkshire dates between 1770 and 1780 and has a single sheet round dial, here seen unrestored with the silvering looking very drab (Plate 356). The case is of mahogany with much intricate marquetry inlay in the form of flowers, urns and swags. The case is rather unusual and it is difficult to deduce much from its styling. Shaped door top, high base, separate hood pillars, ogee mould (here unusually as the trunk-to-base mould) are all suggestive of northern work, as perhaps too are the carved rosettes to the swan-necks.

The twelve-inch eight-day longcase clock by Adam Costen of Kirkham in Lancashire dates from the 1770s (Plates 357 and 358). The maker set up in business there in 1739 and was still working in the 1780s. By now chapter rings tend to be broader than earlier but their width is accentuated by the absence of half-hour markers. Minutes are now marked by dots instead of the earlier minute track and dotted minutes are the predominant style from now on, especially in the North. These probably copy the dotted minute principle from the newly fashionable japanned dials, which were launched in the early 1770s.

The dial centre is now of the polished ground type on to which this example has crisply executed, though naïve, engraving of flowers, leaves and birds centred on a vase. The engraver has worked his winding apertures into the pattern as flowerheads, always a sign of forward planning. The older style of lavish floral engraving on to a *matted* centre, formerly popular in this region, has by this time mostly passed from use. Here the spandrels are of the question mark pattern, more commonly found on larger dials than twelve inches. The blued steel hands appear original, the serpentine minute-hand style now having largely replaced the older straight form.

The oak case of the Costen clock stands 6ft.7in. and is crossbanded in mahogany, as very many northern cases were by this date. It has the *verre*

PLATES 359. AND 360. *Two thirteen-inch longcase dials by David Collier, the left.during his Gatley period of the 1770s, the right during his Eccles period of the 1780s. Little difference in the style. Blued steel hands probably original on both.*

églomisé style panel to the hood, in this instance on a wooden base, more sensible than the glass version it imitates as it is less prone to damage. Northern features include: *verre églomisé* panel, free-standing pillars, shaped door top, complex moulds (especially the ogee top-of-trunk mould), high base. Shaped hood backsplats and quarter pillars to the trunk can occur anywhere but were particularly popular in the North-west. The flat top still occurs here but the majority of square-dial clocks now have some sort of shaped pediment, such as a caddy or a swan-neck.

Two thirteen-inch eight-day longcase dials by David Collier are shown together for comparison in Plates 359 and 360, the first signed at Gatley, the second at Eccles. David Collier was born at Gatley (otherwise known as Gatley Green), just south of Manchester, Lancashire, in 1721, moving in about 1780 to Eccles just a few miles north-west, where he died in 1792. The Gatley dial (Plate 359) probably dates from the 1770s, the Eccles dial (Plate 360) from the 1780s. Both are similar in style, having engraved centres on to a polished ground, the Eccles dial incorporating the two mythical birds often found on Lancashire dials of this period. Both dials have a twelve o'clock moon, the earlier Gatley one having a starry sky background, the later Eccles one having landscape backgrounds. This was the usual stylistic sequence of decorating moon dials, but there are exceptions. This type of moon dial could be knocked on twelve-hourly by a wedge on the hour pipe.

Both dials use the mouth calendar. This saved making the extra wheel needed for a box calendar, but on twelve o'clock moon dials it was especially convenient because the same hour-pipe wedge could be used to drive both moon disc and calendar disc. On both dials the absence of half-hour markers is typical of this period. Both appear to have original blued steel hands using the serpentine minute hand style. The Eccles clock uses the familiar (especially in Lancashire) question mark spandrel, but the Gatley dial has spandrels of a rather unusual rope-swag pattern.

The oak cases of both clocks have considerable mahogany trim and have a

PLATES 361. AND 362. *The cases of the two David Collier clocks, the left one being the Gatley clock, the right one being the Eccles clock. Both in oak with mahogany trim, about 7ft. and 7ft.6in respectively.*

number of features in common – complex moulds, high canted-corner base each with inset mahogany clipped-corner crossbanding, separate pillars and the overall stockiness of a north-western case. The swan-neck of the Eccles case (Plate 362) was an alternative to the flat-top of the Gatley one (Plate 361). The Eccles clock has the typical shaped door top and the distinctively north-western cushion mould above it. The flat-topped door of the Gatley clock is an exception to the general rule, but, despite this, it is still very clearly a northern case. The Gatley case stands about 7ft., the Eccles one about 7ft.6in.

PLATE 363. *Twelve-inch dial of about 1780 by Charles Edward Gillett of Manchester, with typical Lancashire birds-and-branches centre. Blued steel hands may be original.*

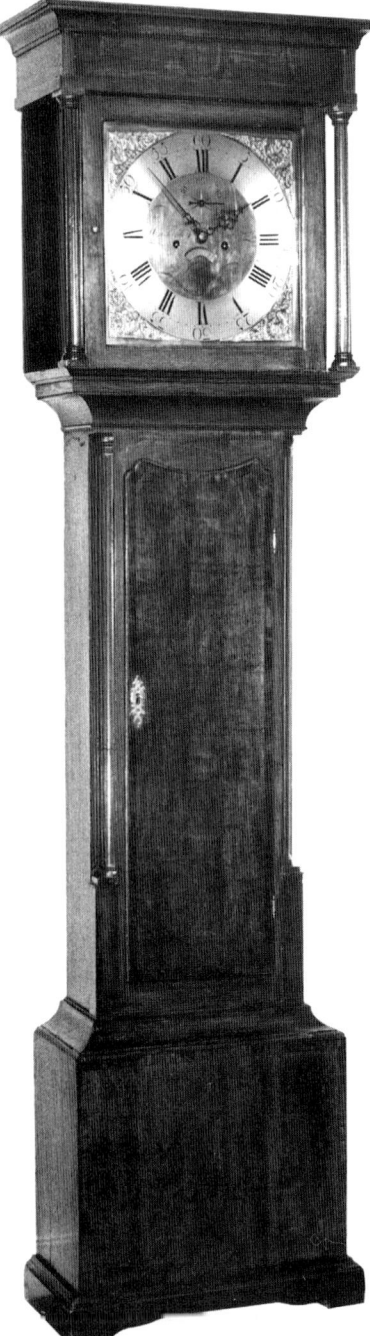

PLATE 364. *Thirteen-inch longcase dial by Winstanley of Holywell dating from the 1780s, the centre here modestly engraved. Spandrel pattern uncommon. Blued steel hands probably original.*

PLATE 365. *The oak case of the Winstanley clock has walnut trim and stands 6ft.7in. This door top shape was popular in parts of Cheshire.*

The twelve-inch eight-day longcase dial by Charles Edward Gillett of Manchester, Lancashire, dates from about 1780 (Plate 363). The maker is known to have worked there from about 1770 till 1800. The dial keeps the scroll-engraved centre style popular in the 1770s, here including the two popular Lancashire (jay-like, pheasant-like) birds. The small fishing boat scene within the calendar ring is unusual for the North, however. The engraving is first class. Dotted minutes suggest post 1770.

The thirteen-inch eight-day longcase dial by Winstanley of Holywell in Flintshire, North Wales, also dates from the 1780s (Plate 364). Here the engraving is also well done but the centre is far more sparsely decorated – compare the compass rose within the seconds area with that of the Gillett dial in Plate 363. The mouth calendar using twelve-hour drive was a cheaper

PLATE 366. *Very unusual example of late marquetry casework on to walnut housing a clock of the 1780s by Holland Farmer of Caernarvon and standing 6ft.6in.*

PLATE 367. *Mahogany longcase of the 1780s by Peter Fearnley of Wigan, showing many features typical of better square dial work in the area including mahogany 'flying saucer' finials. Height 7ft.8in.*

method and the impression is that this clock was kept down to a price. The blued steel hands are believed original.

The Winstanley case is of oak with walnut crossbanding and is typical of the more restrained north-western casework of the period (Plate 365). It stands 6ft.7in. Fluting to hood pillars and trunk quarter columns was more common in the North than South at this time on such simple country casework. The usual northern indicators are present – separate pillars, shaped door top, high base.

The walnut eight-day marquetry longcase clock pictured in Plate 366 is signed Holland Farmer, Caernarvon (North Wales), an unrecorded maker or perhaps a partnership between Messrs. Holland and Farmer. The clock seems to date from about 1770-80, but this puzzling case has some earlier features,

PLATE 368. *Twelve-inch dial of about 1780 by Taylor of Whitehaven with fine engraved centre making a design feature of the winding holes. Blued steel hands probably original.*

PLATE 369. *The Taylor case is in red walnut (normally used in solid form, as here) and has an architectural pediment used by occasional makers in this area. Height 7ft.1in.*

not least being the all-over marquetry. It seems to be an example of a casemaker deliberately working in the old-fashioned style and using some features (D-mould and lenticle glass) more normally associated with the early years of the century. It stands 6ft.6in. The case is unlikely to have been made in the early eighteenth century as size, proportion, balance, pillars, fret and mouldings are all wrong for that period.

The square dial longcase clock was sometimes made in north-west England in the grander mahogany style of the arched dial, as with the example pictured in Plate 367 by Peter Fearnley of Wigan in Lancashire, which has many of the features usually associated with the arched dial style. The dial measures fourteen inches square. This example dates from the 1780s, Peter Fearnley being a prolific and well-known clockmaker, born in 1749 and dying in 1826.

Typical regional indicators are: canted base and complex canted moulds, raised clipped-corner base panel, overall square shape of base, tiny ogee bracket feet, shaped door top, *verre églomisé* panel to hood front, free-standing hood pillars with mahogany caps and bases, original finials in mahogany in what I call 'flying saucer' pattern, being the Lancashire form of the flambeau finial. The height is about 7ft.8in.

The twelve inch eight-day longcase clock by William Taylor of Whitehaven dates from the late 1770s or early 1780s. This prolific clockmaker was born about 1738 and died in 1801. The dial (Plate 368) is exquisitely engraved in the scrollwork-centre style, here working into the design the winding holes as flowerheads, a principle often used in better class work. The blued steel hands are believed to be original. By now the square box calendar is becoming a little old-fashioned but was kept by some makers.

The case of this clock (Plate 369) is made from red walnut, which came from

PLATE 370. *Most unusual dial of the 1780s by John Lees of Middleton having the centre painted with a landscape scene. Blued steel hands probably original.*

PLATE 371. *The oak case of the Lees clock has mahogany trim and stands about 7ft.6in. Original finials and support blocks. Feet missing but were probably small ogee type.*

the Americas and is sometimes called American walnut (Americans call it black walnut). This wood was used mostly in solid form from the 1760s to 1790s, mostly in north-western England and Scotland. Its figuring is less bold than with mahogany. The hood has one form of pitched pediment, used occasionally anywhere but more commonly in the North-west of England. The case is slender (on account of the small dial) but has distinctively north-western features, and stands 7ft.1in.

The John Lees of Middleton (Manchester, Lancashire) clock in Plates 370 and 371 dates from the 1780s and shows an unusual feature which seems to be limited to a few clocks from Lancashire and Cheshire rather than being a national trend, and that is that the central zone of the dial is painted. This was perhaps intended to give the best features of both brass and painted dials, but seems not to have been a very widespread practice. This clock is otherwise quite conventional for the period and region.

The Lees case is of oak with mahogany trim and has many features typical of the time for this area.

The eleven-inch eight-day clock by Reynolds and Earle of Oxford shows a single sheet type of dial (Plate 372). Such clocks were silvered over the entire surface for clarity of time reading. This can be dated very closely to between 1797 and 1799 because of the known brief partnership of these two makers. Single sheet dials were known in the North but were less common there.

The oak case of the Reynolds and Earle clock (Plate 373) stands 6ft.6in. and shows a continuing use of what were very traditional, even by this time old fashioned, southern features – integral hood pillars, flat-top door still of the full-length type.

PLATE 372. *Eleven-inch single sheet dial by Reynolds and Earle of Oxford dating closely to 1798. Single-track minutes were sometimes used at this time but mainly on single sheet dials.*

PLATE 373. *Simple oak case of the Reynolds and Earle clock standing 6ft.6in. This traditional style is little changed from almost half a century earlier.*

The square dial was by this time very old fashioned in London and had long fallen from use. However, there was a revival of the square dial in such clocks as Plate 374 by Vulliamy, which dates from the late eighteenth century. This is a single sheet dial used in square form for this type of precision clock, not a regulator, but a very formal style of precision clock such as might be made for a gentleman's study. The silvered dial has plain (unengraved) corners, being totally devoid of any form of decoration.

The mahogany case is also in a way a revival of style in the architectural top, reminiscent of some of the cases of the late seventeenth century by such makers as Fromanteel. However, the case is otherwise in the same sort of London styling as found on arched dial examples – fluted columns and quarter columns each inset with brass reeds, brass capitals, shallow base, double plinth. The flat top to the door is used here rather than an arched top, probably because it better suits the square dial.

The single sheet dial was quite often used in all areas for regulators, which were precision clocks made for a clockmaker's own use as his shop timekeeper and for the use of others who might want such a special-purpose clock. The one in Plate 375 is signed by Wilson Clare of Preston, Lancashire and dates from the 1830s-40s. The silvered dial carries no decoration and has a separate hand and dial for hours and minutes. Such clocks usually have maintaining power, which by this period was of what is usually known as the 'Harrison' type.

The mahogany case is formal and uses the architectural pediment as on the Vulliamy clock in Plate 374, but this case has decorative carving and other high quality trim as sometimes seen on formal furniture of the day. The height is 7ft.7in. Many regulators had glass doors to allow a view of the pendulum, which was often of a special type – here of telescopic construction in alternate metals to avoid expansion error.

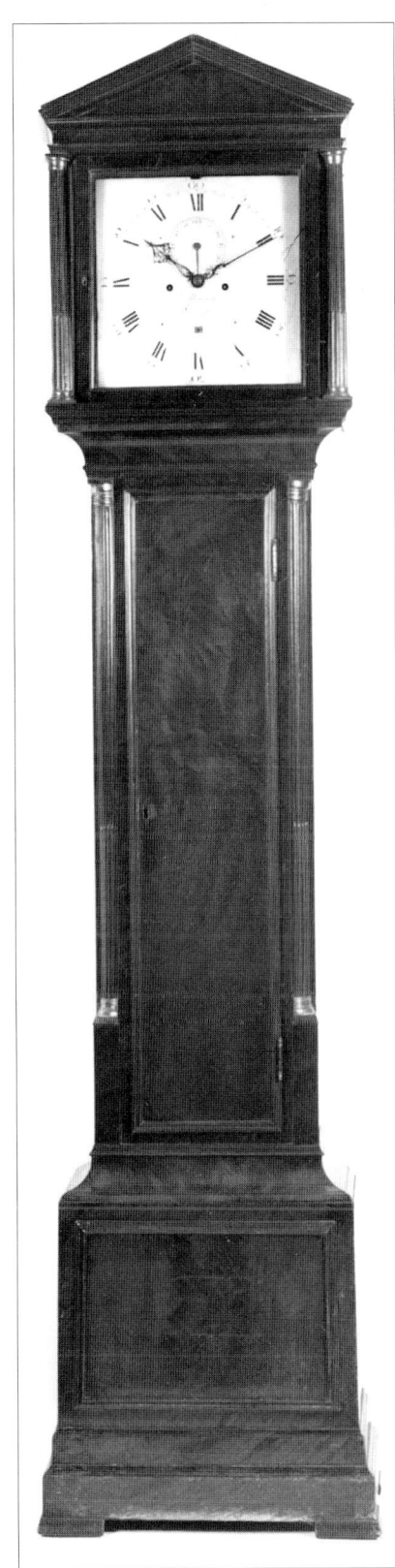

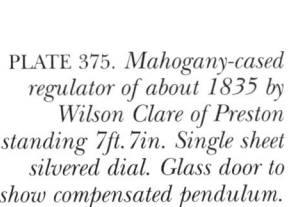

PLATE 374. *Mahogany square single sheet dial longcase of the late 18th century by Vulliamy of London standing about 7ft. A very formal style made only in London. (Photograph by courtesy of Sotheby's Sussex.)*

PLATE 375. *Mahogany-cased regulator of about 1835 by Wilson Clare of Preston standing 7ft.7in. Single sheet silvered dial. Glass door to show compensated pendulum.*

PLATES 376 AND 377. *Two arched dial London walnut longcases of typical 1730s and 1740s styling, left by John Drury, right by William Scafe. Both may once have had a superstructure to the hood. (Photograph of Scafe clock by courtesy of Derek Roberts.)*

CHAPTER TEN

EIGHT-DAY ARCHED DIAL WORK

W̲e now examine what is probably the largest and most diverse group of all, namely eight-day arched dial clocks from about 1730 to the end of brass dial clockmaking. I have broken this group into very simplified regions to try to illustrate better the principles and strength of these regional influences. Naturally there were exceptions to the general trend in all areas, but the distinct polarity of concept is evident between on the one hand London (together with much of southern and eastern England) and on the other hand Lancashire (together with much of northern and western England and Wales). This division of styles influenced by these two major centres of taste became stronger as time progressed and is therefore much more apparent in the 1780s than the 1730s.

1. LONDON (WITH THE SOUTH AND EAST)
This region covers not only London itself but also all southern and eastern England including the East Coast counties as far north as Scarborough on the Yorkshire coast, the South Coast counties as far west as Dorset and Wiltshire, and the inland counties as far as Warwickshire and Nottinghamshire. The boundary could be seen roughly as a line from Bridport in Dorset to Scarborough in Yorkshire. Naturally the influence of London is weakest towards the further reaches of this territory from the capital. Throughout this zone most clockmakers followed the London concept for much of the time, departure from it being only by occasional choice and by occasional clockmakers of individually-minded nature.

The clock by John Drury of London (Plate 376) dates from the 1730s, that by William Scafe of London (Plate 377) from the 1740s. These are typical examples of the London walnut-veneered longcase of the period, some of which had an upper caddy pediment above the flat of the hood. The case styles

PLATE 378. *Classic twelve-inch London dial of about 1710 by Francis Wells – plain centre, diminutive floating half-hour markers, seconds still marked in units of five. Original steel hands (blueing here worn away).*

PLATE 379. *Eight-day twelve-inch longcase dial of about 1750 by John Bottrell of Coventry, based on the London principle, though the work not so fine as London – e.g. matting irregular. Engraved decoration to calendar is more to provincial taste. Original blued steel hands.*

PLATE 380. *Eight-day twelve-inch longcase dial of about 1750 by James Norman of Charminster based even more closely on London taste – note the small floating half-hour markers. Original blued steel hands.*

are highly distinctive and can be seen to have been followed in concept by some provincial makers for their oak cases.

Whilst these particular clock dials show certain unusual features (such as seconds dials in the arch), the general dial styles are typical of the day and the dial of about 1740 by Francis Wells of London in Plate 378 shows most of the stylistic features they embody. Principal amongst them are: matted centre, un-ringed winding holes, square calendar box, floating half-hour markers, signature often on a name-plate within the dial centre or on the boss in the arch.

The twelve-inch longcase dials by John Bottrell of Coventry in Warwickshire (Plate 379) and James Norman of Charminster in Dorset (Plate 380) both show this London influence though made at widely separated places around the same date, about 1750. The Norman clock has particularly obvious London features of style, including the strike/silent switch in the arch which was a feature of a good many London and London-influenced clocks and was uncommon in the North-west. The Norman clock has a five pillar movement, another feature long retained in London and little used outside that area of influence. The Bottrell clock has a hint of engraving around the calendar box, a feature we shall see used to a larger extent as we progress northwards. Of the two the Norman dial has crisper, more confident engraving and more even matting.

Both these clocks are housed in quite simple oak cases of broadly similar design (Plates 381 and 382) modelled on the London walnut examples we saw earlier. London features echoed in simpler manner here include: integral

PLATES 381 AND 382. *Oak cases of the clocks by Bottrell (left) and Norman (right), loosely based on the London walnut model seen in Plates 376 and 377. Some early features still retained such as hood side windows and integral hood pillars.*

PLATES 383 AND 384. *Two twelve-inch dials of the 1760s-80s compared, the one on the left by John Dison of St. Ives being earlier than the one on the right by Joseph Blundy of London. Remarkably similar layout and styling. Both originated from London style though the London dial has retained unchanged a style which by then would be old-fashioned in the provinces.*

hood pillars, hood side windows, simple moulds. The flat-top doors instead of arched are a concession to country joinerwork.

The twelve-inch longcase dial by John Dison of St. Ives in Huntingdonshire has the characteristic London style of about 1760-70, with typical name-plate signing and strike/silent in the arch (Plate 383). By this time half-hour markers have ceased to be used and seconds are marked in units of ten rather than of five. The movement has the London five-pillar construction and also an anti-rattle spring attached to the central knop of the top pillar closest to the hammer, the latter a feature used on many London clocks but only seen in the North on clocks which closely followed the London school. The twelve-inch longcase dial by Joseph Blundy of London (Plate 384) is a closely similar dial though dates from perhaps twenty years later. London styling progressed little from now on and the optimum styling of about 1760 seems to have been continued by many makers there for another thirty years. Both dials are excellently engraved in that very formal and very conservative style. Dial foot ends and calendar ring studs show in the Dison matting, and calendar ring studs and seconds dial backplate rivets show in the Blundy dial centre.

The twelve-inch longcase dial by Thomas Harben of Lewes in Sussex (Plate 385) dates from the 1760s and shows the alternative style of the polished dial centre with 'Chippendale' style engraving excellently done and working the winding squares into the overall design. This was an optional style some makers selected at this period instead of the matted centre, but the preference for this engraved centre was considerably stronger in the North. The dial is

PLATE 385. *Twelve-inch longcase dial of the 1760s by Thomas Harben of Lewes in Sussex showing the Chinese Chippendale engraved centre style which was popular with some makers as an alternative to the London matted centre. Blued steel hands probably original.*

PLATES 386 AND 387. *Two cases of broadly similar styling, the oak case (left) of the clock by Harben of Lewes based on the same principle as the grander mahogany case (right) by Henry Fish of London. Both date from the 1760-80 period.*

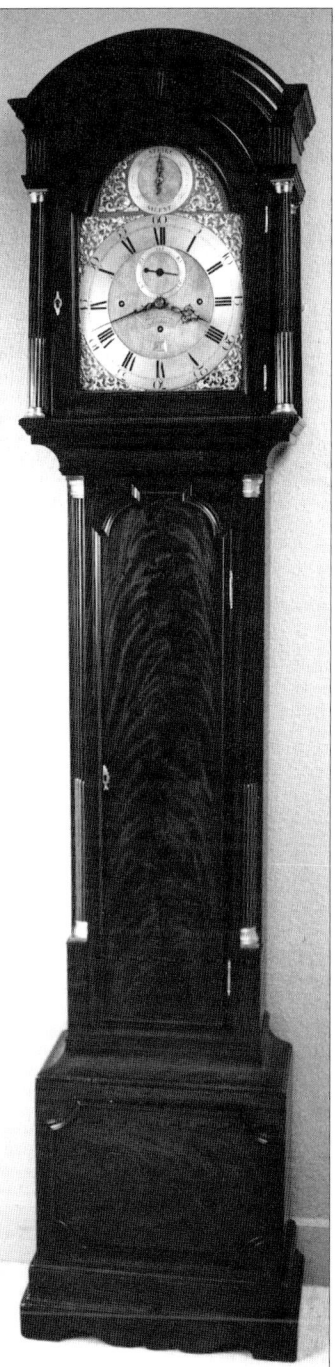

otherwise very 'London', even to its strike/silent arch feature.

The oak case of the Harben clock in Plate 386 has mahogany crossbanding and stands about 7ft.1in. This is the country equivalent of the case illustrated in Plate 387, a very fine quality London mahogany case housing a three train clock of 1760-70 by Henry Fish of London. This represents the best quality case of the day with brass reed inserts in the hood pillars and trunk quarter pillars, both with brass caps and bases. The door top is a characteristic shape on some London cases of this time. The raised clipped corner base panel and the double plinth are also typical London features.

PLATE 388. *Twelve-inch longcase dial of the 1770s by John Wood of Grantham with many features derivative of London, though using the engraved centre. Original blued steel hands throughout.*

PLATE 389. *Twelve-inch longcase dial of the 1770s by Moore of Ipswich in Suffolk, also based on London concepts, but using the polished rather than engraved centre, in this instance purely as a place to engrave the name. Blued steel hands original throughout.*

The twelve-inch longcase dial of the 1770s by John Wood of Grantham in Lincolnshire (Plate 388) is of the same type as the Harben dial (Plate 385), but uses the thirty-hour style of 'mouth' calendar, which saved a wheel. The twelve-inch dial of the same period by Moore of Ipswich, Suffolk (closer to London) shows a more restrained version (Plate 389), using the polished dial centre but with no centre engraving beyond the name. This plainer form was more preferred by London makers, some of whom saw decorative centre engraving as decadent, and presumably the London influence was felt more strongly in Ipswich than Grantham. Both dials have excellent engraving. Both have the London strike/silent system in the arch.

The twelve-inch longcase dial by James Pepper of Biggleswade in Bedfordshire (Plate 390) is of the same general style, though here the engraved centre is more stylised with indeterminate architectural features, the centre somewhat less skilled in design and execution. This clock also has a five pillar movement along the London principles.

Cases in burr walnut veneer continued to be made in London until about 1750, after which time mahogany took over as the newer, stronger, finer wood. The earliest mahogany was very dark in colour and had little figure; this is often called Spanish mahogany, being from the Spanish West Indies, but is sometimes termed Cuban mahogany. Later mahogany (from the third quarter of the eighteenth century) was frequently of paler colour with richer figuring and is often referred to as Honduras mahogany. The terms Spanish or Cuban

PLATE 390. *Twelve-inch longcase dial of the 1770s by James Pepper of Biggleswade, Bedfordshire, based a little more loosely on London and making more use of the engraved centre theme.*

and Honduras are used in a generic sense to describe the type of timber and are not intended to specify actual islands of origin.

This new timber was used in solid form for the construction of most of the case, but the frontal areas were veneered with the most finely figured pieces, sometimes called crotch mahogany or flame mahogany, these being the best figured pieces from the area where a large branch met the trunk. Crotch mahogany was not used in solid form for several reasons, including its sheer cost and its instability. The doors (and often base panels too) of London longcases are almost always of crotch mahogany veneer set on to a door made of solid mahogany of much plainer grain for stability. The same applies to many provincial mahogany case doors, though some are veneered on to oak basewood.

The dome-top case of the Henry Fish clock (Plate 387) is one London mahogany style. Another more popular one was what is today called the pagoda top case. This was made in two basic versions, one having quarter columns to the trunk, the other simpler form being square trunked. The best examples usually had brass reed inserts into the fluting of the columns, a feature which began about 1760. The general case style is often referred to briefly as a 'London mahogany' case. Oddly enough, some examples of what

253

we usually call the 'London mahogany' pagoda style were made in burr walnut, but these petered out by the later 1760s when mahogany became the exclusive timber for these cases in London.

Because of the considerable height of some London pagoda cases quite a number were later cut down by removing the pagoda section to leave what looks rather like a dome top, but this has a much thinner moulding than the true dome top and the whole effect is that the clock looks pinched at the hood.

Lacquer cases continued in popularity into arched dial times, the case style now based on the earlier arched dial walnut forms seen in the Drury and Scafe cases in Plates 376 and 377. By the 1740s the pagoda case style was made in lacquer too but by the 1770s lacquer fell from fashion.

London cases are very distinctive in style as they fall into a limited number of shapes and finishes. Strange as it may seem, London-made cases can sometimes be found on provincial clocks. This would happen where a provincial maker, or his customer, expressed a preference for a case of the newest or most fashionable type made in the capital. The lacquer cases of provincial clocks can often be seen to have been London made. So too can some examples of walnut pagoda style and of 'London mahogany' cases, that is mahogany pagoda examples. Of course some provincial cabinetmakers also copied the London style. The influence of London mahogany casework, especially of the pagoda style, was strongest in the south-east corner of England, that is in those counties immediately adjacent to London itself.

The pagoda-top mahogany cased clock by James Ivory of London (Plate 391) represents the better type of London pagoda with quarter columns and brass reed inserts. Alongside is a lacquer-cased clock by Monkhouse of London (Plate 392) in the same general shape and style, though lacquer cases kept the square-section trunk corners. Both have double plinths, customary in London work, both date from the 1760s and stand about 8ft. high and both have typical matted centre London dials of the day with strike/silent switches in the arch.

The mahogany-cased London clock in Plate 393 has a different type of pediment occasionally found in London work and usually known as a break-arch. Again this has the brass reed inserts to the quarter columns. Hood side windows were retained in London casework to the end of the brass dial period. This clock has the single sheet type of silvered brass dial with plain corners and dates from the 1790s, though would show little variance if as much as twenty years earlier. For comparison we see in Plate 394 a clock of the same general concept, a mahogany break-arch longcase of the 1770s by Martin Hall of Yarmouth in Norfolk. The dial is of the typical matted centre London style, signed on a name-plate, but has a rocking ship in the arch, perhaps to personalise the style to the seaport in which it was made. The case is one provincial variation on the London break-arch theme.

The single sheet silvered brass dial was an alternative form to the normal composite dial with separate chapter rings and was known from the 1760s but was not universally popular. When London makers used such dials they generally left the corners and arch devoid of any decorative features, in other words blank apart from any necessary numbering or lettering. Provincial makers usually included flower sprays or scroll engraved designs. The twelve-inch dial by Evans of London (Plate 395) is typical in its very plain and formal style, with the usual strike/silent arch feature. The dial by Edward Stevens of Boston, Lincolnshire (Plate 396), is from a bracket clock of about 1770 and is seven inches wide. This shows the kind of floral corner/arch decoration typical of single sheet dials made in the provinces. Bracket clocks more than

PLATES 391 AND 392. *Fine mahogany pagoda-style longcase of about 1760 by James Ivory of London seen alongside a London-made lacquer case of broadly similar age and styling. Lacquer cases did not have the fluting to the columns. Both stand about 8ft. high.*

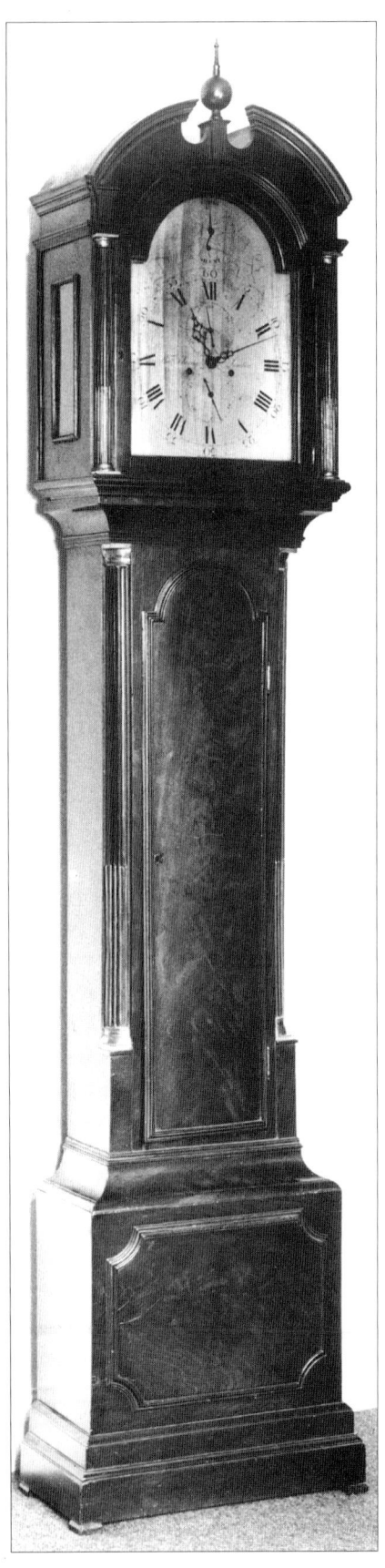

PLATES 393 AND 394. *Fine mahogany London clock with single sheet dial dating from about 1795, here using the break-arch hood form, seen alongside a provincial mahogany version of the same basic style housing a rocking ship clock of the 1770s by Martin Hall of Yarmouth.*

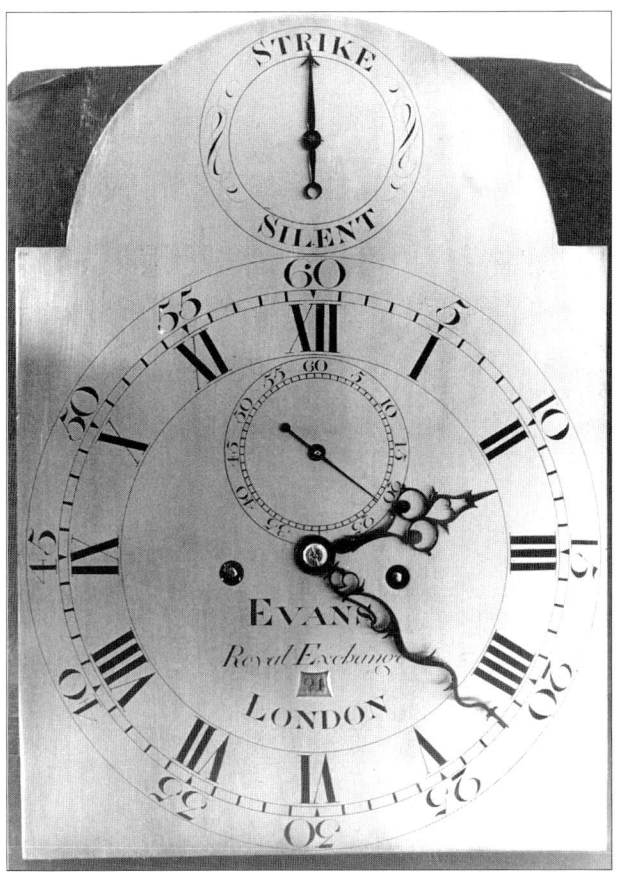

PLATE 395. *Twelve-inch longcase dial of the 1780s by James Evans of London of the single sheet silvered dial type. This style was usually very plain when used by London makers, as here. Blued steel hands probably original.*

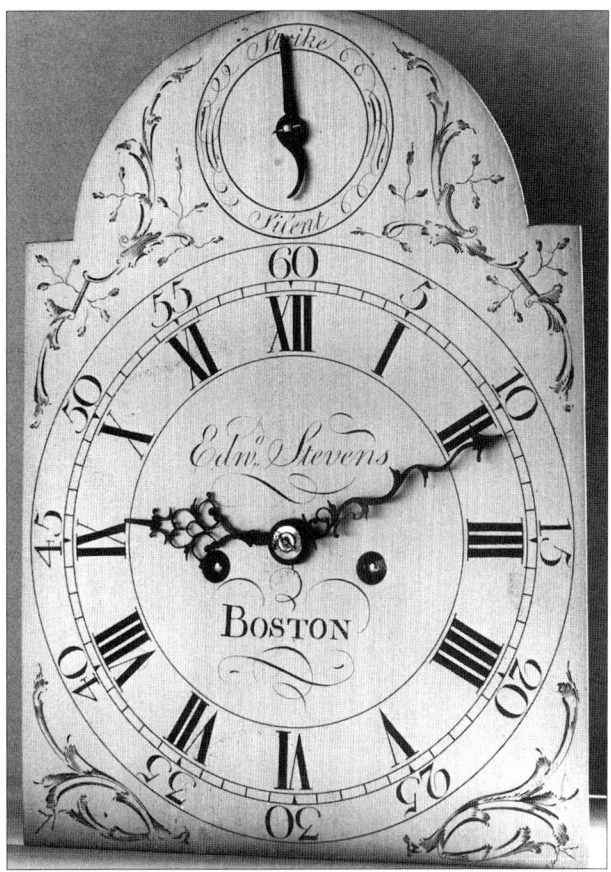

PLATE 396. *Bracket clock dial of the 1770s by Edward Stevens of Boston in the silvered single sheet style, very reminiscent of longcase work, but in a more decorative manner than with London makers. Blued steel hands probably original.*

longcases usually have some hint of London styling about them, illustrated in this example by the strike/silent arch lever. The longcase dial by John Gullock of Rochford in Essex (Plate 397) is a provincial example of about 1775, here with a rocking ship, the latter not a popular feature on London dials. The movement has five pillars, London style.

An alternative form of London mahogany case had a crested pediment as seen in Plate 398 of the 1770s-80s by John Fladgate of London. The square trunk form was made in pagoda-top style as well as with the crested pediment, the 1780s example in Plate 399 being by George Philip Strigel of London. Another alternative to the type with quarter columns to the trunk had reeded canted trunk corners, sometimes with brass reed inserts, as with the example shown of the 1780s by Samuel Farquharson of London (Plate 400).

The longcase clock of about 1790 by John Wilson of Peterborough, Cambridgeshire, is rather unusual. The dial (Plate 401) has a solid centre in the manner of a single sheet dial and the numbers are engraved on to this instead of the usual chapter ring. Otherwise the dial has strong London influence with typical strike/silent switch in the arch. The fine mahogany case of London pagoda style (Plate 402) has inset brass reeds, but the shaped hood backsplats are not typical of London work and it may well be that this is a locally made case following the London style. The original finials are of very unusual embossed drum-shaped style.

PLATE 397. *Longcase dial in the single sheet style made about 1775 by John Gullock of Rochford, Essex. This example has a rocking ship in the arch.*

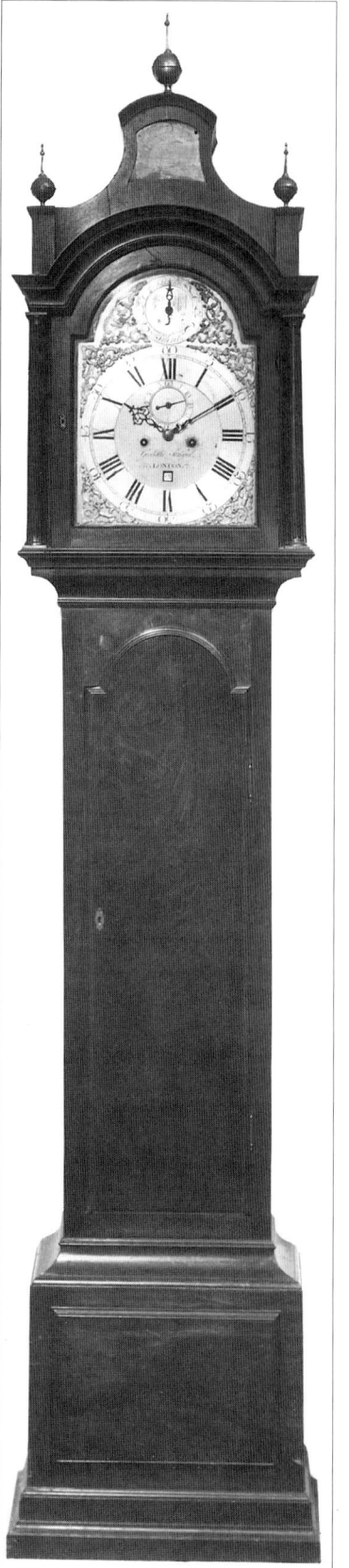

PLATE 398. *Mahogany longcase of the 1770s by John Fladgate of London, the case here being of the crested top style.*

PLATE 399. *Mahogany pagoda style longcase of the 1780s by George Philip Strigel of London, this example with square-cornered trunk. Height about 8ft. (Photograph by courtesy of King & Chasemore.)*

PLATE 401. *Longcase dial of the 1790s by John Wilson of Peterborough, based on London styling but here with an unusual form of separate dial centre with chapter ring as a single piece. The single-track minutes are always a late sign. Original hands by now are of matching pattern in blued steel – seconds hand may be a replacement.*

PLATE 400. *Mahogany longcase of the 1780s by Samuel Farquharson of London, the pagoda style case having canted trunk corners, inset with brass reeds. Height 7ft.5in. (Photograph by courtesy of Parsons, Welch & Cowell.)*

PLATE 402. *Mahogany case of the clock by John Wilson of Peterborough in the London pagoda style, though probably not London made. May originally have had the London-style double plinth. Height about 7ft.10in.*

PLATE 403. *Twelve-inch dial with rocking Father Time made in the 1750s by John Coates of Tetbury, Gloucestershire. The hour hand seems to lack its external loops. The arcaded minute band is not common in this region.*

2. THE WEST COUNTRY

The West Country is the name given to the region of south-west England covering those counties clustered round the Bristol Channel as far south as Cornwall. Bristol was the principal city, its importance as a major seaport making it the prosperous capital of the South-west and *the* most important provincial city in Britain, second only to London.

The eight-day clock by John Coates of Tetbury in Gloucestershire dates from about 1750 and is an example of a clock which shows no particular regional stylistics (Plate 403). The birds-and-basket engraving on the otherwise plain matted centre, the plain winding holes, the rocking Father Time are all features which could occur anywhere in the country. The arcaded minute band on the chapter ring, often called a 'Dutch' minute band, was more popular in north-west England than elsewhere, but could appear anywhere. The tiny floating half-hour markers and the name-plate signature tend to point towards southern styling. What is overall a very handsome dial cannot be localised at all other than by the maker's name.

Mahogany was imported into Bristol from the West Indies from the middle of the century, after which time this timber was extremely popular in better quality clocks in this area. In arched brass dial clocks a style had developed by the 1760s, or even earlier, which we now often call a 'Bristol mahogany', which indicates not simply a Bristol-made clock in a mahogany case, but one in a certain type of mahogany case, which was very popular in this region.

The longcase clock by John Duffett of Bristol (Plates 404 to 406) is typical in

PLATE 404. *Twelve-inch longcase dial with rolling moon and Bristol tides made in the 1760s by John Duffett of Bristol for Edward and Sarah Webster, probably as a marriage clock. The scroll-engraved centre was popular in the West (and North).*

PLATE 405. *Detail of the hood of the mahogany case of the John Duffett clock showing double pediment style (here a dome topped by a swan-neck), double pediments being popular in the Bristol area, though many were later cut down to a single pediment.*

many respects of a Bristol mahogany clock, this example dating from the 1760s. The dial centre bears the names 'Edward & Sarah Webster', this presumably being a marriage clock. The florally engraved dial centre on to a polished ground was the preferred style in this area (and in many western areas) from this time on, though examples of matted centres do occur.

A considerable number of these mahogany Bristol clocks of the second half of the eighteenth century have rolling moon dials and many of them, as here, also have 'High Water at Bristol Key' tidal dials incorporated, obviously on account of the importance of the port to so many in the area. (Quay was very often spelled as Key in the eighteenth century.) The high tide time is read along the outer ring of roman numerals, so that the dial here is showing the 16th lunar day with high tide at VIII (eight o'clock). Tidal times varied by complicated fractions of minutes each day so that such high water readings represent not the actual turn of the tide but an average high tide time somewhere close enough to the turn to indicate high enough water for landing or embarking.

The spandrels are very much a provincial pattern. The blued steel hands are probably original. The floating half-hour marker here still shows a hint of London influence but the dial has otherwise a distinctly non-London look.

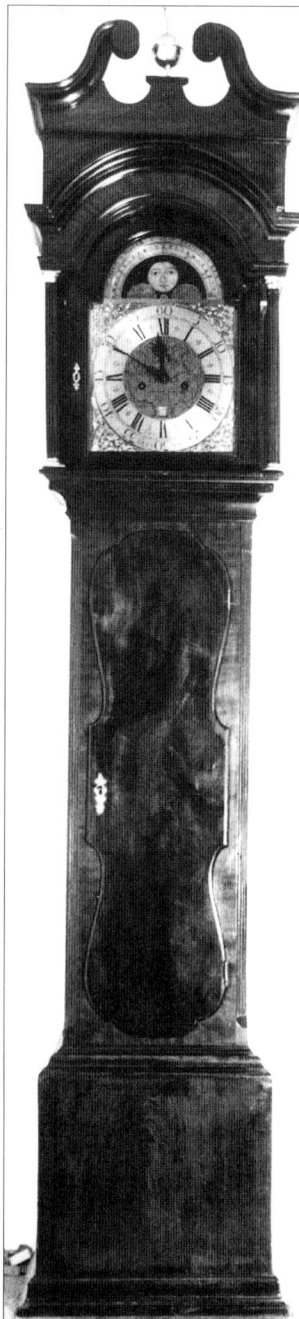
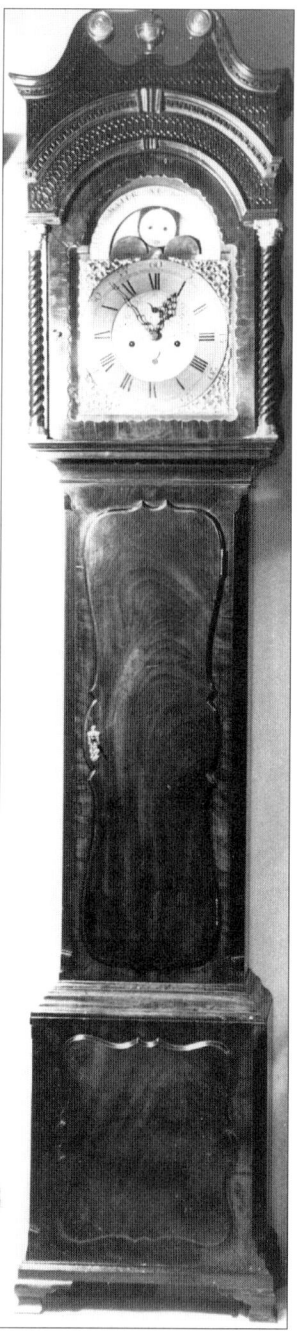

PLATES 406 AND 407. *The mahogany case of the John Duffett clock (left) compares with the mahogany case of an eight-day moons-and-tides clock by Henry Lane of Bristol dating from later in the century. Both exhibit different forms of the serpentine door, a feature popular with some late 18th century Bristol casemakers.*

The case shows certain features typically associated with Bristol cases, some of which of course housed clocks made in other towns in the region. Most obvious is the strange serpentine trunk door seen on this case and that of the Henry Lane clock of the 1790s in Plate 407. On some clocks this shape is repeated in the raised base panel, as in the Lane example. Clocks from this area often have canted corners to the trunk, often fluted, perhaps because fluted quarter-columns would have been thought a bit too fussy with this style of case. Many also have the same canted corners to the base.

A double pediment to the hood is a feature of some such cases. In the Duffett example this takes the form of a swan-neck on top of a box on top of a dome top. In the Lane case the swan-neck sits directly on the dome. It is as though the

PLATE 408. *Twelve-inch longcase dial with moon and Bristol tides made about 1780 by George White of Bristol. An engraved central scene was a feature of some clocks of this region at this time, though this particular domestic scene is perhaps unique.*

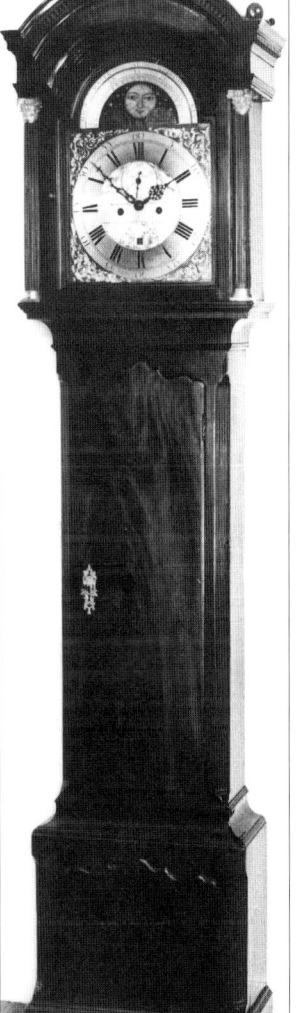

casemaker could not decide when enough was enough. The result of the double pediment is that such clocks are often quite tall, both these examples being about 8ft. high. The double pediment fashion faded towards the end of the century.

Some of these later cases have a wavy inner edge to the hood door, as in the Lane example. Some too have lavishly applied blind fretwork about the hood and some have rope-twist pillars with Corinthian brass caps and bases. All these features are seen in the Lane case. The overall effect of shaped panels and fretwork (some frets of oriental style) gives a suggestion of Chippendale styling.

Another feature popular with some West Country clockmakers (and occasionally in other regions too) was to have the dial centre engraved with a small scene, often a landscape with cottages or a seascape with ships. The twelve-inch longcase dial of about 1780 by George White of Bristol (Plate 408) is an example of this type of dial style, but here the dial centre has a most unusual interior drawing-room scene showing a servant putting on his mistress's shoe or slipper. Just what this particular scene signifies I cannot say, unless it was perhaps done on the instructions of the purchaser to illustrate her status in society.

By this date the half-hour markers have fallen from fashion. The spandrels are of a type particularly popular in the West Country. The two moon 'humps' here show a stylised rising sun and an armillary sphere, merely two decorations on the theme of time used as alternatives to the usual hemispheres to fill the available spaces. The blued steel hands appear to be original, by now with the serpentine minute hand form.

The mahogany case (Plate 409) stands only 7ft. with a dome top pediment. The trunk has fluted canted corners, as also does the base. Here the door top is quite busily shaped and is echoed in the similarly shaped top of the raised base panel. Shaped door tops, other than the simple shouldered arch as used in London, tend to be a sign of western or north-western work.

PLATE 409. *The mahogany case of the George White longcase stands only 7ft. high. The door and base panel have matching top shapes, something which was more a whim of the cabinetmaker than a regional style.*

263

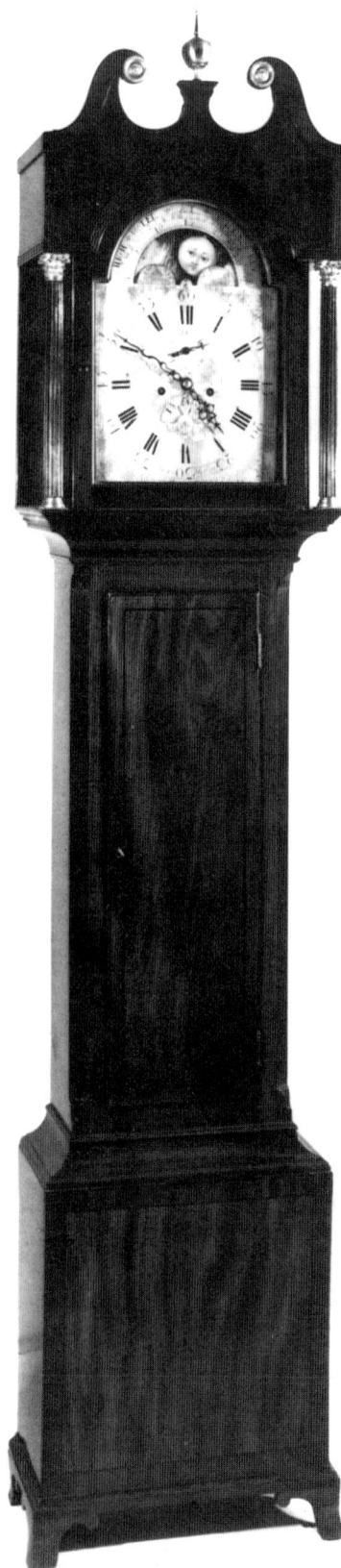

PLATE 410. *Single sheet longcase dial of about 1780 by William Millsom of Bristol, showing moon and tides.*

PLATE 411. *Mahogany case of the William Millsom clock, being simpler in style than the grander versions shown earlier.*

PLATE 412. *Eight-day longcase dial of the single sheet style made about 1780 by James Hine & Company of Exeter, showing moons and high tides at Topsham Bar. The calendar marks every single date in the month, which is unusual and a bit crowded. (Photograph by courtesy of Clive Ponsford.)*

The longcase dial by William Millsom of Bristol is an example of the single sheet silvered brass dial used for a West Country moon dial showing high water at Bristol (Plate 410). In this instance the tidal hours are shown rather unusually in arabic numerals rather than roman. This clock dates from late in the century. The mahogany case of this example is of considerably simpler form (Plate 411).

The single sheet dial example in Plate 412 is by James Hine and Co. of Topsham in Devon and dates from the 1780s or later. The engraved corners show the Four Seasons theme, unusual as late as this. High water is registered at Topsham Bar.

The Thomas Field of Bath clock dates from the 1790s and shows a later form of the double-pediment top, here a swan-neck on to a flat top (Plate 413). This particular mahogany case has less strongly regional characteristics. The shell inlays come into vogue about the turn of the century. The dial is of the single sheet type and has that type of centre showing a miniature seascape (Plate 414).

PLATE 414. *Detail from the Thomas Field of Bath dial showing the central engraved scene. The single sheet dial seems to have been more popular in this region than elsewhere.*

PLATE 413. *Mahogany longcase of about 1790 by Thomas Field of Bath, the dial of the single sheet type. Again the double pediment theme is obvious in the flat top surmounted by a separate swan-neck.*

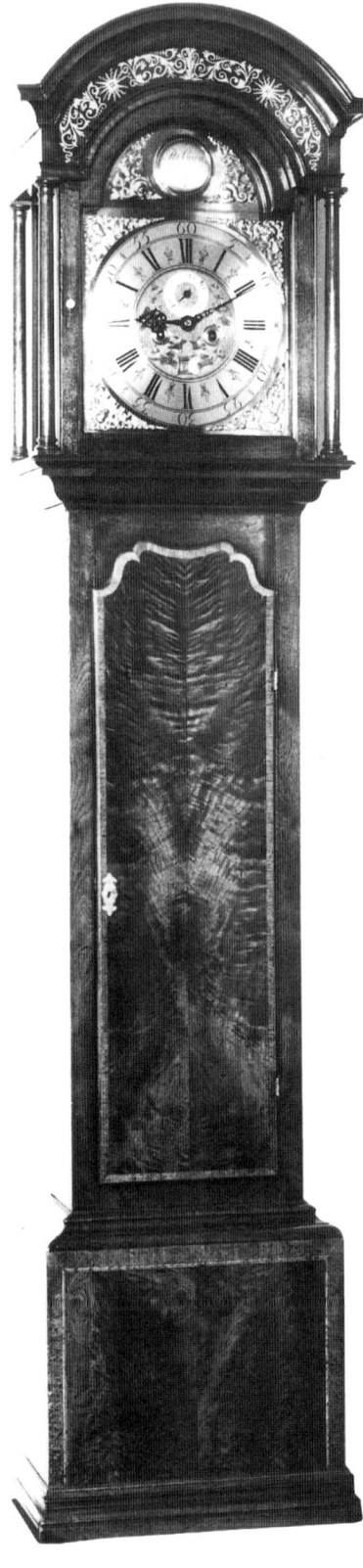

PLATE 415. *Thirteen-inch longcase dial of the 1730s by Peter Green of Liverpool, with separate arch. The centre treatment of engraved work on to matting is very north-western. Original blued steel hands.*

PLATE 416. *Walnut case of the Peter Green of Liverpool clock standing about 7ft.2in. North-western features include separate pillars (especially separate rear pillars), shaped door top, verre églomisé panel for a 'fret'.*

3. LANCASHIRE AND THE NORTH-WEST

For our purposes this area covers not only the North-west as far as Lancashire and Cheshire, but Wales and the Midlands. Lancashire itself was *the* strong centre of influence. This influence also extended to the more northerly counties of Cumberland and Westmorland but we shall examine those later as a separate area.

The thirteen-inch longcase dial by Peter Green of 'Leverpool' in Lancashire (the old spelling of Liverpool) dates from the 1730s and shows several stylistic features which clearly differ from London work (Plate 415). The dial centre is matted and has considerable engraving *into* the matting, here of two birds and two dolphins. This increasing encroachment of engraved ornamentation into the matted centre is typical from now on of those north-western clocks which retain the matted centre principal – the other category being those which used the engraved centre on to a polished ground.

Peter Green's dial also uses the large floral-twist half-hour markers, still attached to the inner quarters ring, and retains large half-quarter markers – both features very different stylistically from a London dial of the day, the former a feature often used in the North-west from this time on. The engraving is of excellent quality with some individuality – the curved 1s in 10 and 15. Ringed winding holes remained popular for much longer in this area than in London, where by this time ringing had ceased. The blued steel hands are original.

A further oddity is that the arch of the dial is separate from the square, something not done in London where the arched dial casting would be in a single piece. The reason for this is that Green's biggest demand, at this time

PLATE 417. *Very bold fourteen-inch longcase dial of 1730-40 by Andrew Knowl(e)s of Bolton. An exceptional degree of engraved work even for a north-western dial, here on to a polished centre. Original steel hands have lost their blueing.*

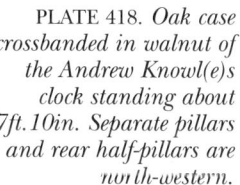

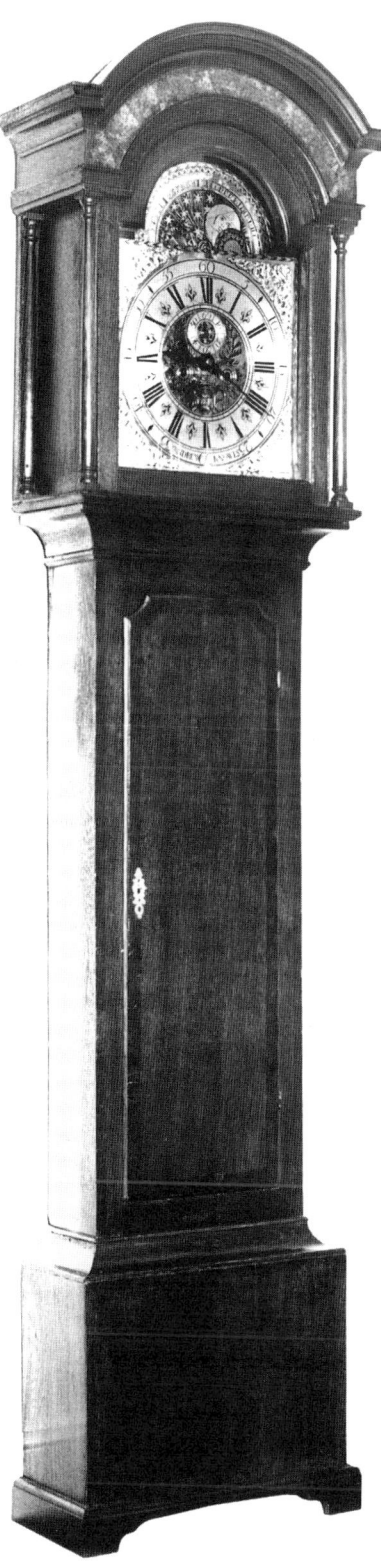

PLATE 418. *Oak case crossbanded in walnut of the Andrew Knowl(e)s clock standing about 7ft.10in. Separate pillars and rear half-pillars are north-western.*

when arched dials were just becoming fashionable, would be for square dial clocks. By adding a separate arch when occasion demanded he could avoid the cost of carrying a stock of two different types of dial sheet. This separate arch idea was largely confined to early eighteenth century makers in the North-west (including Ireland). Within a very few years the complete arch casting took over.

The case of the Green clock (Plate 416) is in walnut and represents the north-western interpretation of the London dome-top walnut case of the day. The principal differences are that the case includes certain north-western features: separate rather than integral hood pillars, crossbanding to the door and base (though examples of these last two features are occasionally found on London clocks), book-matching to the door and base, shaped door top, *verre églomisé* panel (of glass painted with gold scrollwork on the front, greeny/blue on the reverse) beneath the hood topmould where London cases might have a fret. The height is about 7ft.2in.

The fourteen-inch longcase moon dial by Andrew Knowl(e)s of Bolton in Lancashire, seen in Plate 417 before cleaning, is an exceptionally bold style, using the polished ground style of dial centre. The engraving is so strong and extensive that there would have been no room left to show matting. Dials became larger earlier in this region than elsewhere. The engraved moon has very unusual slots around the arch, thereby showing each of the four weeks adjacent to the current lunar day, which is here showing the 15th for full moon. The starry sky background was a feature of most engraved moons. Herringbone banding is here used a little later than normal. The clock dates from the 1730s or 1740s.

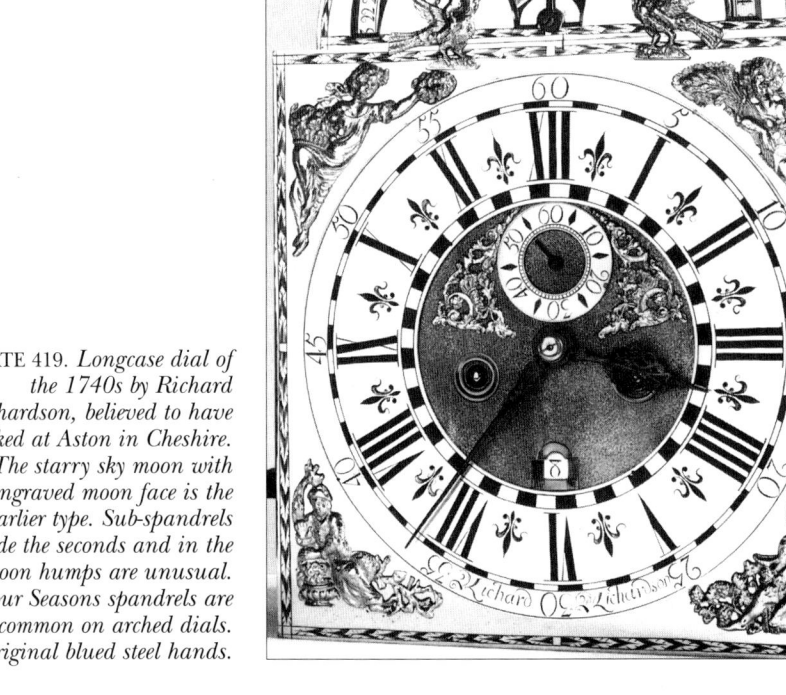

PLATE 419. *Longcase dial of the 1740s by Richard Richardson, believed to have worked at Aston in Cheshire. The starry sky moon with engraved moon face is the earlier type. Sub-spandrels beside the seconds and in the moon humps are unusual. Four Seasons spandrels are uncommon on arched dials. Original blued steel hands.*

PLATE 420. *The Richardson case is of oak and stands about 7ft.4in., based loosely on the London walnut case of the day. The D-mould to the door is an early feature, here retained later than usual.*

The bold and large half-hour markers are typical of the area, as is the late use of half-quarters and the heavy ringing of the winding holes. The engraving is of a very high standard and very distinctive. The steel hands are original, here having lost their blueing from incorrect cleaning. The whole dial is as different from London styling as could be imagined.

The case of the Knowl(e)s clock (Plate 418) is in oak with walnut crossbanding and stands about 7ft.10in. This is the oak equivalent of the Peter Green walnut example in Plate 416, having many features of style in common with it, as far as that is possible with a quite different wood.

The dial by Richard Richardson (Plate 419) is another very bold and strongly engraved example of north-western work dating from the 1740s. The maker is believed to have worked at Aston in Cheshire from about 1735 and to have died there in 1756. The chequerwork effect of the half-hour and minute rings is very unusual but suits well the bold engraved moon and herringbone edging. Ringed winding holes are to be expected though the dial centre here is of the (southern) matted type without any engraving.

Two small spandrels alongside the seconds dial are unusual but are seen on a few clocks of this period, principally by makers from this region. The two small spandrels on the moon humps are seldom seen in this position and are a whim of the maker. Here the main spandrels are the Four Seasons type, more often found on square dial work than arched dials. The blued steel hands are original. Altogether this is the work of a maker who went his own way stylistically whilst retaining certain features which indicate the general region.

The case of the Richardson clock is in oak and stands about 7ft.4in. (Plate 420). Early features include the D-moulded door lip, here unusually on the hood door as well as the trunk door. Glass side windows are retained, though most north-western makers had dropped them by now. This is the north-western equivalent of the oak cases in Plates 381 and 382 by John Bottrell of Coventry and James Norman of Poole, with which it makes interesting comparison.

The three-train longcase clock by Thomas Stones of Heaton (Plates 421 to 423) dates from the 1740s or early 1750s. The place is believed to be the Heaton near Manchester in Lancashire and the matted ground with extensive floral engraving and ringed winding holes is a style indicative of that area. The signature using 'de' and 'fecit' is a little old-fashioned but some northern makers did this into mid-century.

The clock plays a tune on six bells, a hymn called 'York' popular in musical clocks as it needs only six notes. The clock plays three verses every fourth hour, at 12.00, 4.00 and 8.00. The musical barrel is most unusually made of *wood* and still survives in good order despite its age (Plate 422). The iron pegs of the music barrel are wide and strong and were simply hammered in.

The walnut case (Plate 423) is the north-western equivalent of the London walnut case of the day, differing in the book-matching, free-standing hood pillars, crossbanding on the lip of the top-of-trunk mould and in having quarter columns to the trunk (an early use of a feature more usually associated with mahogany cases). The unusual aspect is the early use of blind fretting in what we might now call the Chinese Chippendale manner, but such fretting was in use before Chippendale popularised it. The height is about 7ft.6in.

The thirteen-inch longcase dial by David Collier of Gatley, Cheshire (near Manchester), is highly typical of the mid-century eight-day 'Lancashire' moon

PLATE 422. *The Stones musical movement from the back showing most unusual* wooden *pin barrel controlled by an inside countwheel.*

PLATE 421. *Three-train (musical) longcase dial of 1740-50 by Thomas Stones of Heaton, a dial in many ways old-fashioned, yet with a fully-developed engraved centre on to matting. Blued steel hands original.*

PLATE 423. *The walnut case of the Stones clock stands about 7ft.6in and displays early use of chequered blind fretwork.*

PLATE 424. *Thirteen-inch longcase rolling moon dial of the 1750s by David Collier of Gatley near Manchester. Engraving on to a matted centre was popular in the area. Painted landscape/seascape moon dials have now usually replaced the engraved starry sky forms.*

PLATE 425. *Dome-top oak case of the David Collier clock, typically north-western with the* verre églomisé *style panels and shaped door top. Height 7ft.6in.*

dial, with copious floral engraving on to a matted centre, in this instance incorporating the winding squares into the design as flowerheads (Plate 424). The sunken seconds dial has a scalloped edge, something which happened here and there at the whim of the maker, even in London. The blued steel hands are original.

The oak case (Plate 425) is reminiscent in shape of that by Peter Green in Plate 416, being a dome-top example crossbanded in walnut, the crossbanding on the base being set in from the edge, as was not uncommon in the area. The base has canted sides and canted moulds, a very north-western feature. The *verre églomisé* panels to the hood are here on a wooden ground. Free-standing hood pillars and wavy-topped doors are now more or less standard on cases from this area. Height about 7ft.6in.

The thirteen-inch moon dial by John Taylor of Manchester (Plate 426) dates from the 1750s and has the same type of matted centre with engraving which was very popular in the region, not unlike that of the Collier dial in Plate 424. Again the winding squares are incorporated into the design, which is always a good sign implying forethought. Half-quarter markers are retained here unusually late. Note the quirky top line of the 1 numerals, showing a little individuality on the part of the engraver who did an excellent job here. A motto or legend, here set around the arch, was popular with some makers from this region; almost all concern man's transient state.

The case of the Taylor clock (Plate 427) is seen alongside another of similar style housing a clock of the same period by Nathaniel Brown of Manchester (Plate 428). By now the swan-neck pediment has come into vogue, a style which was to become *the* most popular of all pediments in the North (and was seldom used in London). The *verre églomisé* panel was one method of infill used below the swan-necks – seen as blind fretting on the Brown case. This

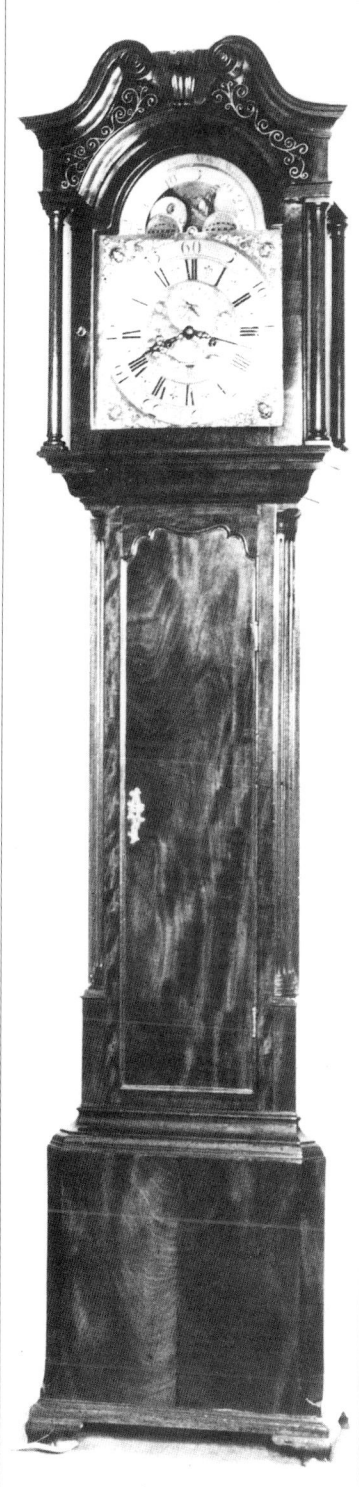

PLATE 426. *Thirteen-inch rolling moon longcase dial of the 1750s by John Taylor of Manchester, very typical of mid-century Manchester area work. Late use of half-quarter markers. Mottoes were popular with some north-western makers.*

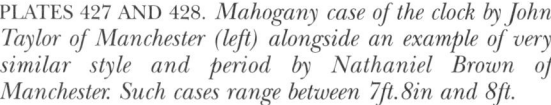

PLATES 427 AND 428. *Mahogany case of the clock by John Taylor of Manchester (left) alongside an example of very similar style and period by Nathaniel Brown of Manchester. Such cases range between 7ft.8in and 8ft.*

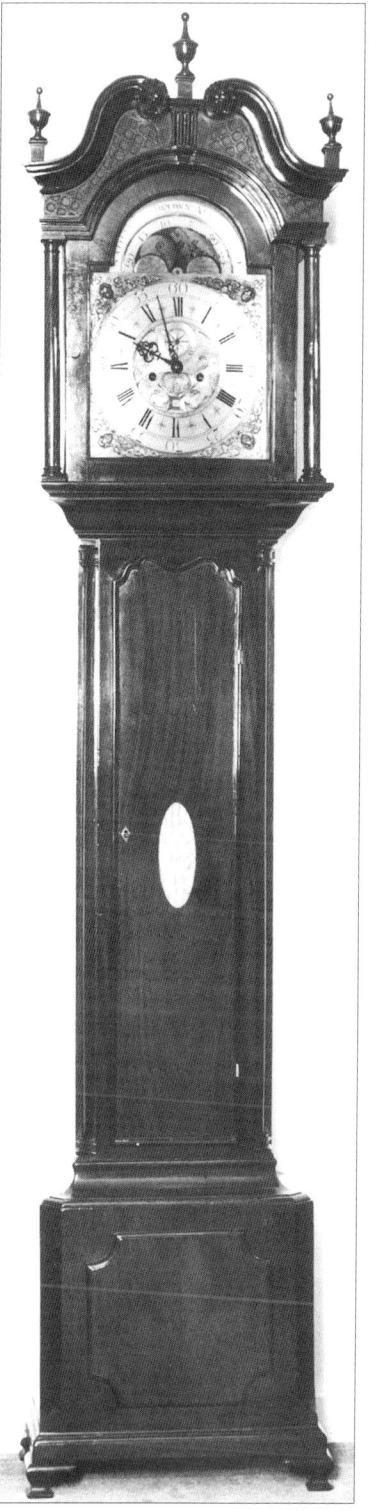

kind of swan-neck is joined at the centre by the infill, whereas the more normal swan-neck has a central open space for a finial. The long wavy-topped doors have a double lip mould which developed from the earlier D-mould. Free-standing hood pillars are standard from now on in this area and many such clocks have an ogee top-of-trunk moulding, as do these two. The bases

PLATE 429. *Twelve-inch rolling moon longcase dial of the 1760s by Smith of Chester, by now lacking half- and quarter-hour markers and therefore plainer in style, though still using the engraving on to matting centre.*

PLATE 430. *Mahogany case of the Smith of Chester clock standing about 7ft.4in. The* verre églomisé *principle is here used below the swan-necks. Typical north-western case of restrained style.*

PLATE 431. *Oak-cased longcase of the 1760s or 1770s by Robert Cawley of Chester, still using the older form of starry sky moon. The* verre églomisé *treatment below the swan-necks has in this example become a fully painted form. About 7ft.4in.*

and base moulds are canted. Some cases have a flush base and some a raised panel, the latter becoming increasingly popular with time. Unexpectedly small ogee bracket feet were popular on these cases. The Taylor case stands 7ft.8in, the Brown one about 8ft.

The twelve-inch dial by Smith of Chester (Plate 429) is in the same general style with a copiously engraved matted centre, but dates from a little later in the 1760s. The dial is therefore now plainer, lacking the half-hour markers and band. Again the use of the motto is to some degree a regional feature.

The mahogany case of the Smith clock stands about 7ft.4in. and has the more typical swan-neck with a wide set centre space for the finial and its support (Plate

PLATE 432. *Twelve-inch three-train (quarter-chiming) longcase dial of the 1760s by Francis Bayley of Uttoxeter, showing considerable London influence.*

PLATE 433. *Unusually ornate version of a single sheet rolling moon longcase dial made in the late 1760s or early 1770s by Major Scholfield of Rochdale. The sporting subjects are unusual, suggesting a custom-made dial.*

430). Fluted quarter-columns and wavy door tops were a regular feature of such cases. The complex trunk-to-base mould (here having an unusually narrow canting, as has the base) is typical of the region, as too are ogee bracket feet.

The oak-cased clock by Robert Cawley of Chester (Plate 431) dates from the 1760s or 1770s and has certain similarities of style with the Smith case. Here, however, the swan-neck has a complete infill which in this example is painted with flowers on to a green ground. This is an uncommon treatment of the infill but was practised by some north-western makers. Otherwise the case has many features resembling the Smith one. The ogee top-of-trunk mould was a popular feature in the region.

The twelve-inch dial of a three-train longcase clock chiming the quarters (Plate 432) was made in the 1760s in Uttoxeter in Staffordshire, which is a county to the south-east of this region. This dial could almost pass for a London dial, as it retains the more formal, matted centre style. However, the ringed winding holes and the floral tracery around the name are not London features. This example demonstrates that not all clockmakers followed the regional trend, which of course weakened anyway in those parts closest to London influence.

The single sheet silvered dial was not hugely popular in the North-west. When it was used there it was usually more busily decorated than its London or south-eastern counterpart. The example by Major Scholfield of Rochdale, Lancashire, dates from the later 1760s and is an exceptionally handsome example (Plate 433). Presumably this was made to order for a foxhunting customer as the corners and the painted scenes of the moon disc show foxhunting scenes. Here Scholfield uses the serpentine Dutch minute band and large floral-based half-hour markers as well as floral centre swags and 'Chippendale' swags on the moon humps to make a particularly spectacular dial. The arch motto reads 'Make use of Time and Remember Death'. The blued steel hands are original.

PLATE 434. *Fourteen-inch rolling moon longcase of the 1780s by Edmund Scholfield of Rochdale, the floral engraved dial centre typical of many north-western makers of the period.*

PLATE 435. *Mahogany case of the Edmund Scholfield of Rochdale clock standing about 7ft.4in. Book-matching was popular in the area at this period.*

The fourteen-inch rolling moon dial by Edmund Scholfield of Rochdale, Lancashire (son of Major Scholfield above) dates from the 1780s (Plate 434). The maker was born in 1730 and died in 1792. Here we see a late example of the brass dial, which by now uses dots for minutes in the manner of the newer fashion of painted dials. The polished centre is lavishly engraved with floral swags, as was the fashion in this region by this time. The maker retains unusually late a form of floral twist half-hour marker. The large question mark spandrels were often used on thirteen- and fourteen-inch dials of this region. The original hands are of blued steel.

The case of the Edmund Scholfield clock (Plate 435) is in mahogany with distinctly Lancashire features which include: book-matched door and base, wavy-topped door, infilled swan-neck, relatively broad proportions (caused by a limited height of 7ft.4in. with a fourteen-inch dial), canted base and canted base-to-trunk moulds, tiny ogee bracket feet. The case has many fine features of quality including cast brass paterae (rather than thin stamped ones), shell inlays above the door (uncommon in Lancashire work), Corinthian brass capitals, dentil moulding along the run of the swan-necks.

The dial of the longcase clock by John Ellison of Halewood, near Liverpool, Lancashire (Plate 436), is another late brass dial example dating from the 1780s, again using dotted minutes, a feature copied from the earliest painted dials. Half-hour markers have generally fallen from use by this time. The spandrels are of the familiar question mark pattern. Here the dial centre is of the matted ground type with floral engraving, used on occasion even as late as this since not all clockmakers turned to the polished centre form. The engraving is of excellent quality and the hands are original in blued steel.

The longcase clock by John Clifton of Liverpool, Lancashire, dates from the 1780s and also has the dotted minute system (Plate 437). The maker died in 1794. Here the dial centre is an unusual type with a chequerwork pattern done by engraving a criss-cross pattern, each square being engraved in a sort of crosshatched manner. Several versions of this design exist and these seem to appear only in Lancashire. The clock has centre calendarwork – a third hand from the centre making one circuit of the dial per month. The centre calendar was little used outside this north-western region. The main hands are of blued steel but the calendar hand is of brass for easy recognition.

The mahogany case (Plate 438) is of distinct Lancashire style, having many of the features we saw in the preceding clocks. However, some Liverpool cases of this time have additional stylistic features seldom seen elsewhere, namely *double* pillars to the hood (two integral with the door as well as two free standing) and what we call a brickwork base (Plate 439), where the base has cornerpieces standing proud rather like stone blocks on the corner of some buildings. These two features seem to be confined almost entirely to clocks

PLATE 436. *Rolling moon longcase dial of the 1780s by John Ellison of Halewood (near Liverpool), here using the engraving on to the matting form of centre which was still a preferred choice of some makers in the area. Original blued steel hands.*

PLATE 437. *Rolling moon longcase dial of the 1780s by John Clifton of Liverpool, using a form of chequered dial centre popular with a few north-western makers of the period. Centre calendar with brass hand for distinction.*

PLATE 438. *Mahogany case of the John Clifton of Liverpool clock, the double front pillars being a characteristic limited to Liverpool and immediate area. The hood on this example has glass infill panels, probably originally with* verre églomisé *decoration, now worn away. Height about 8ft.*

PLATE 439. *Detail of the brickwork base of the Clifton case, a feature confined to the Liverpool area.*

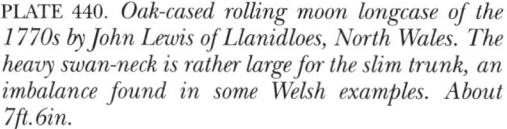

PLATE 440. *Oak-cased rolling moon longcase of the 1770s by John Lewis of Llanidloes, North Wales. The heavy swan-neck is rather large for the slim trunk, an imbalance found in some Welsh examples. About 7ft.6in.*

from Liverpool and close by, though just occasionally the brickwork base will appear elsewhere.

The hood has the infilled swan-neck we associate particularly with this region, here with three glass panels, the third one being in the centre to form a projection to take a finial. The hood also has a box top which stands proud above the corners of the swan-necks, also a feature more or less restricted to this region. The height is about 8ft.

The oak-cased eight-day longcase clock by John Lewis of Llanidloes in North

PLATE 441. *Longcase clock with penny moon and tides in the arch made about 1740 (or a little earlier?) by William Nicholson of Whitehaven. Several features distinctive of the North-west. (Photograph by courtesy of John Carlton-Smith Antiques.)*

Wales dates from the 1770s and stands about 7ft.6in. (Plate 440). The dial has a polished centre with restrained floral engraving and uses a mouth calendar, unusual on eight-day work but cheap to make. Two small spandrels are set on the moon humps rather than the more usual engraved hemispheres. The arched brass dial was not common in Welsh clocks, as most were of square dial form, probably to keep the height and cost down. Those which are arched dials very often have a rolling moon, as here. The dial style in this example is much more related to north-western styling than to that of London.

The case is of very thick and heavy oak, needlessly thick as the hood sides are of one-inch timber! Heavy timbers are a feature of many north-western clocks. In style this case has several features of north-western influence, most striking being the heavy swan-neck with solid infill and the shaped door top. Welsh clocks in general were influenced from Lancashire in the North and from Bristol in the South, hence their stylings are very non-London.

4. THE LAKE DISTRICT

The Lake District is that area of north-west England which encompasses the old counties of Cumberland and Westmorland (today grouped together as Cumbria). The types of clocks made there amount to an extension of the styles of the Lancashire region. In these most northerly counties of western England, however, their relative remoteness from London, and even from the much more populated Lancashire area, meant that many makers developed their own idiosyncratic ways. Of all the regions of England it seems to me that Lake District clocks frequently include the most varied and interesting range of clockmaking and casemaking. There is often something particularly interesting about a Lakes clock.

This longcase dial by William Nicholson of Whitehaven (Plate 441) has limited floral engraving over most of its matted centre, a feature which puts it

PLATE 442. *Twelve-inch longcase clock made in 1750 by William Porthouse of Penrith for William and Margaret Fawcet, probably a wedding clock. This maker tended to keep to the matted centre style.*

PLATE 443. *The oak case of the Porthouse clock is carved all over, but it is difficult to say when.*

PLATE 444. *Movement of the Porthouse clock showing large cheese-head wheel collets, trip pins in the end of his rack lifting piece to take a repeater trigger spring (here missing).*

PLATE 445. *Thirteen-inch rolling moon dial made around 1750 by John Whitfield of Clifton near Penrith, a maker of very varying and sometimes eccentric styles. Note circular grain of matting and arcaded minute band.*

in north-western England. The clock could date from about 1740 retaining a late use of half-quarter markers and quite early use of ten unit seconds registration. Half-hours of the large and bold flowerhead/fleur-de-lis nature and large half-quarter markers are also north-western, as are the busily-ringed winding holes. The arch carries a penny moon feature with lunar date read by a pointer on the outer scale, high tide times on the inner scale.

William Porthouse was a well-known and very prolific clockmaker working in Penrith, Cumberland, from about 1725 till his death in 1790. The greatest part of his output was in thirty-hour clocks, but he did make a considerable number of eight-day examples. The twelve-inch longcase dial by William Porthouse pictured in Plates 442 to 444 is dated on a plaque in the arch to 1750 with the names of the first owners, William and Margaret Fawcet. This was presumably a marriage clock made for their wedding in that year. Porthouse more than any other maker adopted this method of incorporating the names of the first owner(s) on the dial and many of his clocks have that feature, quite often with the year as well. At this period Porthouse often signed his dials on a strap, as here, a system not widely employed in the north-west.

Most of Porthouse's clocks have matted centres, though some of his later clocks use the floral engraving polished centre style. This particular clock has a very formal almost London-like dial centre with plain box calendar and unringed winding holes. On the other hand, his chapter ring has bold, anchored half-hour markers which are clearly north-western in style.

The oak case is in the dome-top style, not common in this region. It is profusely carved all over and stained to a near black colour (Plate 443). The carving may have been done later. Shaped door top and ogee mould above and below the trunk (and even an ogee mould to the dome top) all show north-western styling.

PLATE 446. *The oak case of the Whitfield clock has a caddy top and stands 7ft.2in.*

279

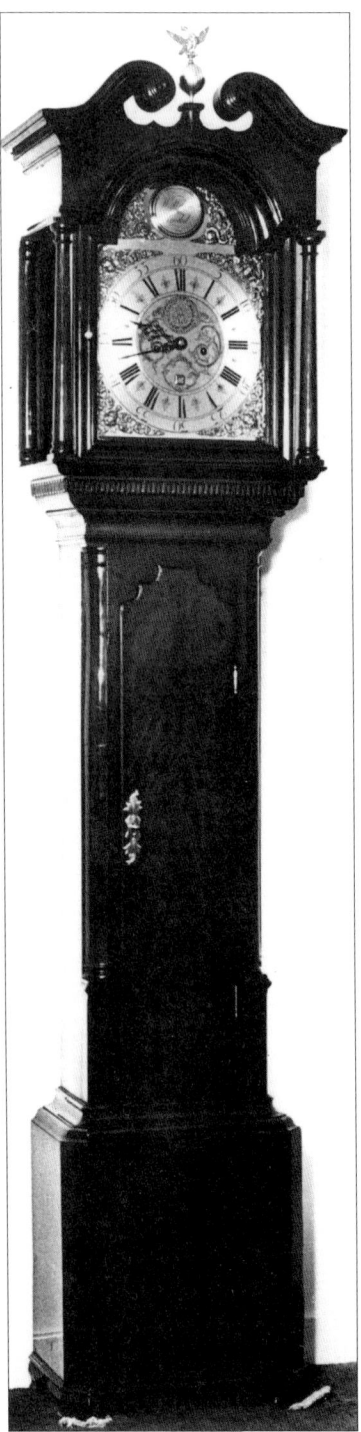

PLATE 447. *Thirteen-inch longcase dial of the 1760s by William Wilson of Kendal, who here uses the engraved centre style incorporating the monogram of the first owners instead of a seconds dial.*

PLATE 448. *The mahogany case of the Wilson clock stands 7ft.10in. and has several features popular in the area.*

The movement is sturdily made though conventional. Porthouse dials are of cartwheel cast form, as we would expect. His pillar forms vary at different times. Those shown in Plate 444 are fairly traditional but sometimes he used ornate double-baluster pillars and sometimes more heavily finned ones than on this movement. He probably changed his style with time.

Not much is known about John Whitfield of Clifton, a village close to Penrith in Cumberland, except that he worked in the 1740s-1760s period and his clocks can be of quirky, even eccentric, styling. The thirteen-inch rolling moon dial in Plate 445 has the upper rim of the arch wider at the sides than the top, an unusual treatment perhaps forced on the maker by the maximum moon wheel size he could cut. The moon disc itself is engraved and silvered and has the starry sky background usual on the earlier type of moon dial.

The dial centre is matted and the circular grain shows quite plainly. Ringed winding holes and mouth calendar features make this quite plainly different from London matted centre styling. Here Whitfield has ceased to use half-hour markers and has adopted an unusual arcaded minute band rather different from the more usual form we call a Dutch minute band (seen in Plate 433). The blued steel hands are original and of slightly odd style. The date is about 1760 or perhaps a few years earlier.

The oak case (Plate 446) stands about 7ft.2in. and has a caddy top above a flat pediment in the manner of some London walnut clocks but, of course ,in a very much more countrified style. Black string lines surround the door and base, the door having a part starburst inlay to the top. Pillars are attached to the hood door in the early manner.

The thirteen-inch dial of the longcase clock by William Wilson of Kendal in Westmorland (Plate 447) dates from the 1760s. The maker was apprenticed in 1735 and was working from about 1745 to 1781 when he died. Here the maker uses the polished dial centre with floral engraving very finely done and

PLATE 449. *Thirteen-inch rolling moon longcase dial made in the 1760s by John Benson of Whitehaven and having centre calendar. Here the maker uses the older starry sky form of moon disc.*

PLATE 450. *The case of the Benson clock in red walnut stands about 7ft.8in. Many features indicative of the region include: ogee top-of-trunk mould, complex trunk-to-base moulds, tiny ogee feet, swan-necks supported by 'keystones', wood rather than brass caps/bases.*

incorporating the initials of the first owner as a monogram, a feature of a number of Lake District clocks of this time. The monogram is so ornate it is hard to decipher but could be the letters RAR interwoven – probably for husband and wife in the manner of a marriage clock. The floral sprays incorporate the winding holes as part of the design.

The mahogany case (Plate 448) stands about 7ft.10in. Pillars and trunk quarter-columns are wooden capped, a feature of many north-western clocks where cabinetmakers often preferred them to brass caps on mahogany cases. Similarly the swan-necks (open rather than infilled in the Lancashire manner) have wooden terminals rather than brass ones. Canted base and base moulds are typical of the area, as is the shaped door top. The top-of-trunk mould is an exaggerated ogee shape trimmed with dentil moulding, another feature which is largely north-western.

The thirteen-inch rolling moon dial by John Benson of Whitehaven in Cumberland dates from the 1760s, perhaps a little later (Plate 449). This well-known maker was working there from about 1750 and died in 1798. The dial has the scroll-engraved polished centre here calibrated to take a centre calendar, the latter a popular feature in north-western arch-dial clocks. The spandrels are of the large 'Lancashire' cherub head pattern, very popular in the north-west in the third quarter of the century for dials of thirteen inches and above. The painted moon disc has the starry sky background, usually being the earlier form.

The case of the Benson clock (Plate 450) is made of red walnut, surprisingly popular in this particular region for better quality eight-day moon dial clocks, though not as costly as mahogany when new. This wood has a quieter figuring than mahogany and a colour verging towards brown rather than the red of mahogany. Cut a sliver from a hidden interior part and its natural colour shows distinctly brown compared to mahogany's red. The wood was often used

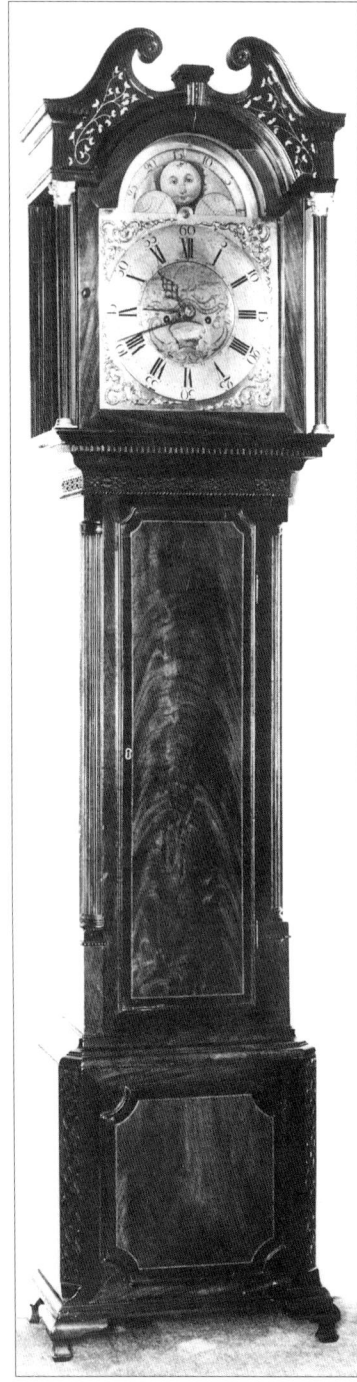

PLATE 451. *Detail of hood and dial of a rolling moon longcase of about 1785 by Taylor of Whitehaven, having centre calendar and visual ebbing/flowing tide dial above VI. Verre églomisé panels below the swan-necks. Considerable 'Lancashire Chippendale' influence.*

Plate 452. *The mahogany case of the Taylor clock stands about 7ft.10in. A combination of many north-western features, only some of which are strictly typical of the immediate area.*

in solid form. Crossbanding, as in this example, can be in red walnut but more often is in mahogany.

The absence of brass capitals on the hood and trunk, the carved rosettes to the swan-necks, shaped door top, ogee top-of-trunk mould, blind fret above the door top, canted base with complex trunk-to-base moulds, tiny ogee feet, free-standing half-pillars to the rear of the hood – all these are distinctive characteristics of the area and period. The height is about 7ft.8in.

The rolling moon longcase clock by Taylor of Whitehaven in Cumberland (Plates 451 and 452) dates from about 1785, an obvious late sign being the dotted minutes. A high and low water feature shows at VI.

This case is in mahogany and has many regional features in common with the Benson example in Plate 450, but this one has brass caps and bases. The swan-neck itself has inset *verre églomisé* panels (an influence from nearby Lancashire) and has dentil-capped 'blocks' at each side of the upper hood from where the swan-necks emerge, such blocks being a feature of many Lakeland cases, including the Benson one. Here the canted base has blind fretting down its cantings. The height is about 7ft.10in.

5. THE NORTH-EAST

This region consists of those eastern English counties adjacent to the Scottish border, principally the old counties of Northumberland, Durham and northern Yorkshire. Here the style of clock dials followed more closely than in the North-west the original London principle of the undecorated matted centre and in some respects it is a continuation of that south-eastern region. There were exceptions, however, and especially in the later eighteenth century when the floral-engraved polished dial centre style became increasingly

PLATE 453. *Twelve-inch longcase dial of the 1730s-40s by Robert Henderson of Scarborough, based on the London style, the matted centre engraved with an extension of the basket-of-fruit theme. Original blued steel hands.*

popular in some parts of this region, as it did elsewhere in provincial England. Dials in this area remained longer at the twelve-inch size (again following London fashion) and, though later examples grew to thirteen inches, the fourteen-inch arched brass dial seen in Lancashire was uncommon here.

There was considerable movement of clockmakers from the north-western English counties into the North-east (especially from Lancashire to Newcastle) and therefore north-western styles would have emigrated to the North-east. Conversely, some north-western clockmakers of the late eighteenth century are known to have had their dials engraved in Newcastle-upon-Tyne. This must mean that there was interaction of ideas between east and west and that clockmakers and engravers in the North-east were aware of the two different stylistic options available. It seems to me that the engraved centre form must have been more costly than simple matting.

Robert Henderson was born in 1678 in Cumberland but moved to work in Scarborough by 1708 to become the first clockmaker there, remaining there till he died in 1756. He was a Quaker, though his clocks show little of what we think of as Quaker styling. However, Quakers would often have had to set aside their own tastes to cater for non-Quaker customers. The fact that he was the first clockmaker there may have had a bearing on his bringing in what were largely London styles and may also have had an effect on later clockmakers copying the local tradition he started.

The twelve-inch dial longcase clock shown in Plates 453 and 454 dates from the 1730s-40s. In the arch is a rocking figure of Father Time, a feature which could occur anywhere in the England (except London itself) probably just on the whim of the customer. The matted centre has ringed winding holes and considerable engraving around the calendar box showing the basket of

PLATE 454. *Blue lacquer case of the Henderson clock, almost certainly London made, standing about 7ft.10in.*

PLATE 455. *Another twelve-inch longcase dial by Robert Henderson of Scarborough dating from about 1740, the dial again based on London principles, even to the small floating half-hour marker.*

PLATE 456. *The oak case of the second Henderson clock is a simple country version based on the concept of the London walnut case of the day.*

'plenty', a sort of extension of the birds-and-basket theme.

If we exclude the basketwork and the rocking figure, what remains is essentially a London dial, even to the floating half-hour markers. Several makers in the Yorkshire part of this area (including Henderson) used five pillars on their movements, London style. The case is a blue lacquer example of the pagoda style and is clearly a London-made case, further evidence that Henderson is trying to pass on the dignified London styling to his clients.

The second twelve-inch longcase dial by Henderson (Plates 455 and 456), dating from about 1740, is also virtually a London dial even down to the name-plate signature, the only *un*-London feature being the ringed winding holes. An unusual feature is seen in the arch boss, which is engraved 'R. Temple, Owner'. This is a provincial feature not really common in the North-east and the inclusion of the word 'owner' is a very unusual feature anywhere.

The oak case of this Henderson clock is the provincial equivalent of the London walnut case of the day without a caddy. Mould shapes, door shape (here with D-mould lip), integral pillars, hood side windows, shallow base (even though this may have been shortened further later) – all these are taken from the London standard.

Henry Hindley was born in 1701 in Wigan, Lancashire, and moved to York about 1731, where he worked till he died in 1771. His work is of the highest calibre and usually, as with the example shown in Plates 457 and 459, of a very formal nature and based very much on the London styling of the day – very different from the styling used by clockmakers in his place of origin.

The twelve-inch dial shown in Plate 457 dates from about 1750 and is in all respects but the name a London dial of very up-to-date nature, without half-hour markers. The arch calendar is typical of Hindley in that all features of the dial are for function, not decoration. The fact that London styling was popular in this area may well be due to his influence.

Hindley's movements are distinctive in that he used his own baluster pillar style, which is virtually his trade mark (Plate 458). He also used a regulator

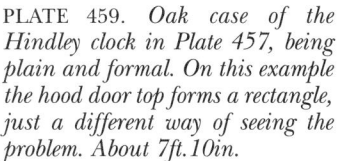

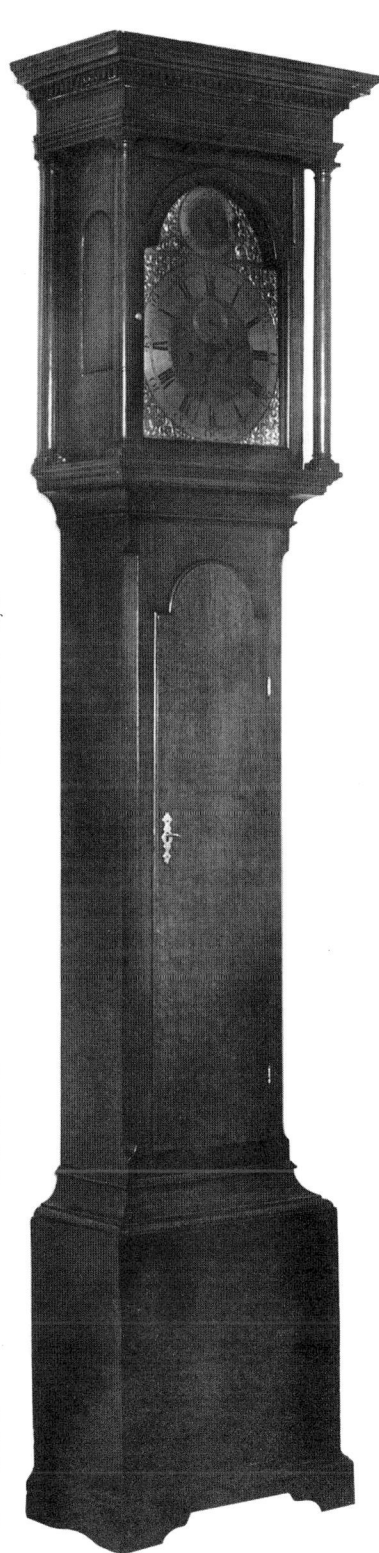

PLATE 457. *Twelve-inch dial of a longcase clock by Henry Hindley of York dating from about 1750. Hindley's style is that of London – plain, formal, absence of decorative features, early dropping of half-hour markers. Original blued steel hands. Arch calendar.*

PLATE 458. *Movement of a different Hindley eight-day longcase, showing his arch calendar drivework and his unique baluster pillars, which are virtually his trade mark.*

PLATE 459. *Oak case of the Hindley clock in Plate 457, being plain and formal. On this example the hood door top forms a rectangle, just a different way of seeing the problem. About 7ft.10in.*

PLATE 460. *Hood detail of a longcase by John Cornforth of Stokesley, c.1750. Dial style originating from the London form but with developments.*

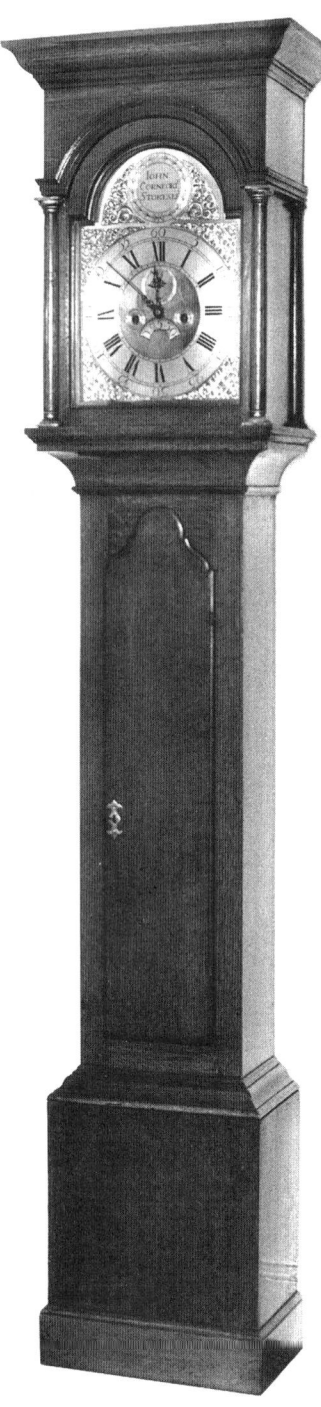

PLATE 461. *Oak longcase of the clock by John Cornforth of Stokesley, the general case style still a simplified form based on the London walnut style of a generation earlier. About 7ft.8in.*

type of pin crutch, fitting into a slot in the flat pendulum rod rather than around it as a normal fork.

Hindley's cases are also very formal. Some are in walnut, but this one in Plate 459 is of oak, very simple with London style moulds, arched door top, and hood side windows. In this example the hood pillars extend the full height of the hood, leaving the hood door as a rectangle in the manner of some bracket clocks. The height is about 7ft.10in

The twelve-inch longcase clock by John Cornforth of Stokesley in North Yorkshire in Plates 460 and 461 dates from about 1750. The maker is known to have been working there during the 1740s. This is a relatively formal dial reminiscent of that by Hindley (Plate 457), but has certain provincial features, namely the heavily-ringed winding holes and the mouth calendar. The oak case too has an echo of Hindley about it, though it has the provincial shaped door top, traditional integral pillars and an ogee hood topmould. The height is about 7ft.8in.

The thirteen-inch three-train longcase clock by John Hall of Beverley (Plate 462) dates from the 1750s – the third train winds through the calendar hand. This plays a choice of seven tunes on eight bells and here the maker uses the engraved centre style with almost a 'Lancashire' look to it. John Hall did use matted centres for some of his eight-day longcases and it may be the fact that this was a special clock which influenced his decision to have the engraved centre. The arcaded minute band of the 'Dutch' style was occasionally used by

PLATE 462. *Thirteen-inch three-train (musical) longcase dial of about 1750-60 by John Hall of Beverley. Floral 'Chippendale' engraving on to a polished centre ground. Moon disc smaller than usual, probably to allow space for tune selector and chime (music)/silent dials.*

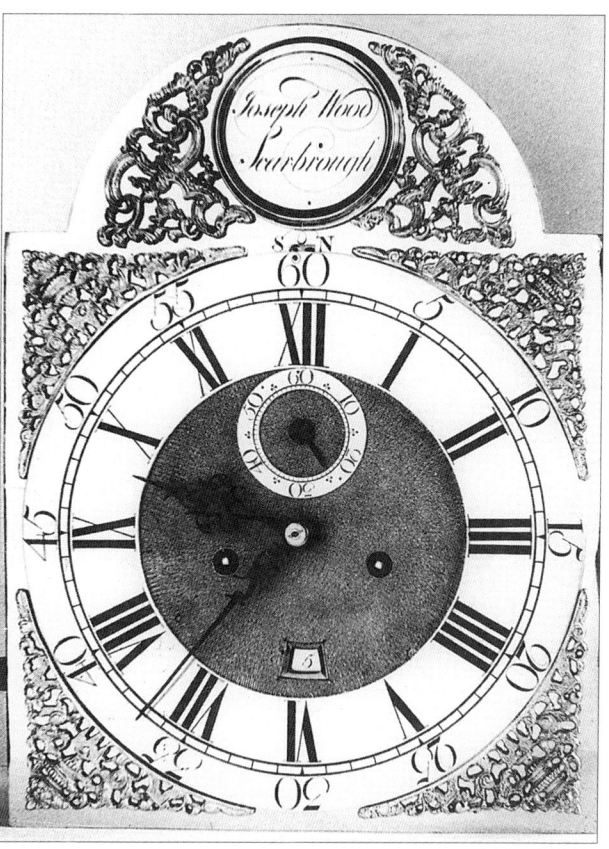

PLATE 463. *Twelve-inch longcase dial by Joseph Wood of Scarborough dating from the 1760s. Dial style almost pure London. S/N switch for strike/not strike (above 60) was a feature common in London work but not usually seen in this area.*

Hall, though seldom by other makers in this area. The engraving is finely done throughout.

The twelve-inch longcase by Joseph Wood of Scarborough in Plate 463 dates from the 1760s. Again this plain matt centre style is very reminiscent of London work. A strike/silent switch is positioned above the 60, marked S/N (for strike/not strike). The dial centre matting is heavier and coarser than on London work, showing some circular grain.

John Walker was a London clockmaker who moved to Newcastle on Tyne in Northumberland in the mid-1750s and worked there till his death in 1773. He was an innovative maker and was working in strongly competitive rivalry with other local clockmakers. Amongst other things he invented a three-wheel eight-day clock, an example of which is shown in Plates 464 and 466 dating from the late 1760s. The saving of a wheel was achieved by using larger wheels with higher gearing.

In 1766 Walker advertised the fact that he had invented a wheel-cutting machine of his own design, which 'rounded up' the teeth at the same time as cutting them – other clockmakers did the rounding up by hand as a separate operation. This saved time. He also redesigned both the striking and going trains to save one wheel in each. These economies enabled him to reduce the price of his clocks by something in the order of £1 below those of his competitors. The movements of these clocks have tapered plates (for economy of brass?). The twelve-inch dial of one such clock is pictured in Plate 464, in

PLATE 464. *Twelve-inch longcase dial of the 1760s with centre calendarwork (brass hand) by John Walker of Newcastle on Tyne. Dial provincial in style, though this maker came from London. Original steel hands, though the hour hand is a style we usually see later in matching pairs.*

PLATE 466. *Movement from the three-wheel eight-day clock in Plate 464, being John Walker's own invention.*

this instance using the engraved centre design, but some of these clocks have matted centres.

The movement in Plate 465 is from a different example of his three-wheel eight-day clocks. The fly for the strike is vertically driven through a worm thread and has adjustable blades to vary the striking speed. The movement in Plate 466 is that of the clock in Plate 464, also having a worm-driven fly but here with larger blades. Although London-trained, Walker used the cartwheel type of dial casting typical of provincial work.

The twelve-inch longcase dial by Archibald Strachan of Tanfield in County Durham (Plate 467) dates from about 1770. The maker worked in Tanfield from about 1755 until he moved to Newcastle on Tyne in about 1774. Here he uses the typical engraved centre style popular with many later eighteenth century Newcastle makers, though the engraving is a little naïve. The clock is numbered after his name in the arch 'No. 518'. The moon dial also incorporates tidal readings for high water on the inner band. With a movable pointer you can set the clock for any tide at will but using it vertically here it is

PLATE 465. *Movement from a different longcase clock by John Walker of Newcastle, showing his unique three-wheel movement, the strikework having a worm-driven fly. Date 1760s.*

PLATE 467. *Twelve-inch longcase dial with moon work and tides made about 1770 by Archibald Strachan of Tanfield (clock No. 518). The maker here uses the floral engraved polished centre. Original blued steel hands.*

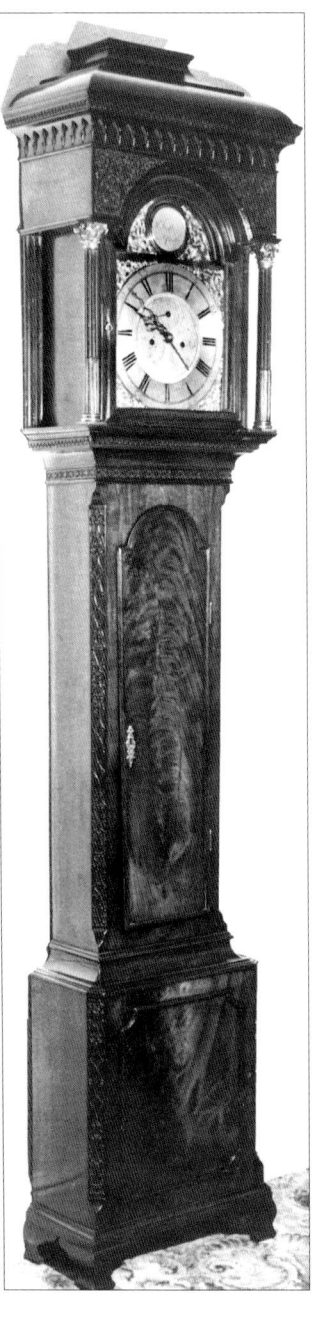

PLATES 468 AND 469. *Mahogany longcase of the 1770s by William Tickle of Newcastle on Tyne (left) with double pediment popular in this area (here a caddy top above a flat) alongside (right) a mahogany longcase of similar age by John Gibson of Alnwick, also with double pediment (architectural pediment above a flat top).*

probably calibrated for Newcastle tide times. Plain moon humps are unusual. The seconds dial is engraved on to the dial centre, which saves the cost of a separate seconds chapter ring. The blued steel hands are original.

Arched dial case styles in this area were very varied, most being in mahogany or oak as walnut was little used here. Many seem to have been designed as a flat-top case with the added option of one of several kinds of upper pediment – frequently a swan-neck, less often a caddy, occasionally even a pitched pediment.

On a great many north-eastern cases this upper pediment has been removed, probably to get the height down. Traces of glue blocks are usually an obvious indication of a removed pediment. Such reduced cases often look very unbalanced, as a flat top is not a happy style with an arched dial clock because

PLATE 470. *Thirteen-inch rolling moon longcase of the 1740s with early form of starry sky moon disc, made by Dollif Rollisson of Halton (Leeds). Unusual in not marking quarter-hour units. Engraved centre reminiscent of a bracket clock backplate of the day. Centre seconds hand.*

PLATE 471. *Oak case of the Dollif Rollisson longcase crossbanded in walnut and standing 7ft.9in.*

of the unused area at each side above the arch. These are known in the trade as 'hammer heads'.

William Tickle (senior) worked in Newcastle from about 1760 till he died in 1798. The clock pictured in Plate 468 dates from the 1770s and has an engraved centre dial. This mahogany case stands a little under 8ft. and illustrates a rather busy mixture of styles. Blind 'Chippendale' frets fill the upper hood front and the canted corners to trunk and base. Canted corners seem to have been popular in this area and so was blind fretting on mahogany examples. The hood pillars have inset brass reeds, London-fashion, but have Corinthian brass capitals, which are very provincial. The caddy is a development from one form of London walnut caddy. Certain features such as the small ogee bracket feet are reminiscent of north-western styling.

The mahogany-cased clock by John Gibson of Alnwick in Northumberland (Plate 469) was made about 1775-80. The maker died in 1844 aged ninety. The case door fits *inside* the door frame cupboard-door fashion instead of overlapping the frame, a method used by some casemakers in this area. It is probably for this reason the door is flat-topped. The pierced (through) cloth-backed fretting is heavier than the blind fretting on the Tickle case. The hood pillars are fluted and counter-fluted. The pitched pediment is unusual.

6. LEEDS AREA

The city of Leeds in West Yorkshire was one of the first provincial cities to have clockmakers. Positioned on the east side of the Pennine mountain range roughly midway between the east and west coasts of northern England, influence was felt here from those strongest trends of fashion from both the north-west and the north-east. What I summarise as 'Leeds' styling was spread over an area about twenty miles wide and ranging fifty miles north and south of that city. Here the fashion came early for engraved polished centre dials and perhaps because of that the variety of such engraved designs and themes was wider than elsewhere. Many Leeds cases follow distinctly recognisable stylistic trends. Casemakers in Leeds sometimes supplied clockmakers further afield and Leeds cases are seen

PLATE 472. *Thirteen-inch longcase dial of about 1760 by John Lawson of Keighley. The arch contains an annual calendar set in rolling moon fashion. Centre seconds, rolling moon at VI, Dutch minute band. A combination of many unusual features, all purely provincial.*

PLATE 473. *The oak case of the John Lawson of Keighley clock stands about 8ft. high. Verre églomisé 'fret'. Many unusual features including starburst inlays and chevron stringing.*

housing clocks by makers from such places as Wakefield and York.

The thirteen-inch dial signed Dollif Rollisson, Halton (Plate 470) dates from the later 1740s. Halton is now a suburb of Leeds but was then a separate village. The maker's unusual first name is believed to be the surname of a female ancestor. This dial would at first sight appear later in date but the maker is known to have died in 1752 and there are also certain stylistic features which date it early. The large dial size enforced the large cherub head spandrel pattern and the large minute numerals, both of which were in more widespread use later.

The finely engraved centre with the stylised head is reminiscent of many an engraved bracket clock backplate of the second quarter of the century and the floral sprays have an echo about them of the old lantern clock dial centre styling rather than the floral sprays on a later Lancashire dial. The unusual engraved 'clouds' on the moon humps and the engraved floral border to the arch are early features, as too is the starry sky background to the painted moon disc (later ones have landscape and/or seascape backgrounds). This clock has a slender counterbalanced centre seconds hand, here indicating 21 seconds past, the counterbalanced tail hidden behind the hour hand.

The case of the Rollisson clock (Plate 471) is of oak with walnut trim and crossbanding (pre-dating mahogany for this use). The unusual shaped door top is echoed in inset crossbanding in the upper door frame and the crossbanding is repeated on the base and hood door. Pillars are still attached to the hood door in the early manner. The clock has the original swept bracket foot. The case stands 7ft.9in. and has the characteristic heavier hood topmould which suggests it is complete and has not had some upper pediment removed.

Colour Plate 16 (page 224) illustrates a twelve-inch eight-day longcase clock by Francis Moore of Ferrybridge in Yorkshire (about fifteen miles south-east

PLATE 474. *Twelve-inch longcase dial by John Gilbertson of Ripon dating from the 1760s. Particular use is made of the engraved-centre styling, even to the seconds and calendar work.*

PLATE 475. *The oak case of the John Gilbertson clock is crossbanded in mahogany and stands 7ft.5in. More slender than many, yet with numerous northern features.*

of Leeds), dating from about 1755. As with many clockmakers who worked in small villages, this man did not usually letter his dials with a place name. The clock has a rocking Father Time in the arch. The matted dial centre shows southern influence, as we would expect at the southern limit of the 'Leeds' area. The ringed winding holes, however, and the bold engraving are very northern, as are the large cherub head spandrels (sometimes called a 'Lancashire' cherub head as they were especially popular there). The half-hour markers are bold ones based on the fleur-de-lis and, though strictly speaking 'floating', they virtually touch the quarters band. The aperture of the mouth type of calendar is unusually large and is made into a decorative feature with a compass rose to match the seconds centre. The original hands are in blued steel.

The thirteen-inch longcase dial by John Lawson of Keighley in Yorkshire (about fifteen miles west of Leeds) has several very uncommon features (Plate 472). First it has a centre seconds hand, counterbalanced as these usually were. The moon dial is placed unusually at six o'clock and this is done because he wanted to use the arch for an annual calendar. This is a most unusual feature, the centre of the disc being painted in four quarters to show country landscape scenes with sporting subjects representative of the four seasons of the year. Around the perimeter is lettered each day of the annual calendar, read against the pointer on the arch, together with the monthly unit and the number of days in each month. Also marked are the major saints' days.

The dial centre is of the polished and engraved type with a little Chippendale decoration but the centre mostly comprises a quotation from the Bible. His chapter ring is of the arcaded minute band type, which we usually call a Dutch chapter ring. All the engraving is sharp and bold with typical half-hour markers of the popular north-western twisted flower type.

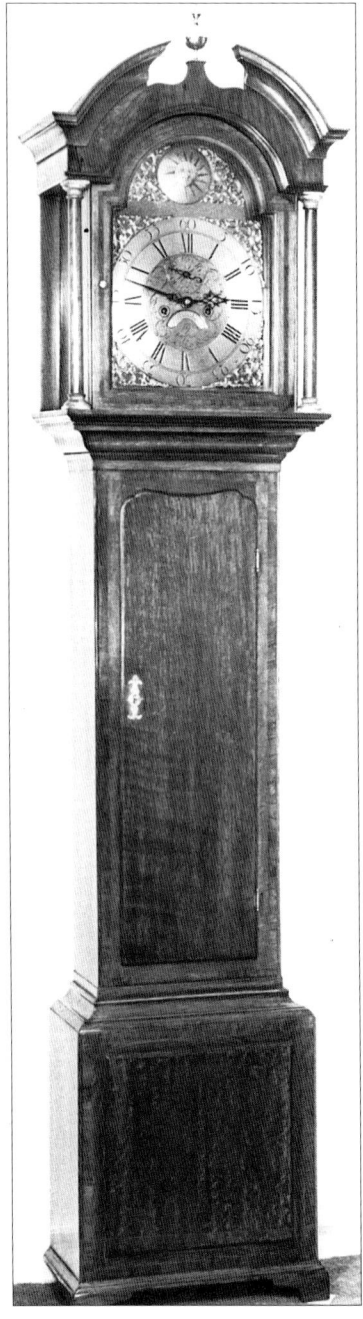

PLATE 476. *Twelve-inch longcase dial made about 1775-80 by Thomas Radford of Leeds using the floral engraved centre style, which most Leeds makers preferred. Winding holes carefully incorporated into the design.*

PLATE 477. *Oak case of the Thomas Radford clock crossbanded in mahogany and in the break-arch style popular in this region. Height 7ft.4in.*

The oak case (Plate 473) is substantial, standing about 8ft. with a hood of the break-arch type, a style popular in Leeds and perhaps influenced from there. Ornate chequerwork inlay of chevron pattern surrounds the hood, door and base panels occupying the position normally taken by crossbanding. Fine starburst inlays feature in the door and base, with ample wavy dentil moulds between hood and trunk. The inset panels beneath the hood topmould are painted on to wood in *verre églomisé* fashion, an influence from nearby Lancashire. The canted base and trunk-to-base moulds also show Lancashire influence. This fine and unusual clock dates from about 1760.

The twelve-inch dial by John Gilbertson of Ripon, Yorkshire, about fifteen miles north of Leeds, dates from the 1760s (Plate 474). This is of the engraved centre type with a finely engraved floral design worked around the winding squares. The calendar is shown by a pointer, a fashion which became increasingly popular as the late eighteenth century progressed. Half-hour markers are still shown but not the quarters, which often dropped first from use.

The case (Plate 475) is of oak with mahogany crossbanding and stands 7ft.5in. The case style is relatively neutral, yet it has northern features such as full rear hood pillars, pillars and trunk quarter columns with *wooden* caps, shaped door top, base higher than wide, carved rosette terminals to the swan-necks. This is altogether a neat and slender case with no particularly localised features but rather a mixture of various northern preferences.

The twelve-inch arched dial longcase by Thomas Radford of Leeds (Plates 476 and 477) dates from about 1775-80. The engraved centre pattern works the winding holes into the design, as the later Leeds dials often did. By now clocks have usually ceased to show half-hour or quarter-hour markings. This has good, sharp engraving and original hands of blued steel.

The oak case has the break-arch hood popular in this region and has mahogany crossbanding which runs down the trunk edges and is inset on the

PLATE 478. *Longcase dial of 1785 by Thomas Radford & Sons of Leeds having the unusual feature of three rocking ships in the arch. Floral engraved centre incorporating the winding squares. Original blued steel hands.*

PLATE 479. *Thirteen-inch longcase dial of the 1780s by William Fletcher of Leeds with rolling moon and centre calendar work (brass hand). Original blued steel main hands. Fine floral centre incorporating flower-head winding squares.*

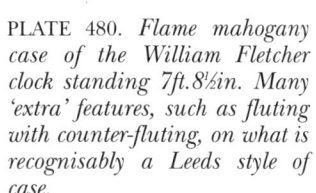

PLATE 480. *Flame mahogany case of the William Fletcher clock standing 7ft.8½in. Many 'extra' features, such as fluting with counter-fluting, on what is recognisably a Leeds style of case.*

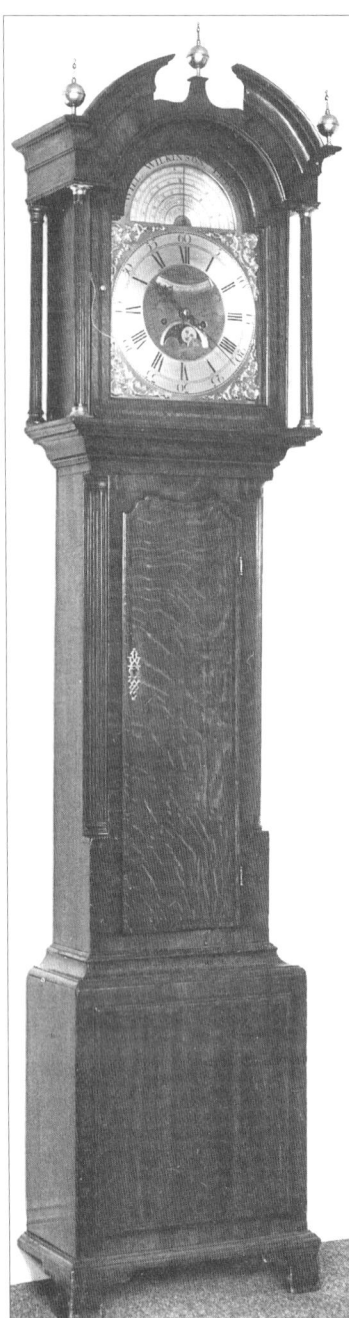

PLATE 481. *Fine thirteen-inch longcase dial with annual calendar in the arch made in the 1780s by William Wilkinson of Leeds. Original blued steel hands.*

PLATE 482. *The oak case of the William Wilkinson longcase has the mahogany crossbanding inset from the edges, a popular traeatment for Leeds cases, on which the break-arch hood was also a favourite form. Height 7ft.8in.*

hood door. This inset crossbanding is a feature on many Leeds cases of this period. The height is about 7ft.4in. The shallow ogee top-of-trunk mould was used on many Leeds cases of this time.

The longcase dial by T. Radford and Sons of Leeds in Plate 478 dates from about 1785, after Thomas Radford's two sons had joined the business. The rocking ship was one form of rocking arch figure which went through phases of popularity, but here the arch most unusually carries *three* rocking ships. Such rocking figures draw additional power from the going weight and are therefore a mechanical disadvantage, but clockmakers accepted this and often used a slightly heavier going weight to compensate.

The dial centre shows a typical well-engraved design, worked around the winding holes and using the later form of pointer calendar. The spandrel pattern is a rather unusual one and is quite a thin casting. These thin castings, often with an edged border, seem to have been popular with some makers here at this time.

The thirteen-inch longcase clock by William Fletcher of Leeds dates from the 1780s and has centre calendar work (here showing the first of the month) and a rolling moon, here showing the 23rd lunar day (Plate 479). The calendar hand is of brass to distinguish it from the blued steel time hands. The engraved centre design has begun to develop stylised swags and incorporates the winding holes. The thin swag pattern spandrels have borders.

The mahogany case (Plate 480) is exceptionally fine with several Leeds features (neutral swan-neck, shaped door top, shallow ogee top-of trunk moulds) but also with several unusual ones not typical of Leeds. The flame mahogany is very spectacular. The hood pillars are fluted and counter-fluted; so are the trunk quarter-columns; so too are the canted base corners. The fluting/counter-fluting theme continues in the inlay beneath the trunk quarter-column supports. Crossgrained veneer to the hood door and around the main

PLATE 483. *Longcase dial of about 1790 by Thomas Dodson of Leeds. Rocking horse soldier in the arch battle scene and various sporting themes within the engraved centre as well as centre calendar work (brass hand). Main hands original in blued steel.*

PLATE 484. *Oak and mahogany case of the Dodson clock showing many Leeds features such as inset crossbanding. Height 7ft.6in.*

door and main base panel is to draw the eye into the main features, that is the dial itself and the door and base panel – this being a principal applied in many better quality mahogany-veneered cases. The height is 7ft.8½in.

The thirteen-inch longcase dial by William Wilkinson of Leeds (Plate 481) dates from the 1780s. This has an annual arch calendar along the lines of the Lawson clock in Plate 472 and probably has its moon dial at six o'clock for that reason. The arch disc rotates once a year and shows the day of the month with saints' days through an aperture below XII whilst showing in the arch area several bands of information such as sunrise/sunset times, an equation of time chart for setting a clock from a sundial, and various zodiacal information.

The finely engraved dial centre includes 'Chippendale' themes. Half-hour and quarter-hour markers have now ceased to be shown. Spandrels are the large question mark pattern. The engraved rim to the chapter ring is a decorative treatment seen occasionally in the area.

The oak case has many typical Leeds features including inset crossbanding and the break-arch hood (Plate 482). Front feet only are a feature found on a number of cases from this area and other northern regions. The height is about 7ft.8in.

The longcase clock by Dodson of Leeds (Plates 483 and 484) dates from about 1790. The dotted minutes and bordered swag spandrels are typical late features. Here the engraved centre is most unusually decorated with scenes of a sporting nature – a fighting cock and a horse race. In the arch is a rocking horse soldier set within a battle scene. Presumably this clock was purpose made with these macho themes for a military and sporting 'blade'. The clock has centre calendar work indicated on a brass hand to contrast with the two main time hands of blued steel.

The oak case of typical Leeds style has inset mahogany crossbanding and trim, with wide crossgrain mahogany veneer being used to frame the door and base panels. The height is about 7ft.6in.

CHAPTER ELEVEN
SCOTLAND

Clockmaking in Scotland was controlled from the earliest times by the guilds of metalworkers, known as Hammermen, each of the major towns ultimately having its own guild designed to keep trade in the hands of local and accredited practitioners and particularly to ban under penalty of the law any intrusion by outsiders. Each guild was known as the 'Incorporation of Hammermen of ...', for example, Edinburgh. The first clockmakers were included within the general category of metalworkers (blacksmiths, whitesmiths, coppersmiths, etc.) but eventually became a separate section within the metalworking trades, the date of that subdivision varying greatly between towns. From the records of some of these guilds we know the names of a number of early clockmakers, but many are known by name only with no example of their work surviving today. From the point of view of studying Scottish clocks we can consider only those by whom work *is* known to survive and for the earliest of *those* clockmakers that work is extremely scarce, rarely more than a handful of clocks being known by any of them.

Domestic clockmaking in Scotland began very largely with the work of immigrant clockmakers, principally from England. This is hardly surprising since the use of the pendulum in Britain originated in, and spread from, London. The earliest clockmakers were based on the major cities, essentially the ports of Edinburgh, Old Aberdeen and Inverness and, a little later, Glasgow, as transport by sea, certainly to and from London, was often easier than by road. Old Aberdeen was then the major city north of Edinburgh, later becoming part of present-day Aberdeen. The Highlands themselves were very sparsely populated.

If we discount the very special instance of royal clockmaker David Ramsay, a Scot who worked principally in London anyway, then the first domestic clockmaker in Scotland by whom work survives is Humphrey Milne (or Mills), who came from Staffordshire to work in Edinburgh in about 1660 till his death in 1695 and whose work is known only by four or five lantern clocks based very much on London styling, at least one in a standing wooden case. Milne seems not to have made longcase clocks at all.

Milne's apprentice and perhaps a relative, Andrew Brown, worked in Edinburgh from 1675 till his death about 1712. Three or four longcase clocks are known by Brown, their work representing the best in London fashion of the day both in the clockwork and the casework, the latter including some marquetry examples. Even in the seventeenth century there were close connections in the clockmaking trade between Edinburgh and London, as evidenced by the tragic occasion when London clockmaker Edmund Appley was taken ill and died on a visit there in 1688, the cost of the church bells for his funeral being paid by Andrew Brown. The result of this close contact is that early Scottish longcase clocks are virtually indistinguishable from London ones in dial, movement and often case too.

Brown's apprentice and successor, Thomas Gordon, worked in Edinburgh from 1703 till he died in 1743, and he too is known to have made marquetry longcase clocks in the London manner.

Frenchman Paul Roumieu, believed to be from Rouen, arrived in 1677 as the first watchmaker to work in Edinburgh. He died there in 1694, succeeded by his

PLATE 485. *Eight-day longcase clock of about 1695 by Thomas Kilgour(e) of Inverness. The styling is very much based on London work of the day. Replacement hands of blued steel. Inside countwheel.*

PLATE 486. *The movement of the Kilgour eight-day clock has five pillars, the centre pillar latched. Solid sheet dial.*

son, Paul junior, who worked there from 1682 till he died about 1711. One or two fine longcase clocks are known by Roumieu junior, again in the London style.

The Kilgour family, Patrick and Thomas, believed to be father and son, worked variously in Old Aberdeen, Inverness and Edinburgh. Patrick seems to have worked mostly in Old Aberdeen, from 1672 to about 1692 or later. It is thought that he was born there and, if so, would seem to be Scotland's earliest native-born clockmaker by whom longcase clocks survive. Where he could have learned his craft is difficult to say, but his work shows signs of London influence. He appears to have made an abortive attempt to move to Edinburgh in 1685/6 but did ultimately move there, being last heard of in Edinburgh in 1702.

Thomas was at Old Aberdeen in 1685 (during his father's Edinburgh spell?), worked at Inverness in the 1690s and Elgin in 1697, married in Old Aberdeen in 1701, then later worked in Edinburgh in 1702. It is very difficult to pin down the wanderings of these two men, but their output of longcase clocks must have been considerable, two or three being known by Patrick and half a dozen by Thomas. Their work is in the best London style (Plates 485 to 488), some marquetry-cased, movements with five finned pillars and usually with

PLATE 487. *Movement of a month-duration longcase clock by Thomas Kilgour of Inverness c.1700. Five pillars. Eleven-inch dial. Outside countwheel striking was retained on long duration clocks.*

PLATE 488. *Marquetry longcase by Patrick Kilgour made just after 1700 with early wide-span arch. (Photograph by courtesy of Sotheby's.)*

inside countwheel striking, two or three being of month duration (the latter retaining outside countwheel striking as was usual with long-duration clocks).

The Scots clockmakers seem to have been very set in their ways as far as style was concerned. This is not surprising when we consider their role model was conservative London, as copied by conservatively fashionable Edinburgh. When the square dial form passed from fashion in London in the early eighteenth century, it passed from use in Edinburgh too and, as most of Scotland at that time followed the Edinburgh lead, it is very seldom one sees a square dial Scottish clock after about 1720. The arch dial form took over absolutely.

Clocks by the very earliest (pre-1720) Scots makers are scarce, probably because of the small number of clockmakers and high destruction rate from their sheer age. However, there is also a scarcity of surviving clocks in the next quarter century, so that it is not until the 1750s and 1760s that we begin to come across them with any frequency. By then, of course, the arched dial had a firm and virtual monopoly of the fashion in Scotland.

Round dial clocks were not popular in Scotland until the early nineteenth century, though an occasional example is seen made in the late eighteenth. Most Scottish round dial longcase clocks date after 1800 and therefore have japanned dials, but occasional eighteenth century examples were made with the single-sheet round brass dial, which had a sort of late revival in Scotland in mid-Victorian times, often giving such clocks the look of a regulator.

It may also be by reason of the strong London influence, filtering through Edinburgh, that the thirty-hour longcase clock is virtually unknown in Scotland. If the eighteenth century Scots public chose to have a longcase clock it was an eight-day, other than in very special instances where long-duration clocks were made, usually having other specialised features. The very occasional thirty-hour clock of Scottish make seems to have occurred close to the English border, perhaps where English tastes overlapped.

The absence of thirty-hour clocks is a great pity and Scottish clockmaking is the poorer for it, since the effect is that in Scotland the longcase clock has only a single origin, based in the London eight-day school. It deprives the enthusiast of an entire sphere of horology, that of the self-taught clocksmith, which in England offers the most exciting, varied and untapped source of study of the development of clockmaking (see Chapter Five).

A few brass dial bracket clocks were made in Scotland, but, as in the English provinces, bracket clocks were very unusual items until the mid-nineteenth century (when they had painted dials), probably on account of their high cost. Those pre-nineteenth century examples that do exist are principally the work of the better Edinburgh makers who catered for the more sophisticated market. Hook-and-spike clocks and hooded clocks appear to be totally unknown in Scotland, probably because of the Scots' aversion to daily winding. *The* major form of domestic clock in which the Scots makers specialised almost exclusively was the eight-day longcase.

By the mid eighteenth century the typical Scottish longcase clock was an eight-day arched brass dial measuring twelve inches in width. The centre of style was Edinburgh and these clocks almost always followed the London fashion of a matted centre with conventional seconds dial. The matted centre was often retained late in those Scottish clocks following Edinburgh influence, as it was in London. However, by the 1760s a different sphere of influence had begun to take effect in certain areas, based on Glasgow, where there was a taste for engraved centres, often on to a very finely matted ground, perhaps a result of influence from north-west England.

Scottish dials, therefore, and to some extent casework too, divide into two broadly separate groups. The larger and more widespread (probably on account of the larger number of cities on the eastern seaboard) was that of east Scotland based on Edinburgh, the lesser that of west Scotland based on Glasgow. The beginnings of these two divisions in Scottish styling are discernible in the late eighteenth century but are not strongly distinctive until after about 1820, by which time the brass dial had given way to the painted dial.

Calendars were usually of the square box type. In fact the mouth calendar was not much used at all in Scottish brass dial work, preference being maintained for the better form, that is the twenty-four hour changing, wheel-driven square-box calendar. In the late eighteenth century some clockmakers used the pointer calendar (see Plate 500, Peddie dial) to match the seconds pointer, principally on single-sheet brass dials, a method often continued into the painted dial period. Centre seconds hands were not popular except in a limited way on single-sheet brass dials of the late eighteenth century. Centre calendars were seldom used in Scotland.

PLATE 489. *Twelve-inch five pillar longcase c.1750 by James Pringle of Dalkeith. The case is of pitch pine.*

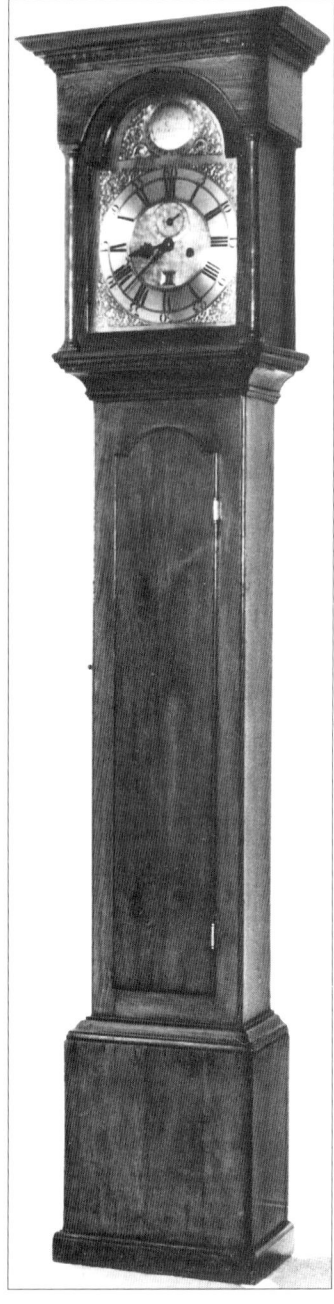

PLATE 490. *The pine case of the James Pringle clock is in remarkable condition with no sign of woodworm damage to which pine is often prone. Height 7ft. Style vaguely reminiscent of the London walnut profile of the day.*

Traditional tastes in Scotland quite often meant that some of those features indicative of period in English clocks were retained later there. Small dial size (twelve inches), matted dial centres, square box calendars, spandrel patterns, hands patterns, and even the use of the brass dial itself after 1770 instead of the 'new' painted one… all these features were retained later in Scotland than in England and might prove misleading towards too early a dating.

Moon dials seem to have been less popular than in England and when they were used it was often in combination with high water indicators, perhaps on account of the particular importance there of the seaports for transportation, the difficulty of road travel being aggravated by the harsh northern climate.

By the mid-eighteenth century most movements had the standard four pillar construction (five pillars being dropped early as unnecessary), and had rack strikework (countwheel striking also being dropped early). Some Scots makers varied their pillar shapes from the English pattern, keeping the central knop but having pillars which tapered up towards the knop, these tapered pillars being more common towards the end of the eighteenth century and into the nineteenth. The pillars on English brass dial movements almost always had a parallel profile, with the exception of a few late eighteenth century makers working close to the Scottish border in such counties as Northumberland, who were perhaps influenced by the Scottish taste. Increasingly often as time went by the central knop itself took on a more squared-shouldered profile than the rounded English style of knop.

Some clockmakers, especially rural ones, seem to have left their work less well finished than we might expect. For example, movement frontplates were often left raw from the casting, sometimes with scribing-out circles clearly visible, without any attempt at cleaning up or polishing. On the other hand some makers took great pains in quite unnecessarily shaping parts such as teardrop-shaped spring tails, just out of sheer pride in their work.

PLATE 491. *Eight-day dial of about 1750 by Robert Melvill of Stirling, the styling very much along London lines. Original blued steel hands.*

The twelve-inch arched dial longcase clock in Plates 489 and 490 by James Pringle of Dalkeith (a few miles south-east of Edinburgh) dates from about 1750. Like many Scottish brass dial longcase clocks, it is a little difficult to date, as the Scots makers sometimes retained features of an earlier style into a period later than we expect them in England. Here Pringle's movement has five pillars, an unexpectedly late use of such a number, as four was the norm in most regions by mid-century. The herringbone edging around and on his name boss is also used here later than was normal in England. However, the dial is 'modern' enough to have dropped the half-hour markers, even though he keeps the quarter-hour divisions and retains the old form of seconds numbering in units of five. The matted centre with square calendar box is a retention of the London principle, as used in most Edinburgh clocks.

The original case is of pitch pine and stands about 7ft. high. The hood door is of oak, however, probably made of this stronger wood to prevent warping. The case was probably painted originally but is now stripped down. The style owes little to London other than in the overall proportion and the integral hood pillars. London cases of this period often had a hood superstructure such as a caddy, which this case never had.

The twelve-inch dial by Robert Melvill of Stirling in central Scotland, seen in Plate 491 in uncleaned condition, dates from about 1750 and echoes the Edinburgh style. Melvill still uses the decorative engraved border to his name-plate. Half-hour markers have now fallen from use and the seconds are marked in the up-to-date style using units of ten. These features changed relatively early in Edinburgh and its areas of strong influence such as fashionable Stirling, following closely on the heels of London.

The oak case (Plate 492) stands about 7ft.3in. and is along the lines of the Edinburgh style with a shallow swan-neck. However, the hood pillars are still attached to the door and the frets below the swan-necks are a little heavier than

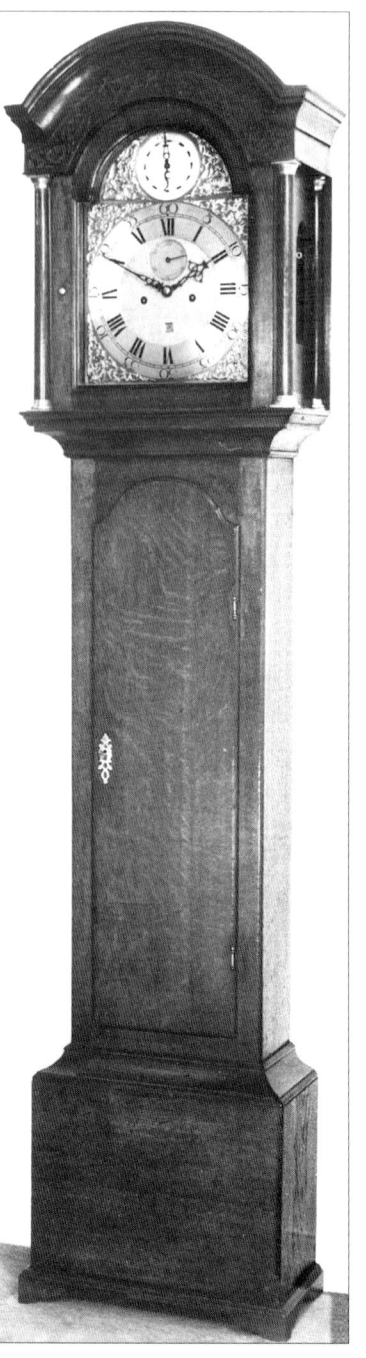

PLATE 492. *Oak case of the Melvill clock standing around 7ft.3in. The style is a simplified form of London work but with a swan-neck pediment, infilled below the swans with fretwork.*

PLATE 493. *Oak cased clock by James Coulie of Montrose dating from the 1760s (see Colour Plate 17, page 313). Height 7ft. The dome top is an alternative pediment form. Retaining hood side windows, London-style.*

PLATE 494. *Twelve-inch eight-day longcase by John Campbell of Campbeltown c.1770. Scroll engraving on to a punch-worked centre is a feature of a number of West Scots clocks of this period.*

we would expect from Edinburgh. Shapes and mouldings are all very simple.

The oak-cased clock illustrated in Plate 493 and Colour Plate 17 (page 313) by James Coulie of Montrose (on the east coast about sixty miles north of Edinburgh) reflects the Edinburgh style, having the formal London-based twelve-inch dial with matted centre and a strike/silent switch in the arch. It dates from the 1760s. The case too has certain echoes of London – dome top, hood side windows, a slight variant of the shouldered-arch door top, simple mouldings.

The eight-day clock in Plate 494 by John Campbell of Campbeltown, remotely situated on the Kintyre peninsula, west of Glasgow, dates from about 1770. Many of the stylistic features are typical of an English dial of the period,

PLATE 495. *Twelve-inch dial of the 1770s by Allan How and Robert Knox of Irvine, again showing western Scots styling in the punch-ground centre. Note the ornate flourishing of the 3s and 5s.*

PLATE 496. *Solid mahogany case of the How and Knox longcase clock with several features indicative of western Scots styling, especially the somewhat heavier shaping of the swan-neck.*

but the centre shows that sort of finely matted centre with floral engraving spreading into it which characterises many dials from western Scotland at this time, the matting being done by a tiny ring-punch leaving a series of dots all over the ground. The engraving seems to have been done before the punchwork.

The dial by Allan How and Robert Knox of Irvine, south-west of Glasgow, dates from the 1770s (Plate 495). This is another example of the floral centre engraved on to a finely matted ground, here even more exuberant than usual. The flowery starburst centre to the seconds dial is an unusual variation and this exotic numbering seems to be the work of either one engraver or a small school of engraving in this area. Note the flourishes on the 3s and 5s. The signature on a boss in the arch was a very common way of signing Scottish dials at this time. No attempt is made to match corner and arch spandrels – dolphin patterns in the arch were an easy option and were used over a wide period in Scotland.

The case of the How and Knox clock stands about 7ft.2in. and is in solid mahogany (Plate 496). When used in the solid form mahogany had to be fairly straight grained to avoid warping, so that the figuring is not nearly so wild as with many veneered clocks. Here the swan-necks have more than usual infill, perhaps a stylistic feature picked up from some north-western English clocks of the day. The pillars are of the three-quarter front and quarter rear type, but here attach not to the door itself but to the door frame, a sort of half-way compromise between the two basic styles. The hood door happens to be a left-hand opener. The tiny ogee bracket feet are also reminiscent of feet on some Lancashire clocks of the day.

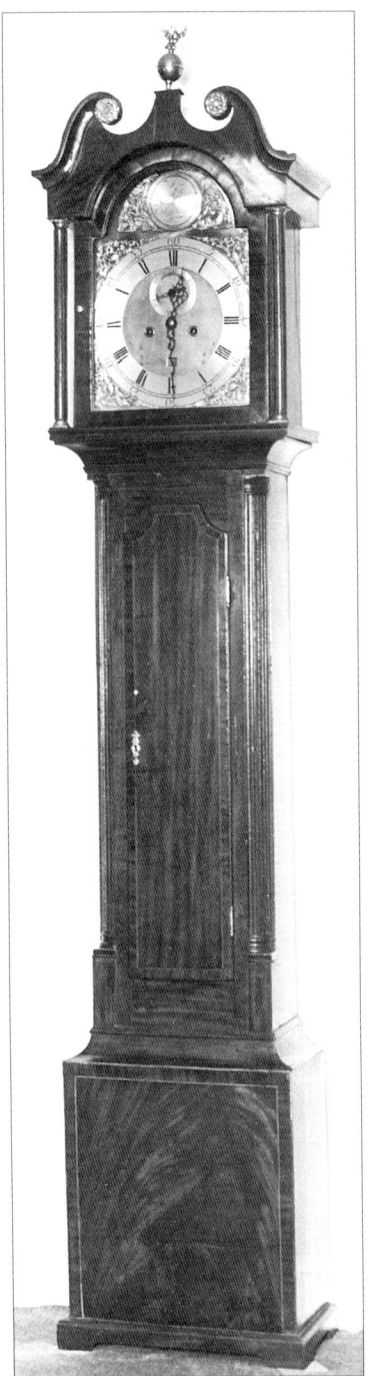

PLATE 497. *Mahogany longcase of about 1770 by Daniel Binny of Edinburgh, the case typical of the small Edinburgh styling of the day, especially in the neat swan-necks. Height 7ft.*

PLATE 498. *Twelve-inch dial of the 1770s or 1780s by William Allan of Aberdeen. The dial centre is lavishly engraved on to a polished ground. Blued steel hands probably original.*

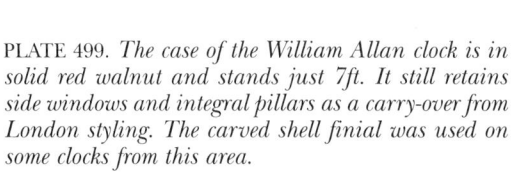

PLATE 499. *The case of the William Allan clock is in solid red walnut and stands just 7ft. It still retains side windows and integral pillars as a carry-over from London styling. The carved shell finial was used on some clocks from this area.*

The clock by Daniel Binny of Edinburgh dates from about 1770 and has the typical Edinburgh style of matted centre dial with name-plate in the arch, very much like a London dial of the day (Plate 497). The neat case is of mahogany, strung with boxwood and crossbanded, and stands only 7ft. high, small height and good proportion being a regular characteristic of the better Edinburgh cases. This case has the typical shallow Edinburgh swan-neck. The fluted quarter-columns were a regular feature on these cases when in mahogany, less common in oak examples.

The twelve-inch dial by William Allan of Aberdeen on the east coast of

PLATE 500. *Single sheet brass dial of about 1780 by James Peddie of Stirling, the engraving typical of this style, the bird-beaked terminals being a regular feature, though varying in execution. Blued steel hands probably original.*

Scotland dates from the 1770s or 1780s and has a most unusual all-over engraved dial centre on to a *polished* ground (Plate 498), unlike the matted centre ground of western Scots dials. The design includes the strange bird-beaked monster heads (at III and IX) which became very popular on Scottish single-sheet dials a little later. The maker has drilled his winding holes through the engraved design, but perhaps this was difficult to avoid in such a densely-woven pattern.

The case of the Allan clock (Plate 499) is in solid red walnut, a wood quite often used on better Scots cases of the 1760 to 1790 period. This timber came from the Americas and was probably seen as a less costly alternative to mahogany. It was usually used in solid form. The hood retains the side windows (influenced from London through Edinburgh) and the integral door pillars, as well as the domed top, here with a split pediment to accommodate the carved seashell motif. Moulds and general styling are very simple. The height is just 7ft.

The single-sheet brass dial by James Peddie of Stirling, in central Scotland about equidistant from Edinburgh and Glasgow, dates from about 1780, maybe a little later (Plate 500). This was a popular style in Scotland and its use continued into the early nineteenth century with little change, even into the mid-nineteenth century with the round dial examples, though these usually have no decorative engraving. At each side of his name the scrollwork terminates in the bird-beaked heads often incorporated into such designs. Sometimes these heads decorate the corners and can resemble birds or sea-horses or serpents. The inclusion in the design of these beaked heads seems to be very much a Scottish theme and often makes them identifiable at a glance from English single-sheet dials.

On this type of dial the calendar is often of the pointer type, as here, to match the seconds. When new these dials were silvered all over to produce a steely effect and one sometimes hears them wrongly described as steel dials.

PLATE 501. *The oak case of the James Peddie clock stands about 7ft. and is typical of many east and central Scots cases of the day.*

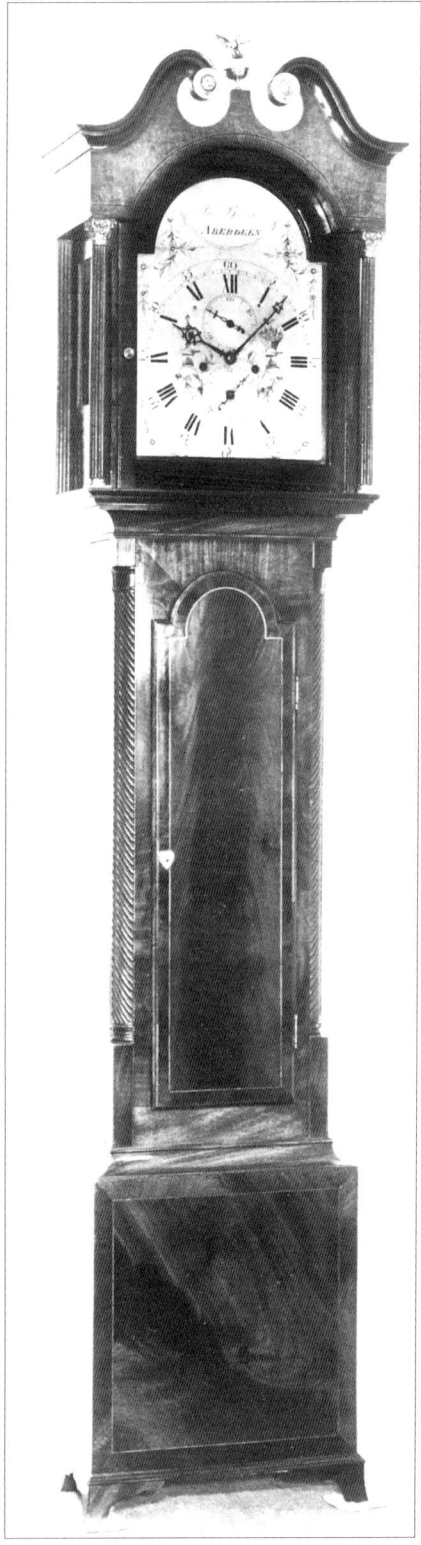

Over the years the silver polishes away to leave an all-brass dial which, of course, can be re-silvered today by any competent restorer.

The oak case of the Peddie clock (Plate 501) stands about 7ft. and is based on the Edinburgh model which spread to influence most of eastern Scotland. The hood still has the side windows and a neat and shallow swan-neck, which has a very characteristic shape to it. Otherwise the case is very simple in style. The double plinth is again a feature originating from the double plinth on many London cases. Door tops are usually shaped, as here, in a relatively simple manner.

On some Scottish cases, this one included, the pillars attach beneath forward-projecting overhangs, a rather strange feature which prevents the door from opening more than ninety degrees. This is more often a feature of western than eastern Scots cases.

The mahogany clock in Plate 502 by John Barron of Aberdeen dates from the 1790s and has a silvered single-sheet brass dial engraved with floral swags. These dials often have pointer calendars to match the seconds hand, as here. The swan-neck case is a version of the typical East Scots swan-neck case, based on Edinburgh, retaining the hood side windows. The rope-twist quarter-columns are a little unusual. Height about 7ft.3in.

PLATE 502. *Mahogany-cased eight-day clock of the 1790s by John Barron of Aberdeen standing a little over 7ft. Single sheet dial. Matching hands of blued steel probably original.*

CHAPTER TWELVE

IRELAND

Domestic clockmaking in Ireland began in Dublin with the work of clockmakers who either came from London or brought with them skills they had learned in London, the relationship of Dublin makers to their London colleagues being very similar to that of the Edinburgh makers. There was considerable contact between London and Dublin and initially the styles of Dublin clocks were virtually identical to those of London. For instance, the famous clockmaker Joseph Williamson, who is often credited with having made most of the equation clocks of his day, was working in London by 1686, yet had served his apprenticeship in Ireland. It is not until about 1700 that we can today trace surviving Dublin work; there must be seventeenth century examples but they are exceptionally rare.

By 1700 the Dublin makers were already beginning to develop stylistic features which were their own. We are speaking principally of eight-day longcase clocks. I don't know of any Irish-made brass dial thirty-hour clocks,

PLATE 503. *Ebonised bracket clock made about 1705 by John Crampton of Dublin 15½in. high including carrying handle.*

PLATE 504. *The dial of the John Crampton clock is seven inches square and very close in style to a London clock of ten or fifteen years earlier.*

PLATE 505. *The engraved backplate of the John Crampton clock showing the verge pendulum resting in its carrying hook, pull-repeat system, three bell nest for the quarters with separate bell for the hourly strike.*

or of more than one or two lantern clocks, and Irish bracket clocks of this age are extremely uncommon.

The bracket clock pictured in Plates 503 to 506 by John Crampton of Dublin appears at first sight identical to such a clock made in London about 1690 or even a little earlier. However, we know from the records of the Dublin guilds that this maker was not allowed to practise his craft there till his freedom in 1704, and that he was still working in 1728. We must therefore date this clock to his earliest working period which is about 1704-1710, which means that the style is slightly old fashioned for the date. The work is of high quality, having a five pillar (London principle) movement. However, the four external (and therefore visible) pillars are decoratively shaped in baluster form, a measure of the maker's individuality of style, whilst the central pillar is of conventional knopped shape, presumably because it was less visible there. The clock has

PLATE 506. *View of the John Crampton movement showing the fusee, standard pattern centre pillar, unusual baluster pillars for the corners, solid dial sheet (London style).*

PLATE 507. *Longcase dial of about 1710 by John Parker of Dublin. Elaborately- engraved treatment in herringbone manner filling almost all available spaces. An unmistakably Irish style.*

pull repeat work on three bells so that it can at will strike the last quarter-hour followed by the last hour – this feature presumably designed for bedside use at night to establish the nearest quarter-hour without having to strike a light.

The longcase dial by John Parker of Dublin (Plate 507) appears to date from about 1710. No facts about this maker are recorded so we cannot check on his dates. The dial can be seen to be closely modelled on London work of the day in respect of the features it displays, yet already by this very early date there are considerable differences in the *way* that information is displayed, amounting to very strong beginnings of Irish styling. Most obvious is the very ornate and lavish treatment of the decorative engraving: herringbone type bordering around the dial edge, outside the chapter ring edge, around the winding holes and both inside and outside the seconds ring.

The matted centre is further smothered in engraved scrollwork so that very little remains visible of the matted surface. This lavish use of engraving on to the matted centre is reminiscent of the general trend which developed much later in the eighteenth century in much of north-west England, perhaps even influenced there by Dublin. Bold half-quarters accompany even bolder anchored half-hour markers, again a feature found in north-west England.

The twelve-inch longcase dial by Thomas Meekings of Dublin (Colour Plate 19, page 317) can be dated to about 1700-1705, as the maker is known to have become free from his apprenticeship there in 1699 and to have died in 1709. This

COLOUR PLATE 17. *Twelve-inch longcase by James Coulie of Montrose c.1765. The dial is very much based on London styling. See page 304.*

dial shows many similarities of style with that by John Parker (Plate 507), especially in the lavish herringbone type of engraving and bold half-hour markers.

The movement of the Meekings clock (Colour Plate 20, page 317) is based closely on London methods using five typically finned pillars and inside count-wheel striking. An obvious difference, however, is that the dial sheet is of the cartwheel-cast (spoked) type, which we know was not the London way, showing that this dial was made locally.

By mid-century this very busy, attractive and presumably also very costly dial treatment had tempered to a much more sober form of matted centre dial, which seems to have become the normal fashion for the typical square dial longcase clock thereafter (Plate 508). The square dial persisted in Ireland well into the last quarter of the eighteenth century (London, of course, had

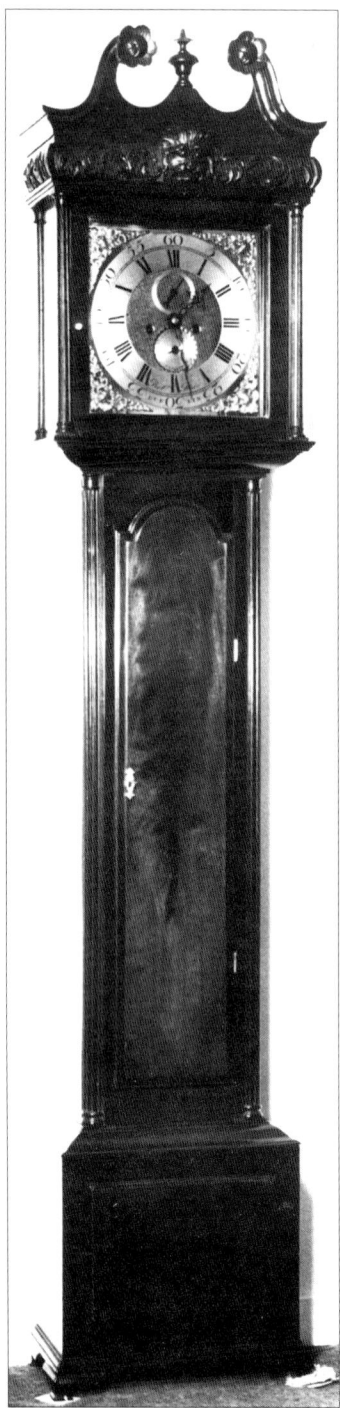

PLATE 509. *Mahogany case of the Robert Rose clock with many features typical of this period, most obvious being the carved frieze and the exceptionally long trunk. Height about 7ft.10in.*

PLATE 508. *Longcase dial of about 1750-60 by Robert Rose of Dublin. Typical plain matted centre and plain chapter ring without half or quarter markers. The large cherub head spandrel was popular on these dials.*

PLATE 510. *Longcase dial of about 1755-60 by Thomas Kennedy of Armagh, the style typical of many such clocks of the third quarter of the 18th century.*

finished with the square dial by the 1730s). The Irish square dial quickly grew to, and settled at, thirteen inches in size, compared with the London version which ended at twelve inches.

Yet oddly the Irish square dial of the mid-eighteenth century returned to some aspects of London styling in so far as it used: the undecorated matt centre (the matting neatly done though often quite coarse-grained); the early dropping of half- and quarter-hour markers; the early use of ten-second numbering. Minute numerals grew large quite early, unlike London dials. This makes the Irish square dial longcase of the second half of the eighteenth century distinctly recognisable.

Dublin was the principal place of manufacture, though by the third quarter of the eighteenth century clockmakers worked in other towns too. An example of a thirteen-inch square longcase dial by Thomas Kennedy of Armagh (Plate 510) shows many typical features of the Irish square dial style, this one dating from about 1755. Casemaking for these clocks at this date was largely confined to Dublin and Belfast. Other centres may have produced casework, but the Dublin model was so strong that we cannot recognise the origins other than as being of the Dublin style.

Such thirteen-inch cases are typically of mahogany, often that very dark and relatively figureless Spanish mahogany which presumably came into Dublin from the West Indies and which seems to have been used at a quite early date here (from about 1750). Many such cases have a swan-neck pediment with carved rosette terminals and have a carved cushion-mould frieze around the upper hood, often with a protruding lion mask or devil's head at the centre, occasionally with a similar mask in the base. Hood pillars have wooden capitals, earlier ones carved in Corinthian style, later ones turned and plain. The door often has a chamfered edge, fielded panel style, and sometimes the raised base panel may be of the same nature. Trunks are exceptionally long,

PLATE 511. *Mahogany longcase of about 1760 by William Betagh of Dublin, who died in 1770. Many characteristic features including carved frieze and rosette terminals. Height 7ft.11in. (Photograph by courtesy of George Mealy & Sons.)*

PLATE 512. *Hood detail of a longcase clock of the 1760s by Thomas Cornwall of Dublin. The fine detail to the carving is typical, as are the carved pillar caps. The inlay is unusual.*

PLATE 513. *Later Irish longcase of the 1780s (note dotted minutes) by an unknown Dublin maker. Certain developments in the style make this case a little more like an English equivalent, though the style is still very distinctively that of Dublin. About 7ft.8in.*

keeping more the long trunk proportion of an English case of half a century earlier. Trunk corners often have shallow canted corners, though some have fluted quarter-columns, these mostly in the later period and sometimes of a smaller diameter than on English cases. These very distinctive features make such cases instantly recognisable. The cases illustrated in Plates 509, 511, 512 and 513 show examples of this style. The features of these cases which compare with English ones relate closest to those of north-west England around Lancashire, though the proportion is very different.

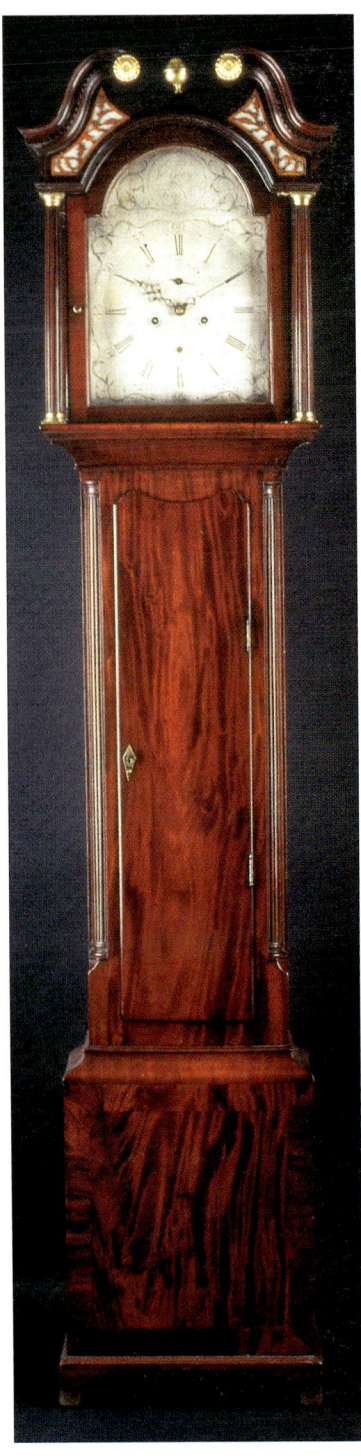

COLOUR PLATE 18.
Mahogany single sheet twelve-inch dial longcase of the 1790s by James Gray of Edinburgh standing about 7ft.4in. Typical Edinburgh styling. (Photograph by courtesy of Sotheby's, Sussex.)

PLATE 514. *Exceptionally fine dial of about 1775 by James Wilson of Belfast (No. 379) from a three-train musical longcase with moon and tidal dial in the arch, equation of time dial below XII and calendarwork at VI. Note the engraved centre style reminiscent of north-west English work.*

Arched dial longcase clocks were made and the example in Plates 514 to 516 is by James Wilson of Belfast, the capital of Northern Ireland, where styles perhaps related more to north-west England. Such dials often had the arch constructed separately from the square and joined behind by riveted straps, as with some of the earliest English arched dials. The arched dial examples would be more inclined to have engraved centres on to polished grounds, similar to the fashion of much of north-west England. Perhaps this engraved centre treatment was on account of the greater cost of such grander examples, which often had rolling moon and tide dials.

Belfast casework for these grander, late eighteenth century clocks was usually

COLOUR PLATE 19. *Twelve-inch longcase dial of about 1700-1705 by Thomas Meekings of Dublin. The elaborate engraved treatment is characteristic of Dublin dials of this period. Fine and costly workmanship. See page 312.*

COLOUR PLATE 20. *Movement of the Meekings clock in Colour Plate 19 showing typical finned pillars, inside countwheel on the main barrel, cartwheel-cast dial sheet.*

PLATE 515. *Movement of the James Wilson of Belfast clock showing musical pin-barrel, twenty-four hammers playing on twelve bells (two per bell for rapid note succession). Serial number 379 visible on frontplate.*

PLATE 516. *The very grand mahogany case of the James Wilson musical clock bears many features reminiscent of north-west England 'Chippendale' styling but a more exaggerated form. Seldom are so many stylistic details incorporated in the one case. Height about 8ft.4in.*

PLATE 517. *Dial of the Barnaby Matthews bracket clock in Colour Plate 21 (page 320), 7½in.wide. Made with verge pendulum, with arch moon and tide work. Unusual form of engraved centre.*

in mahogany and had a considerable affinity with north-west English mahogany casework of the 'Lancashire Chippendale' style, especially in regard to: the absence of brass fittings (wood being used instead for, for example, swan-neck terminals, pillars and quarter-pillar capitals, and finials); overall shape and proportion; keystone supports beneath the swan-necks; complicated and often canted trunk-to-base moulds; tiny ogee bracket feet.

In summary it is fair to say that Irish clocks often have something about them (dials, movements and cases) which is unexpected to the English eye. This is probably a measure of the individuality of styling which is bound to emerge more strongly in any separate region, but the more so where this is an island.

The ebonised bracket clock by Barnaby Matthews of Dublin dating from

COLOUR PLATE 21. *Ebonised bracket clock of about 1750-60 by Barnaby Matthews of Dublin, the case of unusual styling with double pediment. Height about 21in.*

about 1750-60 (Colour Plate 21 and Plate 517) exhibits an interesting stylistic mixture of dial features and a case which, most unusually, has both a swan-neck and a caddy top, two quite different pediments on the same case. Carved and gilded pillar caps and bases embellish fluted pillars (counter-fluted at their bases, London fashion), with side-facing ones at the rear, carved and gilded side-frets as well as glass side windows with additional windows in the upper case sides, five finials all round and a carrying handle. This unusual mixture and duplication of features is probably a mark of Irish individuality.

CHAPTER THIRTEEN

CLOCKMAKING IN AMERICA

The great majority of brass dial clocks in America today were made in Britain and other European countries and were either taken there by emigrants or shipped out in more recent years by dealers. Details of their origins and styles will be found elsewhere in this book. This present chapter relates only to brass dial clocks actually made in America.

Clockmaking in America had barely begun by the year 1700. Records exist of the names of immigrants recorded as clockmakers in the 1680s and 1690s, but no clock seems to be known dating before 1700 and the oldest clock which can be dated with any certainty by any clockmaker working in America is a longcase clock (known in America as a tall clock) dated 1709 and made in Philadelphia by Abel Cottey.

Most of the first clockmakers in America were immigrants from England; one or two came from Ireland, Germany and Holland. They took with them their working practices and styles and the first American-made clocks were made along English principles; a few, a little later, on the slightly different German or Dutch principles.

The first clocks made in America were longcase clocks. Very few examples seem to be known of American-made lantern clocks, even though several of the first clockmakers there are known to have made lantern clocks before they left England. This is probably because by 1700 the lantern clock was already becoming outmoded in England and the thirty-hour longcase was considered to do everything a lantern clock could, but better.

The earliest American-made clocks are very few in number and were made by only a handful of clockmakers, important though these are in marking the beginnings of American work. If we exclude those early clockmakers known there only through records, and count up those by whom clocks are known to survive, it is doubtful whether the work of twenty clockmakers is known before the 1730s, the number growing only steadily thereafter. It was the middle of the eighteenth century before numbers grew to a significant level. Until about 1800 most American-made clocks were longcase ('tall') clocks made in the Eastern seaboard states.

These earliest clockmakers were pioneers in a new land. They had to be as self-sufficient as possible, performing every task associated with the trade including their own casting, forging, wheelcutting and engraving. The alternative was for them to import from England those items they could not personally produce. They could no doubt re-cast brass as need arose once they had obtained the raw material, but it may have been that some felt it easier or more professional to import such highly visible features as dial fittings (spandrels, chapter rings, etc.), especially engraved features. It does appear from the highly competent engraved work of some early dials that some clockmakers may have imported these ready engraved from England. This is especially obvious when such clocks are compared with other early examples where the engraving can be exceptionally primitive. Much of their time was spent in repairing anything mechanical and anything to do with metalwork from church clocks to guns to cooking pans, much as the clocksmith did in England. Clockmaking probably formed only a small proportion of their total work activity.

Early brass dials made in London and in provincial England had solid sheet

PLATE 519. *Dial of a later eight-day longcase clock of about 1730-40 by Peter Stretch of Philadelphia. The dial is very much in the London manner, perhaps even London made. (Collection of Mrs.James Gibbs, USA. Photograph by courtesy of the NAWCC Watch & Clock Museum.)*

PLATE 518. *Eight-day tall clock by Peter Stretch of Philadelphia made about 1715-1720 (see Colour Plate 22, page 330). The case is of American walnut and resembles contemporary English country styling but for the strangely small lenticle glass. (Photograph by courtesy of Graham H. Jeffries.)*

castings. Provincial English dial-casters in some regions soon began to use the cartwheel type of spoked casting and this was in widespread use by the 1730s. Many early brass dials made in America seem to use the cartwheel principle too, but the centre of these is a separate piece from the surround, to which it is joined by riveted straps or spokes. This difference in construction may assist in recognising an American-made dial from an imported English-made one used by an American clockmaker. A further aid to recognition is that, as English-made dials imported for use in America were generally from London, these would have solid dial sheets.

Many of the emigrants from England were Quakers who maintained strong family bonds in both the old and new countries, and the first clockmaking 'schools' were mainly family-based, or sometimes religion-based, as minority sects such as Quakers tended to stick together. The trade was first centred in two major localities: Philadelphia and Boston, with a third centre developing slightly later in New York by makers moving from one or the other.

PLATE 521. *Mahogany case of the Peter Stretch clock standing about 8ft.6in. including wooden finials. The style and proportions are not unlike a provincial English case (though many have lost their caddy top). (Collection of Mrs. James Gibbs, USA. Photograph by courtesy of the NAWCC Watch & Clock Museum.)*

PLATE 520. *Movement of the Peter Stretch clock showing solid (London-type) dial sheet, simpler pillar style consistent with later period. (Collection of Mrs. James Gibbs, USA. Photograph by courtesy of the NAWCC Watch & Clock Museum.)*

PLATE 522. *Eleven-inch dial from an eight-day longcase clock of about 1720-25 by Benjamin Chandlee of Nottingham (Maryland). The chapter ring is very English in style and may even have been made and/or engraved in England. Inside countwheel strike. (Photograph by courtesy of Graham H. Jeffries.)*

PLATE 523 (opposite). *Twelve-inch dial of an eight-day clock by Benjamin Chandlee Junior of Nottingham made about 1755. Spandrels of gilded pewter. Blued steel hands. Unusual dial centre shows a mixture of English influence features. (Photograph by courtesy of Chris H. Bailey, USA.)*

THE FIRST PHILADELPHIA CLOCKMAKERS

The first recorded clockmaker in Philadelphia was Samuel Bispham, an Englishman who bought land there in 1696. At least one clock survives by him, but little seems to be known about him or his origins. It may be significant that English Quaker clockmaker Isaac Hadwen of Kendal, Westmorland, who himself visited relatives in Philadelphia in 1718 (see page 418) took an apprentice named Isaac Bispham in 1727, but this may be coincidence.

The second clockmaker in Philadelphia was another Quaker, about whom considerably more is known – Abel Cottey, born in 1655 in Tiverton, Devon. Clive Ponsford uncovered much of his earlier background when researching for his book *Devon Clocks and Clockmakers*, finding that Cottey worked initially at Crediton, then from the late 1690s at Exeter, and sailed for Philadelphia from Bristol in 1700 under very dubious circumstances. The local Quakers believed that he left Exeter secretly in the night and hid away in Bristol till his ship sailed, leaving shameful debts behind him. Cottey worked in Philadelphia from 1700 till his death in 1711. In England Cottey made lantern clocks but only a single American-made longcase is known by him (and also a loose dial sheet), the clock bearing the initials of his apprentice (BC) on its frontplate and the date 1709. The dial of that clock has engraved corner decoration rather than spandrels, a style known to have been favoured by some groups of English Quakers, who often shunned corner spandrels.

Cottey's apprentice was Benjamin Chandlee, another Quaker born in 1685 in Kilmore, County Kildare, Ireland, moving to Philadelphia with his family about 1702 and working initially as apprentice under Cottey, whose daughter, Sarah,

PLATE 524. *Case of the Chandlee Junior clock in cherrywood, the outline showing influence from London styling of the day, though in a simplified form. The 'lenticle glass' is in fact a carved wooden boss. Height about 8ft.6in. (Photograph by courtesy of Chris H. Bailey, USA.)*

he married in 1710. Chandlee inherited land at Nottingham (now in Maryland) from his wife's mother in 1714, at which date he moved to work there till about 1741, then moving to Wilmington, Delaware, not far from Philadelphia, and dying there not long after. Chandlee's son and grandsons continued to trade in Nottingham till the early nineteenth century (see Plates 522 to 524).

Meantime in 1701 another English Quaker clockmaker had moved to work in Philadelphia, alongside Cottey, though presumably in competition with him. This was Peter Stretch, whose origins are set out on page 62. He worked for something less than ten years at Leek in Staffordshire where a couple of lantern clocks and one or two longcase clocks of the highest quality eight-day type are known. He seems to be an example of that type of early clockmaker who concentrated on eight-day work, whereas Cottey was from the thirty-hour lantern clock background. In Philadelphia, however, Stretch could not afford to be choosy initially and thirty-hour clocks are known by him there as well as eight-day ones. Stretch's shop was on the corner of Front Street, which became known as Stretch's Corner. He is known to have sold clocks to the more prosperous families, which suggests he concentrated principally on the upper market and his domination of it may have been a factor determining Chandlee's move to Nottingham. Stretch died in 1746. In 1932 one writer counted twenty American-made longcase clocks known by him, but these seem to have been dispersed and only two or three appear to be known today (see Plates 518 to 521 and Colour Plate 22).

English clockmaker Anthony Ward was working in Truro, Cornwall, in 1705, where he too seems to have catered for the upper-class market – a longcase

PLATE 526. *Walnut case of the Isaac Pearson clock standing just under 8ft. tall including original caddy top and finials. Not unlike a provincial English case of the period in style. (Collection of Mrs. James Gibbs, USA. Photograph by courtesy of the NAWCC Watch & Clock Museum.)*

PLATE 525. *Dial of an eight-day longcase clock c.1720-30 by Isaac Pearson of Burlington, New Jersey. Very unlike English dial engraving, so perhaps he engraved his own dials. (Collection of Mrs. James Gibbs, USA. Photograph by courtesy of the NAWCC Watch & Clock Museum.)*

clock of month duration is known made by him there and a watch. He arrived in Philadelphia about 1717, but moved to New York in 1724, working there till he died in the early 1730s. The success of Peter Stretch may have been a factor in Ward's move, as it is doubtful that Philadelphia could have supported two clockmakers in the same sector of the market.

From about 1725 to about 1759 Joseph Wills (probably English) worked in Philadelphia, presumably in rivalry for the first twenty years or so of that period with Peter Stretch. Wills made some eight-day clocks but also made thirty-hour longcases of posted movement construction, which would not have competed for Stretch's market (see Plate 527 and Colour Plate 23, page 330).

John Wood senior, working 1736-1761, ultimately purchased Peter Stretch's 'Corner' (see Colour Plate 30, page 339) and his son, John junior, continued trading there till 1791, when he was succeeded by Ephraim Clark.

The above were the major clockmakers to work in Philadelphia during the brass dial period of clockmaking and the point in setting them out is not only for documentation, but to illustrate that there seldom seems to have been more than one clockmaker working there at any one time, except when those contemporaneous makers were seeking different markets.

PLATE 527. *Eight-day longcase dial by Joseph Wills of Philadelphia made about 1725. The bold herringbone-engraved border is very reminiscent of some provincial English work of this time. See Colour Plate 23, page 330. (Photograph by permission of Dr. & Mrs. Elliot L. Shack, USA, and by courtesy of Bernard & S. Dean Levy Inc.,New York.)*

Though Philadelphia was the main centre of clockmaking in this region, and especially the main centre of training, other clockmakers settled in the rural towns adjacent to Philadelphia. The Chandlees moved to Nottingham in what is now Maryland and to Wilmington, Delaware. Isaac Pearson, a Quaker clockmaker and blacksmith, was working in Burlington, New Jersey, from about 1710 to 1749 (see Plates 525 and 526), succeeded by his son-in-law, Joseph Hollinghead.

PLATE 528. *Dial of an eight-day longcase clock dating from about 1760 by Nathaniel Mulliken of Lexington, Massachusetts. The dial is exactly in the London style, but for the engraved eagle, and could possibly be a London dial. (Photograph by permission of Dr. & Mrs. Elliot L. Shack and by courtesy of Bernard & S. Dean Levy Inc., New York.)*

PLATE 529. *Case of the Nathaniel Mulliken clock in cherrywood standing about 7ft.4in. The style has certain features reminiscent of English provincial work, some of them (such as caddy top and side hood windows) of a slightly earlier period. (Photograph by permission of Dr. & Mrs. Elliot L. Shack and by courtesy of Bernard & S. Dean Levy Inc., New York.)*

PLATE 530. *Dial of an eight-day longcase by Gawen Brown, made at Edlingham, Northumberland, England, where he was born, about 1740-45. The style is very different from his American work.*

PLATE 531. *Dial of an eight-day longcase clock by Gawen Brown of Boston, made in the 1760s. The dial is based on London style but is not quite like London work. (Collection of Dr. & Mrs. Elliot L. Shack, USA. Photograph by courtesy of the NAWCC Watch & Clock Museum.)*

PLATE 532. *The mahogany case of the Gawen Brown of Boston clock stands 8ft.3in and has certain stylistic echoes of London work, though is overall very different. (Collection of Dr. & Mrs. Elliot L. Shack, USA. Photograph by courtesy of the NAWCC Watch & Clock Museum.)*

THE FIRST BOSTON CLOCKMAKERS

Boston was the other major centre settled by the first American clockmakers. Here too the earliest makers were mostly of English origin, many claiming to have come from London, and the longcase clocks they made were made along London eight-day principles and styling with some thirty-hour clocks being made for the lower market, principally the country areas.

Clockmaking in Boston began with James Batterson, who arrived 'from London' in 1707, but went soon after to New York. Benjamin Bagnall arrived about 1712, a Quaker believed to be from Staffordshire (home county of Peter Stretch) and probably related to clockmaker Randolph Bagnall of Talk-on-the-hill, near Newcastle under Lyne. It is uncertain whether Bagnall may have learned clockmaking in England or under Peter Stretch in Philadelphia. He is the maker of the earliest surviving Boston clocks. He died in Boston in 1773 aged eighty-four, being succeeded by his sons Benjamin junior and Samuel.

Welshman William Claggett came to Boston in 1708 aged twelve and presumably trained under Bagnall. He was working by 1715 but moved almost at once to Newport, Rhode Island, where he lived on till 1749 as a prolific clockmaker. He is known to have been an engraver too. His son, Thomas, succeeded him.

COLOUR PLATE 22. *Eleven-inch dial of the eight-day longcase clock by Peter Stretch of Philadelphia, dating from about 1715-1720 (Plate 518). The style is very much like northern English work of the same period, using bold fleur-de-lis half-hour markers and ringed winding holes. Inside countwheel strikework. (Photograph by courtesy of Graham H. Jeffries.) See page 322.*

COLOUR PLATE 23. *Walnut case of the Joseph Wills clock standing about 7ft.5in. See page 326. (Photograph by permission of Dr. & Mrs. Elliot L. Shack, USA, and by courtesy of Bernard & S. Dean Levy Inc., New York.)*

(OPPOSITE PAGE)

COLOUR PLATE 24. *Hood and dial of a fine longcase clock of about 1750 or a little earlier by James Wady of Newport, Rhode Island. The clock has a rolling moon in the arch with subsidiary dials for tidal times (high water) top left and strike/silent top right. (Photograph by courtesy of Mr. & Mrs. E.M. Wunsch, USA.)*

COLOUR PLATE 25. *The fine case of the James Wady clock is made in mahogany and stands about 8ft.6in including gilded wooden finial. It has echoes of Bristol (England) styling in the double fretted top pediment but other features, such as the carved shell, are local. (Photograph by courtesy of Mr. & Mrs. E.M. Wunsch, USA.)*

PLATE 533. *Eleven-inch dial of a thirty-hour longcase clock with alarmwork made about 1770 (or perhaps a little earlier) by Augustin Neisser of Germantown, Pennsylvania. Original blued steel hands – note tail to hour hand for alarm setting. (Photograph by courtesy of Chris H.Bailey, USA.)*

PLATE 534. *The case of the Neisser clock is in solid walnut, very sturdily made and stands about 7ft.9in. The style is not unlike that of some English country cases of the same, or slightly earlier, period. (Photograph by courtesy of Chris H. Bailey, USA.)*

PLATE 535. *Eleven-inch dial of a thirty-hour longcase clock of about 1775 by Daniel Rose of Reading, Pennsylvania. The dotted minutes are here within a double track. Blued steel hands. Cup-and-ring decoration between the spandrels looks unusual to English eyes. (Photograph by courtesy of Chris H. Bailey, USA.)*

PLATE 536. *The solid walnut case of the Daniel Rose clock stands 7ft.10in. The style resembles a tall and slim version of some English country casework. (Photograph by courtesy of Chris H. Bailey, USA.)*

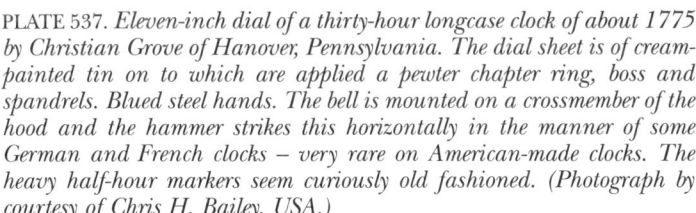

PLATE 537. *Eleven-inch dial of a thirty-hour longcase clock of about 1775 by Christian Grove of Hanover, Pennsylvania. The dial sheet is of cream-painted tin on to which are applied a pewter chapter ring, boss and spandrels. Blued steel hands. The bell is mounted on a crossmember of the hood and the hammer strikes this horizontally in the manner of some German and French clocks – very rare on American-made clocks. The heavy half-hour markers seem curiously old fashioned. (Photograph by courtesy of Chris H. Bailey, USA.)*

PLATE 538. *The solid-walnut case of the Christian Grove clock stands 8ft.1in., its style reminiscent of an English country version of a London walnut case of the 1740s. (Photograph by courtesy of Chris H. Bailey, USA.)*

Several English clockmakers, almost all claiming to be from London, set up in business in the 1720s or 1730s. Little is known about any of them beyond their names and the dates they advertised. Some stayed in Boston only a very short time, probably because Bagnall had cornered the existing market there much as Stretch had done in Philadelphia.

Gawen Brown had arrived in Boston by 1749 claiming to be 'from London'. He was born in Edlingham, Northumberland, in 1720 and spent some time in London, moving there after about 1745. His London period cannot have been lengthy, but he still boasted his 'London' background, probably to impress on his fashionable Boston clientele that he was offering the very best in style and quality. He was a prolific clockmaker over many years, working in Boston till his death in 1801 aged eighty-two. His clocks are very much in the London style (see Plates 530 to 532).

The eight-day longcase clock by Peter Stretch of Philadelphia (Colour Plate 22 and Plate 518) is very English in its styling, not surprisingly since this maker

COLOUR PLATE 26. *Superbly engraved single sheet dial of a three-train centre-seconds longcase clock made about 1780 by John Fisher of York, Pennsylvania. The clock plays seven named tunes on thirteen bells using eighteen hammers. In the arch appears to be a tidal dial. (Photograph by courtesy of Patricia Tomes, Collection of The Historical Society of York County, Pennsylvania, USA.)*

COLOUR PLATE 27. *Cherrywood bracket clock, signed SM for Samuel Meyli of Lebanon, Pa., a Swiss immigrant clockmaker. Thirty-hour fusee timepiece with verge escapement and passing strike (strikes one at each hour). Moon dial in arch. Centre calendar. c.1760. Height 16½in. (Collection of EFL and VAL. Photograph by courtesy of Richard Mones.)*

PLATE 540. *Engraved single sheet longcase dial of about 1785 by Daniel Burnap of East Windsor, Connecticut. Note the screw heads, the normal method by which the dial pillars were attached on single sheet dials. (Photograph by courtesy of Bernard & S. Dean Levy Inc., New York.)*

PLATE 539. *Silvered single sheet dial longcase clock with rolling moon made about 1780 by Peter Schwartz of York, Pennsylvania. (Photograph by courtesy of the Museum of the National Association of Watch & Clock Collectors, Inc., Columbia, Pa.)*

PLATE 541. *Engraved single sheet longcase dial with rolling moon made in the late 18th century by Lewis Curtis of Farmington, Connecticut. (Photograph by courtesy of Bernard & S. Dean Levy Inc., New York.)*

PLATE 542. *The cherrywood case of the Lewis Curtis clock standing about 7ft. 7in. Like many American cases the style has several features reminiscent of English work but from varying regions. In this example these features most approximate to cases from south-eastern England. (Photograph by courtesy of Bernard & S. Dean Levy Inc., New York.)*

COLOUR PLATE 28.
Mahogany bracket clock made about 1765 by Joseph Pearsall of New York, the case pediment shape very typical of New York work. Double fusee, eight-day. Strike/silent to the arch. American work made to resemble the current English style. Original blued steel hands. (Collection of EFL and VAL. Photograph by courtesy of Richard Mones.)

COLOUR PLATE 29. *Walnut eight-day timepiece made about 1730-35 by William Stretch of Philadelphia, with unusually early use of centre seconds and bolt-and-shutter maintaining power, the latter feature exceptionally unusual and probably used to avoid damage to the escapement. (Collection of RM. Photograph by courtesy of Richard Mones.)*

COLOUR PLATE 30. *Eight-day longcase clock dating from the 1730s by John Wood, senior, of Philadelphia. Calendar date in arch, day of the week below XII, centre seconds hand (counterbalanced). Very fine mahogany case with unusual arched side mouldings. Original brass finials. (Collection of EFL and VAL. Photograph by courtesy of Richard Mones.)*

PLATE 543. *Thirty-hour two-handed longcase clock of about 1760 by Augustin Neisser of Germantown (now part of Philadelphia). Repeating facility. Compass rose centre theme engraving reminiscent of some English provincial work. (Collection of EFL and VAL. Photograph by courtesy of Richard Mones.)*

was newly arrived from Leek in Staffordshire in northern England when he made this clock about 1715-1720. We do not know whether Stretch made and engraved his own dials, but we can deduce that he either did this or he ordered the dial from England through some trade contact back in the Leek area.

Some early American brass dials are so similar in style and quality to the English equivalent that the likelihood exists that some of these dials were actually made in England (probably London) and shipped direct. Some early American clocks, therefore, have English-made dials whilst some were American made closely following the English style, and it is exceptionally difficult to tell one from the other.

The solid walnut case of the Stretch clock is not dissimilar to an English country case of the day with integral hood pillars and very simple slender lines. The unusual feature to an English eye is the very small lenticle glass, a feature retained later in some American casework than in England.

Cottey's apprentice, Benjamin Chandlee, made the longcase clock in Plate 522. The dial is very much in the English style with a London-style chapter ring with small floating fleur-de-lis half-hour markers. I don't know whether Chandlee engraved his own chapter rings, but this is so similar to London work that it may have been imported from London. The herringbone edging divides the arch from the square in the early arched dial manner and the heavier engraving around the name boss is unlike London work, which may suggest he did this himself. The spandrels are also believed to have been cast locally.

The twelve-inch eight-day longcase clock by Benjamin Chandlee junior of Nottingham (Plates 523 and 524) dates from about 1755 and shows considerable progression in styling from the earlier clock by his father. The

PLATE 544. *Superb eight-day longcase with moon work dated 1783 by Frederick Heiseley of Hanover, Pa., with centre seconds and centre calendarwork. Heiseley trained under his father-in-law, German-born George Hoff. The dial centre has an engraving of the western hemisphere and the spandrels are of pewter. The movement has two racks to strike hours and ting-tang quarters from the same train. Fine walnut case in Chippendale styling. (Collection of EFL and VAL. Photograph by courtesy of Richard Mones.)*

spandrels are of pewter (gilded), a common feature of his work. The dial shows far more signs of local workmanship now, the matted centre having engraving around the calendar box based only very loosely on the English birds-and-baskets theme, and with cup-and-ring turning an echo of English provincial work.

Towards the end of the brass dial period, the single sheet silvered brass dial was popular in America, as it was in England. American single sheet dials mostly date between 1775, when the first painted (or 'japanned') dials appeared, and 1800. Single sheet silvered dials were principally popular in New England, particularly in Connecticut and also in New Hampshire and Maine. By about 1785 the painted dial increasingly took over, but some Connecticut makers retained the single sheet dial for another twenty years or so. Brass was always costly and sometimes in short supply and this type of dial required less material than the conventional 'composite' dial with chapter rings, etc.

American single sheet dials are often thinner than the English equivalent. Some are of high quality work and richly engraved whilst others are sometimes sparsely decorated with rather amateurish engraving. Some dial sheets are so thin that cracks have appeared at the edges and occasionally the engraver's tools would pierce right through the dial sheet so that some very thin examples were backed by a wooden or iron sheet for strength.

COLOUR PLATE 31. *Three-train eight-day quarter-chiming longcase clock dated 1774 by Peter Schutz of York, Pennsylvania. Solid silver chapter ring. (Collection of EFL and VAL. Photograph by courtesy of Richard Mones.)*

(OPPOSITE) COLOUR PLATE 32. *Hood of the Peter Schutz clock in walnut, the high scrolls of the swan-necks being typical of the region. (Collection of EFL and VAL. Photograph by courtesy of Richard Mones.)*

(OPPOSITE) COLOUR PLATE 33. *Eight-day clock with moon work and centre seconds made about 1765 by John Ferguson of Philadelphia. Fine mahogany case in the Philadelphia Chippendale style. (Collection of EFL and VAL. Photograph by courtesy of Richard Mones.)*

CHAPTER FOURTEEN

SPRING-DRIVEN CLOCKS

Most of this book deals with weight-driven clocks for the good reason that the great majority of provincial clocks were of that nature. The spring-driven clock differed from these in one very important aspect – the spring was an inconstant and therefore unreliable power source. When fully wound a spring pulls strongly and, as it runs down, it pulls ever more weakly, a factor which plays havoc with timekeeping, making the clock run fast when fully wound and slow when nearly run down. The weight, of course, was a constant power source.

The fusee (pronounced fus-ee) gear was a device used in British clocks throughout our entire period of study in an attempt at counteracting this variability in the drive power. The word is of French origin, even though the fusee was seldom used in French clocks! The principle was that the fully-wound spring pulled against the narrow end of the fusee, thereby using more power and, as it ran down, pulled against the increasingly wider end, which needed less drive power. The fusee went some way towards solving the problem, but even with a fusee a spring clock could not compete with the constant pull of a weight.

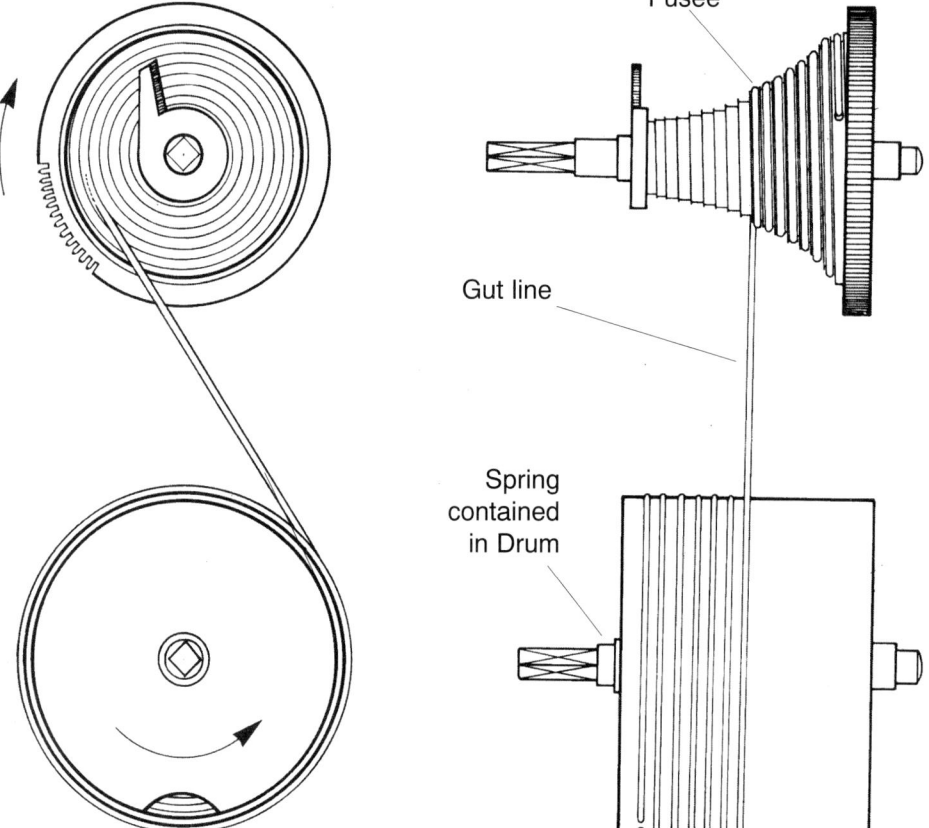

FIGURE 15. *The fusee principle. The fusee itself is the cone-shaped gear attached to the winding square. The arrow indicates the direction of winding. As the spring runs down it pulls against the wider end of the fusee, requiring less pulling power, an attempt at reducing the effect of power loss of an unwinding spring.*

OTHER INADEQUACIES OF SPRING CLOCKS

There were other problems with spring clocks which did not apply to those driven by weights. The spring itself was a difficult item to make – try hammering a piece of iron till it is six feet long, then coiling it so that it will fit into an eggcup in such a way as to be able to wind and unwind thousands of times, and you will see why. Even when such a task was performed, the problem remained that a spring would sooner or later break from metal fatigue. The clockmaker could not make springs, even though he may have learned the theory of this as an apprentice. Instead he had to buy them from a specialist springmaker, this being always a separate specialised craft.

The smaller movements of spring clocks meant that the movement components and the short pendulum itself (the clock's timekeeping regulator, short to fit within a small case) were more prone to expansion and contraction from temperature change, something which would cause serious inaccuracy of timekeeping in days when the contents of chamber pots were known to freeze over during the night!

Being smaller and more delicately made, the movements of spring clocks were more readily prone to problems arising from neglect, lack of oiling, dirt and rusting in damp conditions. The making of a spring clock involved more expense than a weight clock, not only in requiring more components (obvious extras were a fusee, fusee drive chain and a spring), but also in needing a case built in finer woods by a sophisticated cabinetmaker rather than a common joiner. Spring clocks were therefore much more costly than weight clocks.

Spring-driven table clocks employed the verge escapement with the short 'bob' pendulum, which itself was less accurate than the anchor escapement. The verge was retained in table clocks long after the availability of the more accurate anchor escapement because it was not so fussy about the level of the surface the clock sat on, thus enabling a table clock to be carried from room to room if desired, such as to the bedroom at night – which is the purpose of the carrying handle that most earlier ones have.

The anchor escapement was increasingly employed in some spring clocks from the end of the eighteenth century, but for most of our period of study spring clocks used the verge escapement. By the mid-nineteenth century the better timekeeping powers of the anchor escapement were so highly regarded that many older verge clocks were converted to anchor escapement. This is a pity, as an original verge escapement version is more prized today, being unchanged, but it is a measure of the esteem the Victorians had for the better escapement.

Two kinds of spring-driven clock were made in Britain: firstly those which sat on a table or mantelpiece or wall bracket, which are known as bracket clocks even though only a small proportion ever had a bracket; secondly those which hung directly on the wall as a wall clock. The British never took greatly to wall clocks during the period under study and spring wall clocks are seldom met with before the third quarter of the eighteenth century. Spring-driven table clocks were a more regularly produced item, but at all times of this period these clocks were made for a quite different clientele from weight-driven clocks.

THE MARKET

With so many problems and inadequacies involved in bracket clock manufacture, it may seem surprising that there was any market at all for this type,

PLATE 545. *London-made ebonised bracket clock of about 1700 by John Bushman with subdials for pull-wind alarm (top left) and strike/silent (top right), also with pull quarter repeating on six bells. Mock pendulum bob showing in the dial centre. Height 15in. including handle.*

PLATE 546(RIGHT). *Ornately engraved backplate of the Bushman clock showing alarm and pull-repeat mechanisms, the verge pendulum at rest on its carrying hook.*

which was less reliable, less accurate and more costly than a weight-driven clock. Yet the earliest London makers delighted in making them, seeing it as a challenge to overcome these mechanical difficulties and to vie with one another at getting ever more complicated movements into ever smaller cases. But the problem remained as to who would want to buy clocks which cost more than weight clocks and performed less well. The answer was that the demand came from the upper echelons of society, a wealthy clientele who probably had several clocks in the house, including weight-driven clocks. For such customers the novelty, the prestige and pleasure of ownership, and maybe the convenience of portability outweighed questions of accuracy or cost. The wealthy competed with each other to own more complicated or grander or more interesting or smaller or more accurate bracket clocks.

In London such a clientele abounded and in the early days of spring clock making (the late seventeenth and early eighteenth centuries) the best London clockmakers made exceptionally fine and often complicated spring clocks for just that market. The cost was always beyond the average householder and to some owners it was the fact that these clocks were costly which was part of their attraction. Those who can afford a Rolls Royce don't go to work by bus.

THE MAKERS

In the seventeenth century very few provincial makers produced spring-driven clocks. Those who did were virtually exclusively of that London-trained group

PLATE 547. *The dial of the Ferrer clock (Colour Plate 34, page 353) has a very shallow arch, typical of some early arch examples. The signature is latinised, 'formavit' meaning 'created (this)'.*

PLATE 548. *Backplate of the Ferrer clock showing external countwheel for strike control. The fine engraving is typical of many bracket clock backplates of the period.*

PLATE 549. *Arched dial bracket clock in walnut made about 1725 by Barnaby Dammant of Colchester. Strike/silent switch in the arch. The dial design is very similar to that of a longcase clock of the same age.*

PLATE 550. *Movement of the Dammant bracket clock showing the two fusees, verge escapement and bob pendulum. Solid dial sheet, London fashion.*

who first took the upper market (eight-day) trade into provincial cities and towns, and even then these clocks form a minute proportion of their work. Bracket clocks of this period are known from such places as York, Bristol and Leeds, where their makers might conceivably draw trade from the local aristocracy. Bracket clocks by rural makers are unknown at this time – I cannot think of a single example. Some rural clockmakers had the knowledge, skills and often the equipment to make bracket clocks, but they rarely if ever made them simply because there was no rural market for them. Those clients with large country houses would have bought such clocks, if they bought them at all, from makers in the nearest city or from London itself, which they always saw as the prestige shopping centre.

PLATE 551. *Bracket clock of about 1725 in lacquer case and having calendarwork in the arch. Lacquer was an alternative to the more usual ebonised or walnut case.*

In the first half of the eighteenth century a few more provincial clockmakers produced bracket clocks, but a bracket clock by any provincial maker is unusual enough to be noteworthy before about 1770. London makers continued production during this entire period (i.e. pre-1770) and London-made examples will account for the vast majority of bracket clocks met with. By this time a few London makers had begun to specialise in the making of spring clocks and concentrated exclusively on that type of clock, supplying the trade in London and in the provinces. The name on the dial was that of the retailer who passed as the maker, but internal serial numbers or marks will often show that a London-made bracket clock was supplied by, rather than made by, the name on the dial. Oddly enough, the few early provincial examples met with often prove to have been made by the maker himself rather than supplied through him via the London trade.

After about 1770 the London specialists took an even firmer hold and the majority of bracket clocks bearing London or provincial names on the dial will prove to have been made by one of the London specialist makers of spring clocks. One London specialist established in the 1740s was Aynsworth Thwaites, later to become John Thwaites, then Thwaites and Reed. Some of their account books survive today and the serial numbers of some of the clocks they made from 1761 are published in Ronald Rose's book *English Dial Clocks*. From these serial numbers the year of manufacture can usually be established. Rose also publishes a long list of the names of London clock 'makers', some of them very prestigious names, who are known to have bought their actual clocks from this concern and therefore acted only as retailers of spring clocks. The reason provincial makers less often bought their spring clocks from London specialists was probably purely one of price; they had to make the clock personally as the profit lay in the making of it rather than the mark-up.

PLATE 552. *Mahogany bracket clock in what is known as the bell-top style of case dating from the 1760s, maker Benjamin Ward of London. Strike/silent switch in the arch. The dial is very reminiscent of a contemporary longcase dial.*

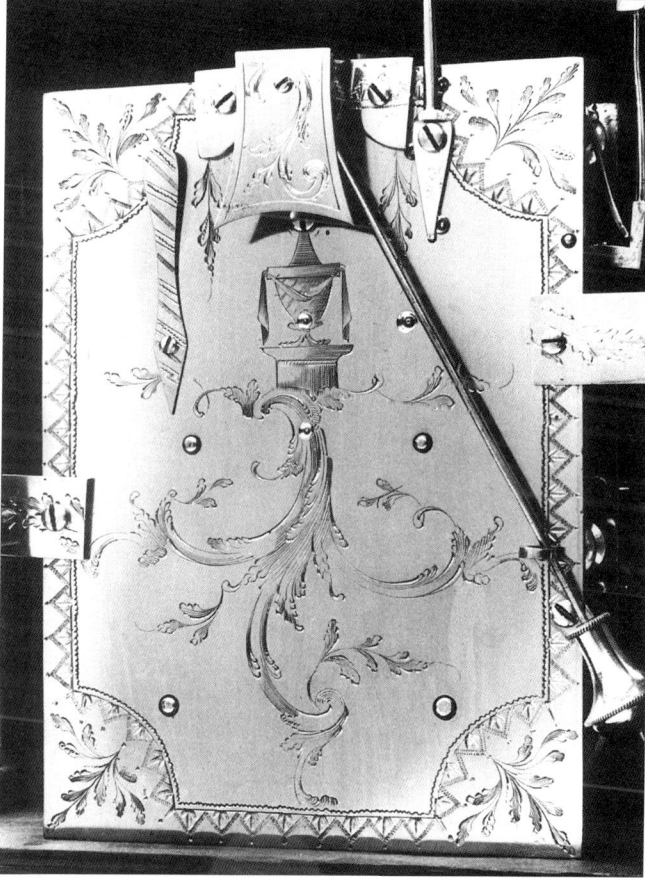

PLATE 553. *Engraved backplate of the Ward clock showing verge pendulum on its carrying hook. The engraving is now of a much different style from earlier periods, often, as here, with a defined border.*

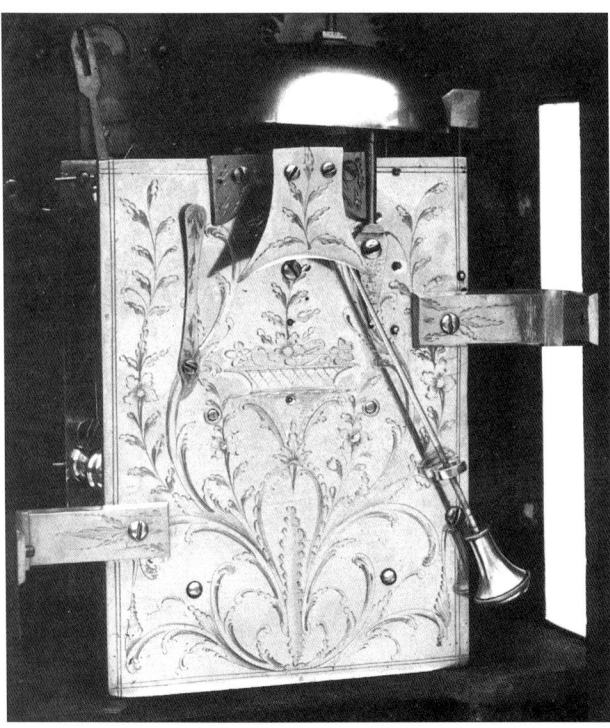

PLATE 554. *Ebonised bell-top bracket clock made about 1800 by James Simpson of Lincoln, the dial of the single sheet type but here engraved with arabic hour numerals, typical of the period. Arabic hours were less popular on bracket clocks than on longcases. Original steel hands of matching pattern as was now the fashion. Height 20in. plus handle.*

PLATE 555. *Mahogany bracket clock of about 1800 by Richard Ward of Winchester in Hampshire, having subdials for fast/slow regulation and strike/silent. Here the dial is part-way towards the single sheet style, yet still having corner spandrels. Single-track minutes appear at the turn of the century instead of the earlier dotted minutes.*

PLATE 556. *Engraved backplate of the Richard Ward clock showing the verge pendulum on its carrying hook. The engraving is typical of many of this time, but before long backplates were left undecorated.*

COLOUR PLATE 34.
Tortoiseshell-cased bracket clock dated 28th June 1707 by William Ferrer (usually Farrer) of Pontefract, Yorkshire, inscribed with the owner's name, Buckley Wilsford. About 15in. See page 348.

WALL CLOCKS

Bracket clocks were sometimes made with a matching wall bracket on which they sat, though these were always in the minority. The alternative form of spring-driven wall clock was that type which hung directly on the wall, such as we were familiar with until recently in the form of a railway clock or a school wall clock. Wall clocks were never really popular for household use in Britain during the period under study, though one type of simple wall clock was widely used in offices and business establishments.

Spring-driven wall clocks exist in two forms. Firstly there was the type known as a cartel clock, which was a simple spring-driven single fusee timepiece (i.e. non-striking) with single sheet brass dial, housed in a very ornate wooden case

PLATE 557. *Three-train musical bracket clock of about 1800 with the dial removed to show the complexities. The clock plays a choice of four tunes. The third hand is for centre calendarwork. Maker Taylor of London.*

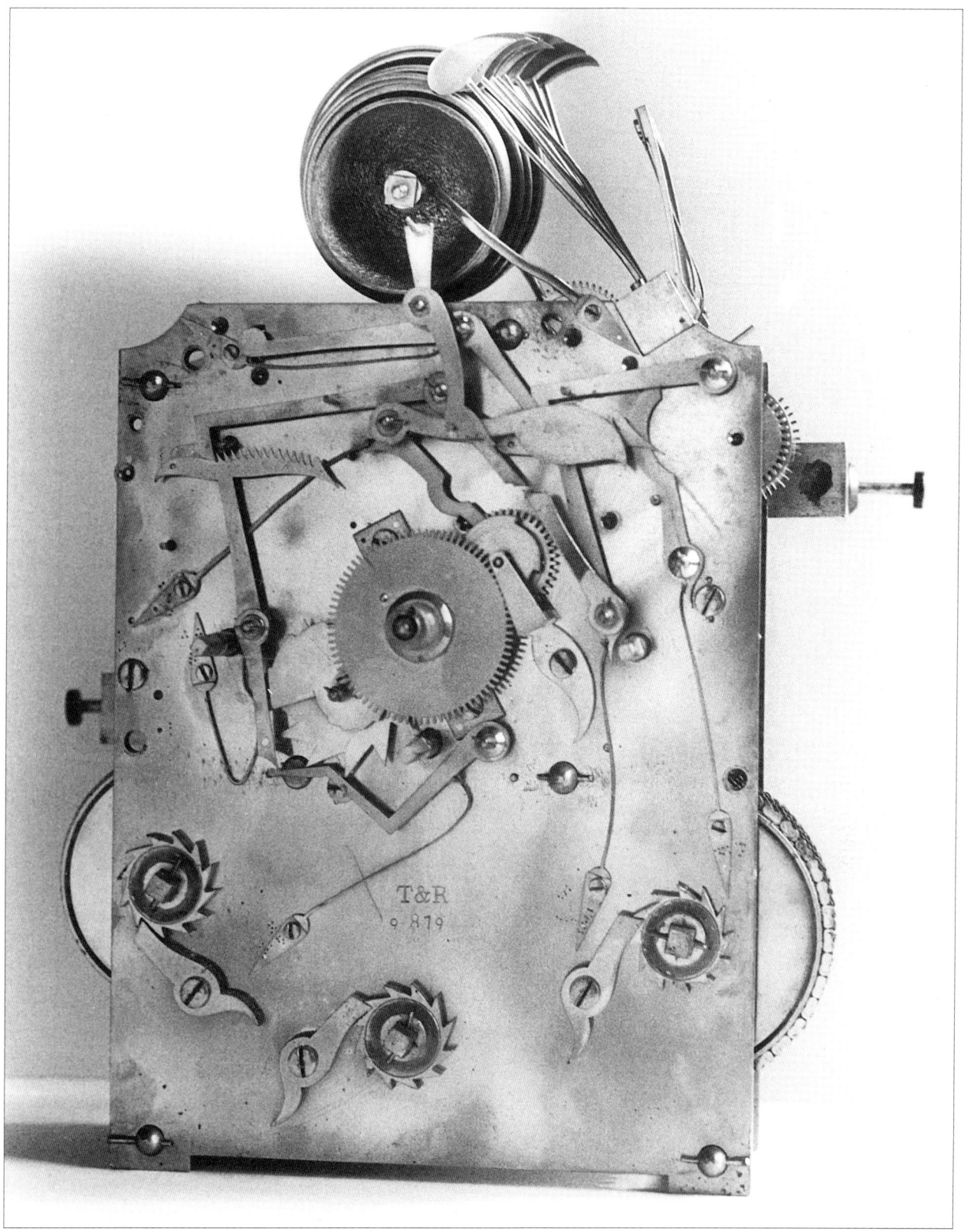

PLATE 558. *Two-train quarter-chiming bracket clock with dial removed signed on the dial by Grant of London. The initials T&R show that this was actually made by specialist London spring-clock makers Thwaites and Reed. The number 9819 dates this to 1837.*

PLATE 559. *Giltwood cartel clock made by Storr and Gibbs of London and dated 1741. Single fusee verge escapement timepiece showing a mock pendulum bob. Height about 2ft.6in.*

COLOUR PLATE 35. *Walnut-cased bracket clock of about 1780 by Thomas Moss of Frodsham in Cheshire. The dial is of the single sheet type. Here the case is clearly provincial, being very different from the London bell-top case of the period. Height about 15in.*

carved into fanciful shapes encompassing birds and branches, and then gilded. This sort of clock began around the middle of the eighteenth century and was clearly designed for grand houses furnished with gilded furniture in the French taste. The actual term cartel clock derives from the French, though in France cartel clocks had gilded metal cases. Cartel clocks had a verge escapement with bob pendulum and were made almost exclusively by London clockmakers.

The other form of spring-driven wall clock is known as an English dial clock or a wall dial clock. This type also began about the middle of the eighteenth century and had the same type of single sheet brass dial with verge escapement, single fusee movement, but was housed in a plain circular

PLATE 560. *Single fusee wall dial clock with single sheet silvered brass dial made about 1770 by Hedge of Colchester. Diameter about 15in.*

COLOUR PLATE 36. *Carved ebonised bracket clock with matching wall bracket made in the 1830s by Adams of London. This is a late example of a brass dial clock, as this type of single sheet dial was used sporadically by London makers until the mid-19th century. Eight-inch dial. Height 2ft.3in.*

wooden rim (usually mahogany) for absolute simplicity. These clocks appear to have been designed for business use – offices, banks, public buildings, or maybe even for the servants' quarters of grander houses. By the early nineteenth century they turned to anchor escapement, though by this date they had japanned (painted) dials. Some earlier examples built with verge escapement were later converted to anchor escapement for greater accuracy.

Cartel clocks virtually always bear on their dials the names of London clockmakers. In this period too the great majority of dial clocks also carry London names on their dials. In any event almost all such clocks of both types were made by the London spring-clocks specialists, supplied by them to others in the trade.

CHAPTER FIFTEEN

Is it Genuine?

WEAR AND TEAR

Over the centuries clocks, like any other machines, will have needed repair; broken parts will have been mended or replaced, dirty parts will have been cleaned. There is no shame in a repair, even though repairers of the past may sometimes have been less skilled and less fastidious than some of today's specialist restorers. After all, the repairers of the past were seldom concerned with niceties of style but more with the attempt to coax life into an ancient clock that was broken or worn out and virtually worthless, and they often did this as cheaply as possible. At least the fact that they did not usually bother to try to 'match' a repair to the style of the original does mean that we can often recognise such repairs quite easily.

Normal wear and tear will have enforced the periodic cleaning of the movement, and probably the dial too, along with re-silvering of the engraved areas, replacing of ropes or gutlines and perhaps of the more brittle parts such as suspension springs and drive springs in bracket clocks. With clocks of greater age we may expect some replacement of more substantial worn-out parts, the likeliest being steel pinions. With verge escapement clocks (or those made with balance wheel escapement) we may also expect that many were later converted to anchor escapement for greater accuracy and such conversion is now usually seen as a measure of the clock's progression through time. Clearly a clock with its original pinions or original escapement is to be preferred, but honest repairs or modifications done with the best of intentions do not constitute a fake, though they may reduce the value of such a clock over one not so affected.

SERIOUS ALTERATIONS AND MARRIAGES

A serious alteration may constitute what we now regard as a fake, even though it may not have been done to deceive at the time. Such alterations mainly consist of two basic types, the results of which we now describe as 'marriages'. A marriage can be one whereby an unrelated dial and movement have been fitted together later or one where a movement together with its dial (genuine or married) has been fitted later into an unrelated case. Coming within the same category but less often seen are examples where a new dial or a new case has been fitted long ago to an old movement – the 'new' part may itself now be old, but much less old than the rest of the clock, usually having been made within the last fifty years.

Such marriages are almost always very easy to spot, largely because they were usually done for expediency and not with intent to deceive, hence no attempt was generally made to conceal the fact that they are marriages. Some marriages have been performed in recent years (the last thirty years?) and some are still being done today, with deliberate intent to deceive by camouflaging the alterations, and these can be very hard if not impossible to detect unless the clock is completely dismantled for a very thorough examination – something which is not possible for the average buyer and probably not for even the most expert buyer when viewing at an auction.

PLATE 561. *Thirty-hour movement with dial removed showing two different sets each of four dial foot holes, evidence that this movement has had at least two different (four-footed) dials in the past. Even with a dial in place four spare holes would be obvious at a glance.*

CHANGED MOVEMENTS

Many thirty-hour longcase clocks were 'converted' to eight-day duration by the fitting of a totally unrelated eight-day movement. This is particularly obvious if the dial is calibrated for a single hand as the present two hands attempt to display minutes which the dial does not show. Occasionally a single-handed thirty-hour clock may have had its movement changed for a replacement two-handed thirty-hour movement with the same incongruous result. The converting of thirty-hour clocks to eight-day duration was carried out until quite recent times as an expedient to avoid what was seen as the nuisance of daily winding. To a purist such a clock has been ruined.

Eight-day clocks too are sometimes found with a non-original movement. If the movement was in a very worn or damaged state, it might have been seen as preferable in the past to fit a different but sound movement rather than repair the original one. Usually any old and 'suitable' movement was used, which might have fitted conveniently.

Evidence nearly always exists of this kind of movement conversion. Tell-tale signs include:

1. One or more spare (now unused) holes in the movement frontplate where the feet of the former dial fitted (Plate 561).

2. Such holes now plugged with brass to disguise them.

3. Spare holes visible in the back of the dial, where the dial feet may have been moved to fit the available holes in the frontplate of the present movement (Plate 562).

4. Such holes now plugged with brass to disguise them.

5. The smudging of the dial back (or the movement frontplate) with artificial dirt, usually a black paint-like substance, to disguise the filling of the old dial feet holes.

6. Poor alignment of the movement features to the dial front, i.e. winding-squares not central to their holes; winding holes elongated to accept non-central winding squares; winding squares protruding proud of the dial surface; hands pipe protruding unusually far beyond the dial sheet; seconds hand not central to its hole; calendar not working or sealed up because this movement cannot drive it; a calendar built for twenty-four hour drive now working twelve-hourly (or vice versa).

7. Movement features out of period with, or inconsistent with the age of, the dial. For example: movement pillar style; number of pillars; method of striking; wheel collet type; lack of wear in pinions.

8. Incorrect hands *can* be an indicator. Often a replacement movement was used *with* its original hands, since they fitted the movement exactly without requiring modification. These hands would almost certainly be out of period with, and/or the wrong size for, the present dial.

9. The presence of a falseplate (which could only have come from a white dial clock) behind a brass dial is a certain sign of a marriage of some sort, as original brass dial clocks did not have falseplates. Sometimes a white-dial movement complete with its falseplate was fitted as a replacement movement to a brass dial.

Sometimes a movement was constructed later specifically as a replacement for a badly abused or damaged one. In this instance there would be a perfect fit of movement to dial with no alterations. Here the only signs of the marriage would be those stylistic and constructional features of the movement which were evidently much later (or occasionally much earlier) than the period of the dial – as outlined in point 7 above. This is very difficult for the novice to detect, but fortunately is an uncommon occurrence.

Occasionally a brass dial was made to replace a neglected or disliked white dial. This seems to have happened in the 1900-1940 period, where some owners had their perhaps shabby white (japanned) clock dial changed for what they thought was a 'nicer' brass one. In these instances too the fit of dial to movement will be perfect with no alterations. Often the movement will have its original falseplate, which is an instant giveaway on any brass dial clock. Fortunately such dials almost always betray a total inconsistency in their stylistic features, having engraving, spandrels, numbering layout, half- or

PLATE 562. *Thirty-hour dial sheet of about 1730 signed Will Kay, seen here with the chapter ring removed to reveal the 'cartwheel' casting method, typical of most clocks made outside the London sphere of influence. Tiny holes show where the chapter ring was held at its inner edge by small rivets at X and II. The original three dial feet were riveted in the larger holes at XII, III and VII, but were later moved to different positions nearby to make the dial fit a different movement. Even with a movement in place this would be apparent as one set of empty dial foot holes would be visible from behind.*

quarter-hour markers all of strongly-conflicting periods such as could never have appeared together on the one dial (Plate 563). The dial sheet will be a solid one of rolled brass sheet instead of cast brass, thinner than a cast one and showing no casting marks, sometimes so thin as to be flexible with the fingers. Any engraving will be flatter or more shallow or of more uniform depth than true period engraving and will fail to copy the nature and feel of the true style it imitates.

The marriage of a wrong movement to a dial still in its original case will probably also reveal itself by the fact that certain features no longer fit where they once did. For example the weight(s) may hang in a different position from where the original weight(s) rubbed the case. The pendulum may rub the backboard in a different spot from that indicated by the original marks. The pendulum knock marks (or cuts to prevent knocking) on the case sides may fail to line up with the position of the present pendulum. In other words, some of the indicators listed below under the married casework section may

PLATE 563. *Old 'reproduction' brass dial made around 1900 (probably to replace a worn painted dial) for a longcase clock made by Matthew Parker of Dunfermline, who worked between 1786 and about 1800. Easily recognisable by many features of incongruous date: style of arch spandrels c.1780; corner spandrels c.1700; minutes 1710; lack of half- and quarter-hour divisions post 1760; half-hour markers 1740; centre engraving style 1770; seconds and calendar dials simply of incorrect layout and style!*

well also apply to reveal a married movement.

The above pointers apply principally to longcase clocks, which were the commonest type of domestic clock and therefore most readily offered parts for such marriages. Bracket clocks were far less often altered and a marriage of dial to movement in bracket clocks is very unusual. A marriage of a bracket clock dial/movement to case is also unusual but was a more feasible option. Many of the same features apply for the detection of married bracket clocks as for longcase clocks, but this is often a much more difficult area for the beginner because their movements are less easily accessible.

Marriages of hooded clocks and of hook-and-spike clocks are detectable by the same features as described for longcases, but marriages in these categories of clock are uncommon. Marriages in lantern clocks are a much more difficult area and outside our scope here.

The attitude taken towards a marriage will vary with different owners and today we are perhaps inclined to take a more lenient view than even a very few years ago. The sheer scarcity of early clocks, especially the very earliest clocks of the first thirty or forty years of provincial clockmaking, can mean that they are so rare that if we insist on total originality we may never find one. With the very earliest provincial makers, men of an exceptional pioneering nature such as John Ogden of Askrigg (page 383), Isaac Hadwen of Sedbergh (page 418), Jonas Barber of Bowland Bridge (page 436), hardly any of their early clocks are in their original cases. A collector today having the chance to buy such an early clock may well take the view that he will accept one in a non-original case, especially if the case appears to be compatible in the sense of being of the

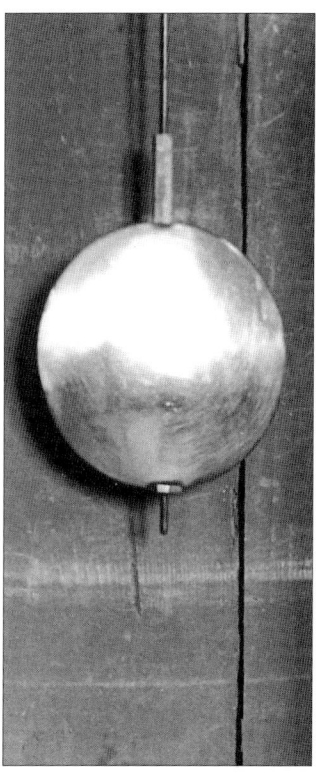

PLATE 564. *Single weight of a thirty-hour longcase clock showing its position during descent and how its rubbing can cause a wear mark in the lower door frame member. Wear marks which fail to match the weight positions can be a sign of a married case.*

PLATE 565. *Longcase pendulum showing how it can often leave two rub marks on the backboard, one caused by the back swell of the bob, the other by the rating nut.*

PLATE 566. *Longcase pendulum which does not hang where the rub marks have previously worn, suggesting that this case is not original to the clock it now contains.*

correct period and style. This is a valid enough approach, but such a buyer still needs to know how to establish whether or not the case *is* original.

Occasionally an old clock was fitted with a replacement movement, as mentioned earlier, and so may be a marriage of movement to dial even though the dial and case are original to each other. Again some collectors would take the view that such a clock was acceptable, despite its shortcomings, but again the buyer needs to be able to tell the genuine from the altered.

CHANGED CASEWORK

Longcase clocks are often found to be in non-original cases. There are several reasons for this. Sometimes a past owner might have wanted to update the fashion or quality of his clock by having a new case made for it, re-casing it at the date in question. In these examples the case would have been purpose-made for the clock and would never have contained any other clock, so that there are no signs of any alteration and recognition can only be through a discrepancy of style/age between clock and dial-with-movement. A very obvious example would be a seventeenth century clock housed in a mahogany case since mahogany was unknown in clocks in seventeenth century England.

Sometimes a clock was re-cased because of damage or worm in the original, sometimes because the owner or dealer thought it deserved a 'better' case. Often this was (and still is) done by dealers to smarten up the clock to help it sell better. A movement-with-dial *not* original to its case can almost always be detected by examination of those features which have had to be altered to make it fit into a case not designed for it, even if the replacement case has the

PLATE 567. *The 'cheeks' of a longcase clock showing its original seatboard fixed in position with its original nails. The scribe line for the original sawing of the cheeks to length can be seen clearly.*

same (i.e. correct) size of dial. Pointers towards recognising a non-original case include:

1. Replaced seatboard (the wooden board on which the movement sits). This is one method used to conceal the fitting of a non-original clock to a case as the thickness of the board can be made such as to cause the movement to sit at the right height. A new seatboard also removed the tell-tale evidence in 2.

2. The nail/screw holes in the ends of the seatboard do not align with similar holes in the cheeks, which are the case side extensions inside the hood across which the seatboard rests.

3. Replaced cheeks, which disguise the altered height at which the seatboard now sits. Touches of the original sawcuts on the backboard may help indicate replaced cheeks. (Plate 567.)

4. Packing pieces, usually slivers of wood, inserted between the top of the cheeks and the seatboard to raise its height.

5. Modern sawcuts in the tops of the cheeks, which may have been cut to reduce the height of the seatboard.

6. Alteration of the mask (the wooden frame which surrounds the dial inside the hood door). The mask could be cut back to enlarge the opening or built up to reduce the opening, or could be replaced completely or even removed totally, all to avoid showing signs of alteration. Such alterations are usually disguised by stain – original masks did not normally need staining when new.

7. A case with a dial aperture (the hood glass size) of the wrong size, altered to make the aperture appear to be of the right size, usually evident by the cutting back of the corner joints. Occasionally these were left without alteration in which case the incorrect size speaks for itself.

8. The pendulum rubs (or could potentially rub) the backboard at a point different from earlier rub marks (Plate 566).

9. The weight or weights rub against the casework at a different point from that where former rubbing has occurred. This is usually particularly apparent if an eight-day clock has been housed in a thirty-hour case since two weights now rub in different places from that where one weight rubbed formerly – and the converse for a thirty-hour clock now in an eight-day case.

COLOUR PLATE 37. *Sophisticated basket top timepiece bracket clock c.1710 signed 'Gabl Smith Barthomley', five pillar single fusee movement with verge escapement pull quarter repeating on two bells. (Photograph by courtesy of Christie's.)*

CHAPTER SIXTEEN

CASE STUDIES OF EARLY PROVINCIAL MAKERS
Illustrative of the beginning of provincial clockmaking

GABRIEL SMITH

Gabriel Smith was born about 1654, the son of a weaver at Audley in Staffordshire, close to the Cheshire border. He lived to the considerable age of eighty-nine, if we are to believe his gravestone, being buried in 1743 at the village of Barthomley in Cheshire, eight miles east of Nantwich. He was a carpenter, bellfounder, engineer and clockmaker, a man of great talent and industry, probably the most important and prolific early clockmaker in the region and the founding member of a three generation family of clockmakers. That he must have learned this craft is obvious, as the clocks he made are almost all of a high degree of sophistication, very different from the primitive work of a rural clocksmith.

His work is based on that of the London style eight-day school, most of his clocks being eight-day longcase examples of high quality including a three-train quarter-chiming clock on eight bells made for William and Margaret Kellsall and dated 1713 and an eight-bell *four*-train longcase dated 1711 chiming quarters and playing a tune every every two hours. However, he is also known to have made lantern clocks and thirty-hour longcase clocks, though these seem to have formed a very small proportion of his output. Three bracket clocks are recorded by him, one of which is illustrated in Colour Plate 37, and he is known to have cast bells and made turret clocks for several local churches. Clearly Gabriel Smith aimed his sights on the upper end of the clock market, as we might expect from someone whose work shows a knowledge of London-school style and methods, yet at the same time he did not turn away cheaper thirty-hour work when that was available to him.

His eight-day clock pillars were finned in the London manner and were usually five in number but even six on some of his earliest examples. However, his pillars were *screwed* into the backplates, a very unusual practice for any clockmaker anywhere, whereas London practice was to rivet them. This very

PLATE 568. *Twelve-inch dial of an eight-day longcase clock signed 'Gabrl Smith Barthomley'. Limited engraving on to a matted dial centre and ringed winding holes are typical of his early clocks. Date about 1710. Note the thin and rather weak engraving of the name compared with the bold, confident work elsewhere on the dial, which suggests two different hands.*

PLATE 569. *Twelve-inch dial of another eight-day Barthomley clock, again with weak signature. Very similar in style to but slightly later than the previous example at about 1710-15.*

PLATE 570. *Six pillar movement of the clock in Plate 568. Note the screwed pillar ends protruding from the backplate. Solid dial.*

PLATE 571. *Twelve-inch dial of another slightly later Barthomley eight-day clock of about 1715. The signature here is more confident. Herringbone engraving around the winding holes and dial centre rim are unusual but seem to be a popular feature with this maker. Original steel hands.*

PLATE 573. *Movement of the clock in Plate 572. The maker by now has turned to the five pillar system, still having screw-thread ends (one pillar later removed in this example). Inside countwheel strikework with trip lever for re-setting sequence. Solid dial sheet.*

PLATE 572. *Early arched dial (twelve-inch, solid sheet) signed at Nantwich and dating about 1725. The herringbone border and penny moon in the arch are pleasant 'extras', with Smith's favourite dial centre herringboning too. The sun face in the dial centre seems to have been a motif Smith enjoyed. Unusual 'Gothic' signature. Original steel hands.*

unusual and distinctive characteristic also appears on the work of his son and of his former apprentices. His earlier clocks have inside countwheel strikework. I have examined a number of his eight-day clocks, though I've not personally seen an example of his thirty-hour longcase work. His earliest longcase clocks (signed at Barthomley) seem to have solid dials but he turned

PLATE 574. *Dial of an eight-day clock signed 'Gabl. Smith Namptwich', using familiar themes such as herringboning and the sun face. About 1725-30. He has now dropped his half-quarter markers. Blued steel hands believed original.*

PLATE 575. *Five pillar movement of the clock in Plate 574. Smith is now using the cartwheel type of dial casting.*

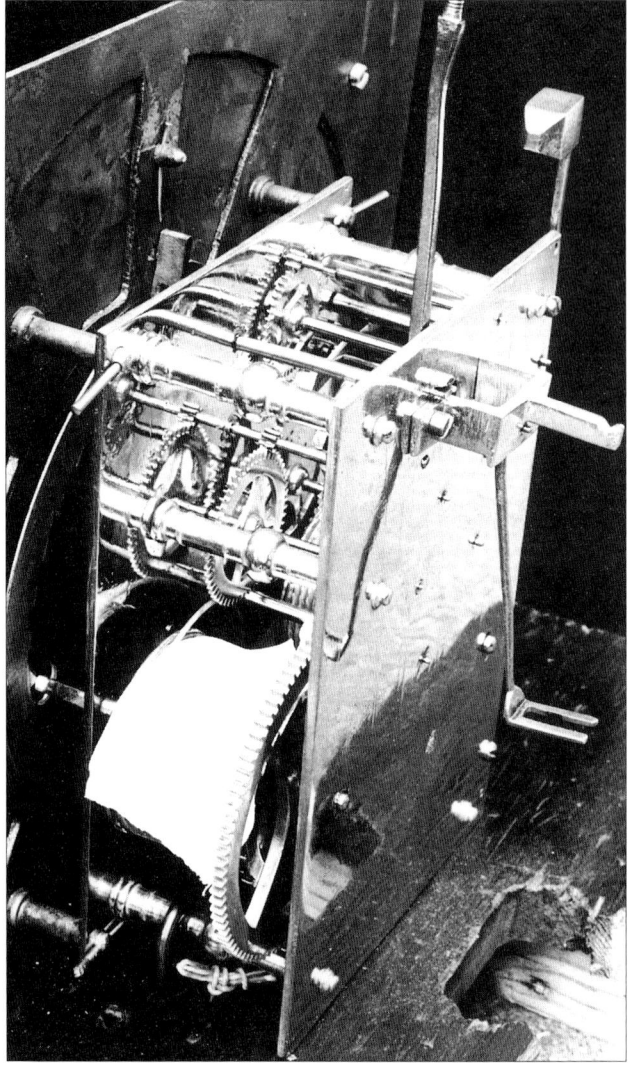

to the cartwheel form on some of his Nantwich clocks, which is what we would expect as a natural progression of method. His dial engraving is mostly bold but his earliest clocks have the name rather weakly lettered. This might imply that the engraving was done out of house and that he engraved his own name on the chapter rings. If so, then he changed this practice later as his Nantwich dials have his name lettered confidently by a professional.

Horological researcher Alan Treherne has dug painstakingly into Gabriel Smith's life and it is through his research that we know details of Smith's biographical dates and movements. He seems to have lived at (or very close to) Audley till the late 1680s and to have moved to Barthomley about 1690, where he worked till about 1722 when he moved to the town of Nantwich. It may be that he saw an opening at Nantwich following the death there in 1717 of clockmaker Thomas Talbot, by whom very few clocks seem to be known, though they too are of the high quality eight-day type. Smith's earliest clocks are signed at Barthomley, later ones at Nantwich, sometimes then spelt Namptwich. No clock seems to be known signed at Audley; it is not known whether this is because his earliest examples have not survived or because he signed all his pre-Nantwich clocks as at Barthomley.

In summary we can see from his work that this is not simply a case of a

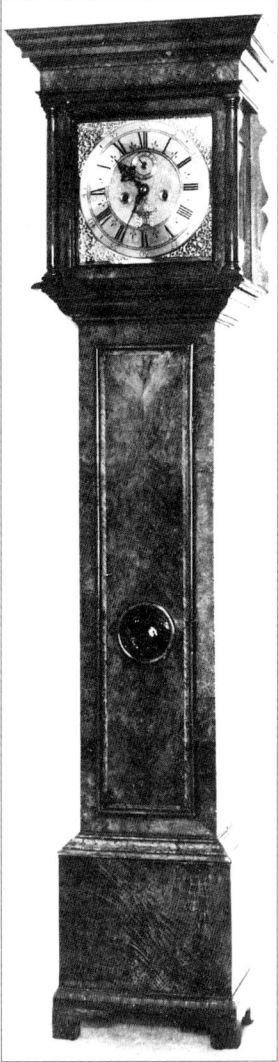

PLATE 576. *Oak case of the clock in Plate 568, standing 6ft.4in. Hood side windows and lenticle glass. Double-D mould to the trunk door.*

PLATE 577. *Walnut veneered case of the clock in Plate 569, standing about 6ft.8in. Many early features typical of the day.*

PLATE 578. *Oak case of the clock in Plate 571, standing 6ft.7in. This example has a caddy to the hood and unusually shallow moulding above and below the trunk.*

PLATE 579. *Walnut-veneered case of the arched dial clock in Plate 572, still using the glass hood side windows and lenticle glass.*

PLATE 580. *Oak case of the clock in Plate 574, standing about 6ft 8in.*

London-trained clockmaker moving into the provinces to introduce London work there – such as we shall see with John Williamson (page 371). Smith was born in the locality, understood a variety of trades in the engineering and smithing lines, thought for himself and adopted his own distinctively individualistic styling by taking what he wanted from both the London school and rural clocksmithing schools. Where he learned all these skills remains a mystery.

COLOUR PLATE 40. *Detail of the hood of the marquetry clock in Colour Plate 39, showing fine detail and quality of workmanship. Slender pillars integral with the hood door, glass side windows, convex moulding, and small caddy top are all features of the period.*

COLOUR PLATE 38. *Eleven-inch solid dial eight-day longcase clock signed 'Jno. Williamson at Leeds', very similar styling to Plate 582 but with slightly larger minute numbers and larger seconds ring. Both movements have five finned pillars and inside countwheel strikework. Original hour hand, later minute hand. Late 1690s.*

COLOUR PLATE 39. *Walnut case with arabesque marquetry in panels housing the eight-day clock in Colour Plate 38. A London-made case standing 6ft.11in.*

JOHN WILLIAMSON OF LEEDS

John Williamson is an example of one of the very earliest provincial clock-makers of that group who had London training and who took with them their London-learned skills into the provinces. Williamson was the first domestic clockmaker to work in Leeds by whom any clocks are known to survive. My researches into his origins proved inconclusive but certain very complicated circumstantial evidence suggests he may have been born in 1651 at Guisley, a few miles north of Leeds, and that he was almost certainly related to the Barber family of blacksmiths and clockmakers who worked at nearby Otley.

John Williamson joined the Worshipful Company of Clockmakers in London in December 1682, not as an apprentice but as a clockmaker already qualified in his trade. He was then not less than twenty-one years old. The Company controlled the clock and watch trade in London and membership of the Company was obligatory for any clockmaker hoping to work in the capital without prosecution. It is not specified where he learned his craft. However, the fact that he joined on the very same occasion as thirty-year-old Jonas Barber from Menston, near Otley (later to work at Ratcliffe Cross in London), and the fact that he later took his skills to Leeds, tend to add to the likelihood that he originated from Guisley. He must have known or perhaps worked in association with (the first) Jonas Barber, who himself worked in London for at least ten years before (being forced into?) joining the Company (see page 436).

The records show that Williamson paid his subscriptions to the Company for only a month. At some time early in 1683 he returned to Leeds where he married in December of that year and where he seems to have worked for the

PLATE 581. *Exquisitely-engraved nine-inch dial of a possibly unique four-wheel train longcase clock incorporating seconds and quarter-chiming on four bells with strikework on a fifth bell. Minutes numbered within the track. This clock is known to date from 1682. The plate-framed movement has four finned pillars. The hands are modern replacements.*

rest of his life. Barber remained in London till his death there in July 1698.

Exactly how long Williamson worked in London is not known, but his style and techniques are those of London work, so if he did not learn his craft there he certainly learned London methods. Only one clock has so far been recorded by him made in his London period, the thirty-hour longcase clock illustrated in Plate 581. This is far from typical of London work in general but is one of a small group of pull-wind short duration longcase clocks known by a handful of London makers of the day including Thomas Tompion, Joseph Knibb and Andrew Prime (brother-in-law of Ahasuerus Fromanteel who introduced the pendulum clock to England in 1658).

This two-handed nine-inch dial plate-movement clock has a four-wheel going train to show seconds and runs for about three days on a tiny 4lb. weight. It has a beautifully engraved tulip-centre dial with a floral seconds feature and is signed within a drapery swag 'John Williamson Near Temple Barr Londini Fecit'. At this early date (1682) the minute numbers are engraved *inside* the minute track, as was the fashion. The blued steel hands are modern replacements.

The strike train incorporates a quarter-chime feature on four bells at each of the three quarters as well as striking the hour on a single separate bell. This is controlled by a single (outside) countwheel notched for hours and quarters. The same wheel triggers the quarter hours by wedges on its circumference, then shunts forward to trigger the hour bell by pins in its side. This is an ingenious system, perhaps not invented by Williamson, but certainly unusual enough to demonstrate that he was a talented maker of considerable initiative. Such a clock would have been made as a special order rather than a mainstream conventional eight-day. Those other known early examples of this type of short-duration clock made in London were few in number, were very much a minority of the output of their makers, did not chime the quarters and none of them was better made than this one, which is the only short-duration longcase clock yet known by this maker.

Over the years I have come across about a dozen longcase clocks made by

PLATE 582. *Eleven-inch solid dial eight-day longcase clock signed 'John Williamson in Leeds Fecit', made about 1695. Fine original blued steel hands. Tiny minute numerals outside the track. Every bit a London dial in style and quality.*

PLATE 583. *Year duration longcase dial signed 'John Williamson in Leeds fecit', c.1690. The clock has a strike train and a ting-tang quarter pull-repeat mechanism. Original hour hand and possibly minute hand. The chapter ring is of the skeleton type.*

PLATE 584. *Month duration longcase dial signed 'John Williamson in Leeds fecit', c.1695. Hour hand original, minute hand later.*

PLATE 585. *Eight-day longcase dial signed 'Wm. Tipling in Leeds fecit', late 1690s or just after. The style is very much like Williamson's but the signature is now on the chapter ring itself. This maker is believed to have been trained by Williamson.*

him at Leeds. One bracket clock is known by him (Colour Plates 41 to 43), signed without any place name, as is a single lantern clock, the latter much abused. The absence of a place name might here suggest these were products made during his London working period but at a time when he was *not* a mem-

COLOUR PLATE 41. *Double fusee ebonised bracket clock signed on the backplate 'John Williamson Fecit' without any place name, dating from the 1680s or early 1690s. The case is very much of the London pattern, but the gilding of the fret front and sides is unusual. This must be one of the oldest known provincial bracket clocks.*

ber of the Clockmakers' Company (membership of which was compulsory by statute). The omission of the London place name may have been to help avoid detection by Company officers. Two watches are also known bearing his name.

From this we can deduce that his main output was in longcase (and occasional bracket) clocks for the upper market, that is those with a duration of eight days or longer. Most were eight-day clocks but two are of month duration and one quite exceptionally runs for twelve months. Year-duration clocks of any kind are very rare, but this particular one is exceptional in having

COLOUR PLATE 42. *The dial of the Williamson bracket clock is very much like a miniature longcase dial. Fine engraving. Superb original blued steel hands. N-S switch above XII offers a strike or non-strike option.*

COLOUR PLATE 43. *Beautiful engraved backplate of the Williamson bracket clock. The absence of a place name may indicate that he is no longer working in London but has not yet decided to sign with his Leeds base.*

strikework (most do not, due to the problem of storing enough power for twelve months of striking) and additionally incorporates a pull-repeat ting-tang quarter chime feature.

This is exactly what we would expect from a man with London schooling, trained to make clocks of the highest quality. Williamson left thirty-hour work to his country cousins, quite literally in such clockmakers as his relative Jonas Barber of Winster (nephew of the Ratcliffe Cross maker mentioned earlier), a man who made clocksmith work of a very different nature from London clocks (see page 436).

Some of Williamson's clocks are illustrated in Plates 581 to 584, 586, 587 and in Colour Plates 38 to 43. He used a London pattern of dial – twelve-inches square, solid dial, matted centre with Tudor rose type central motif, ringed winders and seconds, square box calendar with a hint of engraving around it, signature on the dial base, engraved leaves between cherub head spandrels, tiny minute numbers positioned *outside* the minute track, seconds calibrated in fives *inside* the ring, meeting arrowhead half-hour markers – in these examples at least he did not use half-quarter markers. His hands are based on London hands of the day. His movements have five finned pillars and inside countwheel strikework. These clocks to all intents and purposes are London clocks signed at Leeds. Some of his longcase clocks have lost their original cases. Those cases pictured here, however, are original to the clocks and these are clearly London-made cases, which he must have ordered to be shipped from the capital. His time in London would have meant that he had contact with casemakers there. All in all Williamson's work demonstrates ideally how the first city clockmakers brought the London craft to the provinces.

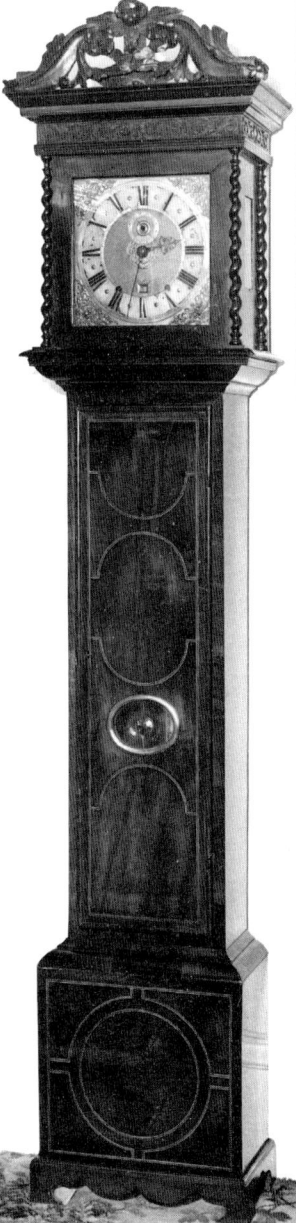

PLATE 586. *Walnut case of the Williamson year clock in Plate 583, London made and with many typical London features. The carved cresting survives, lost from many cases, which is an embryonic swan-neck.*

PLATE 587. *Walnut case of the Williamson month clock in Plate 584 with panelled flower marquetry. Again the carved cresting piece survives, here showing twin cherubs, a popular theme for such crests.*

PLATE 588. *Walnut case of the William Tipling eight-day clock in Plate 585. This too may be London made and has all the typical faetures of a late 17th century London case.*

Williamson's date of death is uncertain. The fact that we do not yet seem to know any clocks of his dating after 1700 suggests he may not have lived long, but he is believed to have still been alive in 1712. His successor in Leeds was William Tipling, who was apprenticed about 1685, and the only person who could have trained him in Leeds was Williamson. Tipling married in 1692, by which date he was probably working as his own master. He died prematurely in 1712, so had a relatively short working life and not many clocks seem known by him. I have come across maybe three or four clocks of his over the years and one is illustrated in Plates 585 and 588. Tipling's clocks are clones of Williamson's, except that he signed them on the chapter ring rather than the lower dial sheet, as was the fashion by then. It can be seen, therefore, that not only was the London school of clockmaking taken into the provincial cities by the first longcase clockmakers there, but that that discipline was passed on to their successors.

PLATE 589. *Ten-inch eight-day dial from the earliest dated Cumberland clock, signed 'Aaron Cheasbrough 1689'. Minutes are numbered* inside *the track at this early period. The matting shows vertical and horizontal (rather than circular) grain. Note the teardrop half-quarter markers.*

AARON AND JOHN CHEASBROUGH OF PENRITH

The oldest dated domestic clock made in Cumberland is the eight-day longcase clock pictured in Plates 589 to 592. It is probably the oldest longcase clock which we can date with certainty known from anywhere in the North of England. It is signed (*below* the chapter ring, as was the early fashion) 'Aaron Cheasbrough 1689'. The surname is spelled erratically; he himself normally used that form but John Cheesbrough used a double e.

Aaron Cheasbrough worked in Penrith in Cumberland from 1699, when his first known child was baptised there. However, the existence of this dated clock shows that he was making clocks at least ten years earlier. He is believed to have been born in the 1660s at the nearby village of Ousby, where Cheesbroughs had lived for generations and where he married in 1694 Catherine Winskill, a local girl from the next village of Addingham. He is next heard of living in Penrith from 1699 until at least 1713, and probably till 1725, when his wife died. A man of this name buried there in 1749 has been assumed to be him, but might just have been his son (Aaron II, born 1702 and not known to have made clocks). Where he lived and worked between 1689 and 1699 is unknown. It may have been at Ousby or at some other place as yet undiscovered, or it

PLATE 590. *Detail showing the signature and engraved work. The scruffiness of the chapter ring is the result of worn silvering. Note the very early form of 3 and 5 numerals.*

377

PLATE 592. *The case of the 1689 clock is veneered in walnut, based on London styling but has provincial touches such as the heavy swell to the barley sugar pillars and the very shallow convex top-of-trunk mould.*

PLATE 591. *The movement of the 1689 clock has five finned pillars, outside countwheel strike control and a solid sheet type of dial, all to be expected at this early period.*

may have been at Penrith, though only *some* of his clocks are engraved with the name of that town on the dials.

How and where Cheasbrough came to learn clockmaking remains unknown. An earlier clockmaker called John Washington is known to have worked in Penrith before Cheasbrough, though no clock is known to exist today bearing his name. Washington was born in Sedbergh in north-west Yorkshire in the late 1620s or early 1630s, the son of a blacksmith. He was a locksmith and clocksmith and moved to Penrith just after the Civil War, working there repairing local church clocks and presumably making domestic clocks from at least 1663 till his death in 1713. Other members of his family remained in the metalworking trades in the Sedbergh area, including his second cousin, Richard Washington who was a whitesmith and maker of the earliest known clocks at Kendal.

PLATE 593. *Ten-inch dial from a two-handed thirty-hour clock signed 'Aaron Cheasbrough in Penreth Fecit' probably dating from the 1690s. Here the half-hour markers are the trident form. Ring-turned circles combine with cup-and-ring drill marks to form an interesting combination of dial decoration, interrupted abruptly by the severe calendar box. This dial was made for a posted movement.*

It seems quite possible that Cheasbrough learned his clockmaking at Penrith under John Washington's tutorship. Since no John Washington clock exists we cannot make comparisons, but one or two clocks do exist by Richard Washington and they bear certain strong resemblances to Cheasbrough's work, which tends to add to this possibility (as it is believed John Washington trained Richard).

It is very difficult to draw conclusions as to whether or not Cheasbrough worked in Penrith all along. The absence of Penrith on the dials of some of them may suggest he was not working in that town when he made them. If John Washington trained him, would he have set up in Penrith in competition with his former master? Or was it perhaps that Washington concentrated on his locksmith work, leaving the clockmaking field clear for Cheasbrough. We may never know, but the fact that all (or almost all?) of Cheasbrough's known clocks date from the period *before* Washington died (1713) and the fact that no clocks survive by Washington may strengthen this view. Thirteen clocks are so far recorded by Aaron Cheasbrough, all of them longcase examples; only two of eight-day duration. No bracket or lantern clock is known by him.

Oddly enough William Porthouse, founder of a very prolific clockmaking family, arrived from parts unknown to set up business in Penrith within a year or two of John Washington's death. Maybe he heard of a gap in the clockmaking market there. No clocks by Aaron Cheasbrough appear to date later than Porthouse's arrival and it is possible that Cheasbrough worked as a journeyman for Porthouse or died before Porthouse's arrival there. Unfortunately records of this sort of relationship virtually never exist.

John Cheesbrough (as he spelt his own name) was buried in Penrith, but did not engrave the town on the only two clocks known today by him so that it is

PLATE 594. *Ten-inch (solid) dial of a thirty-hour longcase clock signed 'Aaron Cheasbrough'. A seconds dial is very unusual on a north-western thirty-hour clock, which must therefore have a four-wheel train. Cup-and-ring turning is again used here liberally. Hands believed original. Date difficult but about 1700.*

PLATE 595. *Posted movement of the clock in Plate 594 with four-wheel train to show seconds, fancy filed hammer stop and square brass posts held by brass nuts top and bottom. This is a distinctive form of posted movement unlike those from West Cumberland (see Sanderson page 395).*

PLATE 596. *Later thirty-hour (cartwheel-cast) dial signed 'Aaron Cheesbrough Penrith Fecit' dating from perhaps 1710-20. The style is now much more formal, the engraving more professional. Note the very bald finish to the severe calendar box.*

PLATE 597. *Detail of the dial in Plate 596 showing the quality of the engraving and much more sophisticated styling of the day.*

PLATE 598. *Exceptionally large (fourteen-inch) dial of month duration longcase clock signed 'John Cheesbrough'. Date difficult but perhaps 1710-20. Many ornate and unusual features suggest this was made as a very special clock, perhaps to impress a particular customer. Flowerheads around the winding holes are most unusual. The legend above the moon reads 'Fugit Irrevocabile Tempus' (Time stops for no man).*

PLATE 599. *Close-up of the engraving of the signature on the dial of Plate 598. Note asterisk half-quarter markers.*

PLATE 600. *Backplate of the John Cheesbrough month clock showing outside countwheel strikework, by now very old fashioned. Engraved flowerheads around the main arbors and on the countwheel and the engraved vase are probably unique on a longcase clock and more akin to bracket clock backplate styling of the day.*

PLATE 601. *Five pillar month-duration longcase clock by John Cheesbrough. The complexity of the movement shows that he was capable of such involved work, even though it was hardly ever called for.*

possible he may have worked elsewhere. One of these clocks, illustrated in Plates 598 to 601, is of month duration and was made in an exceptionally ornate way using both decorative and mechanical features which were by then old fashioned in a deliberate attempt at ostentation. A pull-repeat function allows repeating and brings into play a third train chiming the quarter hours on a separate bell. On pulling, the clock chimes one blow at quarter past, two at half past, three at quarter to, after each of which it repeats the last hour. This repeat is governed by a separate snail and fly, and amounts to a simplified form of grand sonnerie striking, working only by command (i.e. by pulling the trigger, presumably for a customer who wanted grand sonnerie striking only *some* of the time!).

John was born about 1686 and died in 1771. He may have been a younger brother of Aaron or just possibly his son. It seems possible that John worked much of his life for Aaron (or perhaps some other clockmaker such as William Porthouse) as his output seems remarkably small.

JOHN OGDEN OF WENSLEYDALE

John Ogden was a most remarkable clockmaker, born in 1665 into a clockmaking family at Soyland, near Halifax in West Yorkshire. His father was a clockmaker of the clocksmithing tradition; so too were his brothers, as were their sons and his own sons. As tended to happen in families originating in the blacksmith/clocksmith tradition, the nature of their skills and products improved with the years, so that some members of the family (Thomas Ogden of Halifax, for example) ultimately concentrated on the upper market in high class eight-day clocks.

John Ogden was a Quaker and he moved out into the rural dales of Yorkshire about 1690, supposedly to escape persecution in the towns. The 'Fawcett' clock, pictured in Britten and dated 1681 and supposedly made by him after his move from Halifax, has now been seen to be made up from an old Ogden dial, and in any case is impossibly early. He lived initially at Bainbridge, a hamlet of only a handful of houses within the parish of Askrigg in Wensleydale. At Bainbridge there was a Quaker meeting house, of which he became a trustee by the year 1697. He married about 1700 into a well-to-do local Quaker family and played a prominent part in Quaker life in the area.

He was the first clockmaker to work in rural Yorkshire and one of the very first to work anywhere in rural northern England. Many of those other early rural clockmakers who followed him over a wide area of northern England were either trained by him or copied his style or methods – such men as John Sanderson and John Ismay, both of Wigton in Cumberland, Robert Brownless of Staindrop in County Durham, Isaac Hadwen of Kendal in Westmorland, Jonas Barber of Winster in Westmorland...

From the point of view of the student of early clocks John Ogden is an especially important clockmaker, for not only was he the first in the rural North, but he moved residence at short, regular and known intervals, which makes it possible for us to ascribe a reasonably close date to his work and to observe the changes in his constructional and stylistic methods in a way seldom possible with early makers.

As a pioneer trying to sell clocks to a rural population, who had always managed their lives without such luxuries, he had to keep his products within the financial reach of his public. Therefore, like all the earliest rural clockmakers, the great bulk of his output was in simple thirty-hour longcase clocks, even though he was capable of making high quality eight-day work which he did produce in greater numbers later in life. It is for this reason that surviving work by such an early rural clockmaker predominantly consists of thirty-hour longcase clocks and an eight-day clock by any of them is a comparative rarity.

Quakers had their own set of beliefs, not only in religious terms but in their manner of conducting their lives and business. They were trying to sell their wares into a populus often hostile to their beliefs, and they were able to do so only by offering sound quality goods at a keen price, through which practice they gradually acquired a reputation for honest dealing. Their religious beliefs were often in some degree evident in their work and different groups of Quakers in different regions observed varying work ethics. As the first rural Quaker clockmaker in the North, Ogden set a pattern which others followed. In his case he usually avoided using corner spandrels, probably because they depicted a human face (a graven image of man, who was made in the likeness of God?) and perhaps also because it was cheaper! His dial corners, and those of some clockmakers who copied him, are sometimes plain but more often have an engraved verse, his favourite one being 'Behold this hand, Observe ye

PLATE 602. *The dial in Colour Plate 44 (page 393) turned on its side with the (riveted) chapter ring removed. The sheet can be seen to have no cut-outs. Practice engraving shows Ogden experimenting with how to spell 'behold', trying it first as 'behould', further evidence that this might be his first attempt at a dial with this verse. The dial centre is strangely plain, in polished brass with no decorative engraving other than the signature.*

motions tip, Man's pretious hours, Away like these do slip'. The earliest forms of this verse have the word 'tap' rather than 'tip'; he seems to have preferred 'tip' later because of its rhyme. This verse was copied by Hadwen and Sanderson. Occasionally he did use conventional spandrels, especially in his later period, perhaps because he sometimes had to satisfy a customer's taste rather than his own personal inclination, and in one example I know he made his own extraordinary 'spandrels' from matted flat brass sheet (Plate 611).

It seems that when Ogden first lived at Bainbridge in the 1690s he signed his clocks at Bainbridge, which he sometimes spelt 'Bainbrigg'. What appears to be his very first clock (Colour Plate 44 and Plates 602 to 605) is signed 'near Bainbrigg' ('near' is a very strange way for any clockmaker to sign though he used it again on at least one eight-day clock) and another of this period is signed 'of Bainbrigg'. The birth of his first child, listed in 1701 in the Quaker registers, records his abode as Bainbridge, whilst those of his later children from 1702 to 1707, and other records till 1711, stipulate this as Bowbridge Hall, and it seems that he built or acquired Bowbridge Hall (still within Bainbridge hamlet) about 1701. Relatives of John lived later at what is now called Colby Hall in Bainbridge, but whether that is a later name for Bowbridge Hall I do not know. From 1701 to 1711 he changed his signature to 'Jno Ogden Bowbridge'.

Between about 1711 and 1715 he moved to the nearby town of Darlington and his clocks thereafter are signed with this place name. He died in Darlington in 1741. His sons John (born 1704) and Bernard (born 1707) also became clockmakers. John junior moved to work elsewhere in 1729 and Bernard seems to have worked mostly under his father and died about 1745, which means that clocks signed at Darlington (sometimes just 'Ogden Darlington') seem to be by John senior.

The fact that his location can be tied down to short, defined periods means that we are able to date his clocks relatively closely, and this is especially

PLATE 603. *Simple birdcage movement of the 'near Bainbridge' clock (Colour Plate 44) seen in dirty condition as found. The square cast brass posts have swelling capitals and bases. Decorative filing to the hammer stop. Curved front cock holds anchor arbor, lantern clock fashion.*

PLATE 604. *Top view of the same movement showing curved lantern-type front cock, strange semi-circular backcock, typical Ogden shaped fork to crutch, decorative score lines to the brass plates, sturdy iron bellstand.*

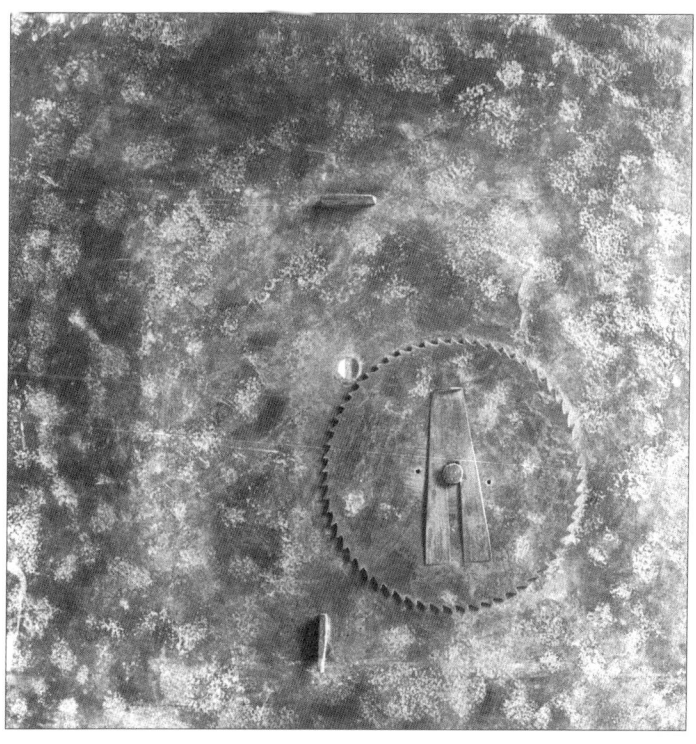

PLATE 605. *Rear view of the same dial showing rough pockmarks typical of casting with some hammering. The vertical lower lug locates into the baseplate and the flat top lug pins into the top plate. Note the* solid *calendar disc offset to one side to show square-box (eight-day-style) calendar reading from the front but using only twelve-hourly knock-on, often termed a 'Westmorland' calendar wheel. The two small holes in the calendar disc were to pin it down to a bed during engraving and are unseen in use.*

PLATE 606. *Solid eleven-and-a-half-inch dial of a unique two-handed six bell quarter-chiming longcase clock dating from the 1690s. A more conventional dial with tulip-engraved centre, reminiscent of a lantern clock, and with conventional cherub head spandrels, both unusual for Ogden. Signed 'Jo^n: Ogden of Bainbrigg'. Chapter ring riveted in place. Arrowhead half-hour markers. Original blued steel hands.*

PLATE 607. *Posted movement of the quarter-chime longcase clock. The going and strike trains are on the right of the picture with large countwheel strike control; the smaller countwheel controls the quarter-chime on the left. Note original large square-headed brass screws.*

instructive at this early time at the very beginning of rural clockmaking. His earliest clocks made during his Bainbridge and early Bowbridge periods are single-handed thirty-hour clocks of posted movement construction – I know of only three posted examples, two of them signed at Bainbridge, including the unique quarter-chime clock pictured in Plates 606 to 609. That particular clock is very important in the history of provincial clockmaking, being the earliest quarter-chime longcase clock that I know of made outside London by a non-London-trained maker. At some time shortly after 1701 he changed to the plate-framed construction, and kept to it thereafter, a change of approach apparently followed sooner rather than later by all northern makers apart from occasional eccentrics such as John Sanderson of Wigton (see page 392).

His earliest dials had solid dial sheets, but within a very few years he had changed over to the cartwheel type of casting. This change took place during his Bowbridge Hall period (1702-1711) and that in Plate 610 is one of his first

PLATE 608. *The quarter-chime movement seen from the chiming side showing chiming barrel with trigger pins, hammer linkage and graduated nest of bells with large separate bell for the hourly strike. Here Ogden is still using a solid dial sheet. This clock is known to have been made for an English convent of Carmelite nuns located in Belgium.*

PLATE 609. *The primitive oak case was made for the clock in Belgium, probably to match wooden wall panelling. Certain features are not dissimilar from an English case of the day. The clock has to be inserted through the hood door for assembling, resulting in the door being much larger than the dial, especially vertically to admit the nest of bells. The clock stands 9ft. high including a 9in. plinth not shown in the photograph.*

cartwheel dials and probably one of the oldest of that type anywhere in the land.

Eight-day longcase clocks from his pre-Darlington period are very rare, and I know of only two. One has the verse corner engraved 'motion's tap' and is signed 'nr. Bainbrigg', suggesting it pre-dates his Bowbridge period. The other is illustrated in Plates 612 and 613. Once established in Darlington his reputation and the pockets of his customers seem to have both increased, as more eight-day clocks are known.

No lantern clocks appear to be known by John Ogden, though one or two are recorded by other members of the Ogden family at Halifax. No spring-driven (bracket) clock is known by him, nor do I recall having ever heard of one by any of the Ogdens. However, when Jane Ogden, widow of John's clockmaking son, John junior, left her will in 1785, she owned more than one spring clock, presumably family made.

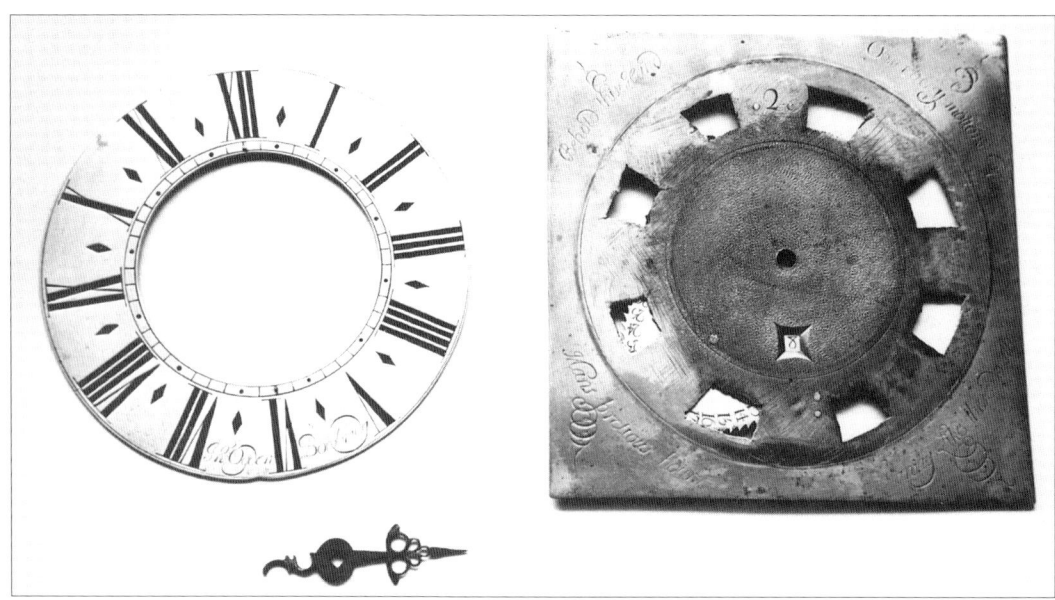

PLATE 610. *The dismantled 8½in. dial of a posted movement thirty-hour longcase clock signed 'Jnº. Ogden Bowbridg', showing several changes in style and construction. He is now using a cartwheel cast dial and also a matted centre and diamond half-hour markers, the latter much liked by Quakers. The Westmorland calendar wheel can be seen in position. Original blued steel single hand. The numeral 2 behind the XII number cannot mean the second clock he ever made, but may mean the second clock of a batch. Two rivets beside the 2 and two more below the calendar are to hold his dial-attachment lugs. His verse corner now has all four lines the right way up and a squiggle now fills the corner centre. Date a year or two after 1700.*

PLATE 611. *Ten-inch dial of a thirty-hour longcase clock signed at Bowbridge, now spelled with an 'e', and using the matted centre and diamond half-hour indicator. Now, however, Ogden has changed to using a plated form of movement. Original blued steel single hand. The 'spandrels' are here of unique nature being cut out from flat matted brass sheet in the form of twin tulips. Date 1705-10.*

PLATE 612. *Twelve-inch (cartwheel-cast) dial of an eight-day longcase clock signed 'Jnº: Ogden Bowbridge', using the diamond half-hour marker, the verse corner with the squiggle, and the matted centre, the same styling as in Plate **610**. Here, however, he rings his winding and seconds holes. The engraving is sharp and professional – presumably done by Ogden himself, as if not I cannot imagine who would engrave for him in such an out of the way place. Date 1705-10.*

PLATE 614. *This ten-inch thirty-hour longcase clock is signed 'Jno. Ogden Darlington and dates from about 1715-20, shortly after his move there. His dial style is almost unchanged from his Bowbridge period, using the same corner verse and squiggle.*

PLATE 613. *The movement of the eight-day Bowbridge clock has four finned pillars (London style), the lower two tapped through for seatboard bolts. The clock has rack striking, being quite an early use of this. Note the cartwheel casting of the dial, the high-domed semi-circular wheel collets, the double-position backcock, and the better eight-day form of twenty-four hour calendar working with an inner-toothed calendar ring riding on two rollers. Note also the filed shaping to the upright hammer spring and hammer tail, and the fancily shaped hammer stop, the latter unusual on eight-day work, not used on plated thirty-hour work, and really a carry-over from posted movement principles.*

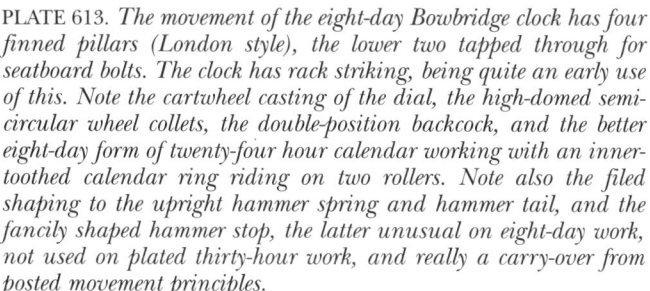

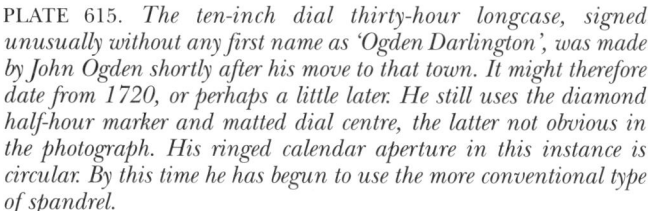

PLATE 615. *The ten-inch dial thirty-hour longcase, signed unusually without any first name as 'Ogden Darlington', was made by John Ogden shortly after his move to that town. It might therefore date from 1720, or perhaps a little later. He still uses the diamond half-hour marker and matted dial centre, the latter not obvious in the photograph. His ringed calendar aperture in this instance is circular. By this time he has begun to use the more conventional type of spandrel.*

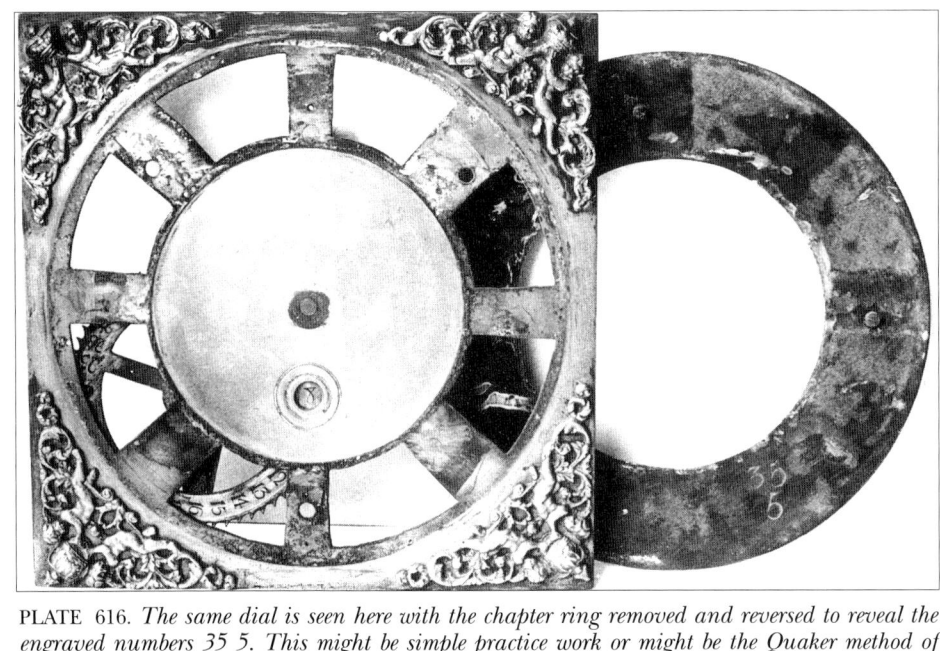

PLATE 616. *The same dial is seen here with the chapter ring removed and reversed to reveal the engraved numbers 35 5. This might be simple practice work or might be the Quaker method of dating, meaning the 5th month of the year 1735, which would be July as the year began in March at this date. Quakers called their months by numbers, refusing to use names chosen after heathen gods. Note the spoked Westmorland calendar wheel.*

PLATE 617. *The plated movement of the 5/35 Darlington thirty-hour clock shows the cartwheel dial with offset Westmorland calendar wheel (by this date a larger spoked wheel rather than a solid disc). The wheel collets are now of a later style than his half-round Bowbridge ones, being of the dome-and-collar type. The pillars too have simplified somewhat.*

PLATE 618. *The oak case of the Darlington (5/35) thirty-hour clock is very simple and crude with tiny, pinched mouldings and the wide topmould overhang characteristic of many early country cases. There are no hood restraints, no internal hood mask, no pillars to the internally-hinged hood door. The trunk door fits inside its frame rather than on to it. Height about 6ft.*

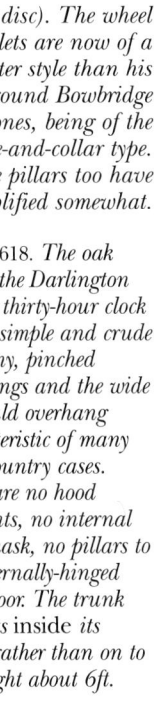

PLATE 619. *This eight-day dial with penny moon in the arch consists of a square dial with separate arch riveted on, as early examples often had. The matted centre and corner verses are similar to his Bowbridge eight-day, as is the diamond half-hour marker, but the wording of his corner verse now reads 'Observe the motion's tip' rather than 'ye', being the more up-to-date form. Decorative score lines are reminiscent of those outside the chapter ring on his first clock (Colour Plate 44, page 393). Date 1720s to early 1730s.*

PLATE 620. *This later Darlington eight-day clock, also with arch penny moon feature, has more conventional styling, and his Quaker traits have now almost disappeared. His minute numbers are now larger and engraving occupies the dial centre. The half-hour diamond is now engraved differently. This dial says more about clocks of the 1730s than about John Ogden.*

PLATE 621. *The pendulum bob of the quarter-chiming longcase clock in Plate 606 is covered in cup-and-ring turn decoration, a most unusual and perhaps unique use of this device.*

PLATE 622. *Nine-inch solid dial of a single-handed thirty-hour clock signed 'John Sanderson Of Wigton Fecit', made in the late 17th century, seen here in tarnished condition – probably his earliest surviving clock. Original blued steel hand of eccentric style. Memento Mori in the top corners, his favourite verse to the centre. The chapter ring marks hours, half-hours, quarter-hours and half-quarters, the latter being exceptionally unusual on a single-handed clock.*

PLATE 623. *Movement of the clock in Plate 622, seen before cleaning. Note nuts to hold pillar tops. Threads below pillars may once have held ball feet. Curved cast support for anchor arbor. Bent iron 'backcock'. Trip lever for re-setting strike sequence.*

JOHN SANDERSON AND THE WIGTON SCHOOL OF CLOCKMAKING

John Sanderson was a prolific clockmaker whose work has long been known but whose life and working dates were until recently a mystery. Over many years I have been involved in detective work into his origins and we now know that he was born in 1671, the son of Robert Sanderson, a blacksmith. The family lived at a small hamlet called Tiffinthwaite, near Wigton in Cumberland, the father dying in 1684 when young John was only thirteen. Robert's widow was left with four young children to support and her own parents and parents-in-law were all dead. The family probably lived at the Grand House at Highmoor with her late husband's elder brother, also named John Sanderson, uncle of young John Sanderson, the clockmaker-to-be. Uncle John died in April 1690 leaving a young widow, at which time his nephew John would have already been five or six years into his apprenticeship. Less than a year later this widow had an illegitimate child fathered by John Sanderson the apprentice clockmaker, nephew of her late husband and then aged only nineteen. This major scandal in the family had considerable bearing on John Sanderson's later life, which seems to have been obsessed with the repenting of sins and the spreading of the word of God.

John Sanderson may well have been apprenticed about 1684 under Quaker clockmaker John Ogden, then newly established in business at Bowbridge in

COLOUR PLATE 44. *Perhaps the first longcase dial Ogden made, with nine-and-a-half-inch solid dial sheet, signed 'John Ogden near Bainbrigg', the corner verse written 'motions tap', changed on later clocks to rhyme as 'motions tip', and with the lower lettering inadvertently engraved upside down, changed to right way up on later clocks. On later dials he also improved the spelling of 'houres' to 'hours', and ultimately 'ye' to 'the'. The decorative incised ring outside the chapter ring is seen on early work by several northern makers. Iron hand original. Note most chapter ring markers consist of arrowheads. Date 1690s. See page 384*

North Yorkshire. Certain features of his style seem to be derived from Ogden and there were precious few others around who could have taught him the trade at that time. Certainly there was no domestic clockmaker then working in Cumberland with the possible exception of John Washington of Penrith, by whom no domestic clock survives today.

Quakers held to a strict moral code and it seems that John Sanderson was expelled from the Ogden household on account of his sins, his apprenticeship cut short before its completion. He was working on his own account at Tiffinthwaite, Wigton, in 1691, where he married in that year a local girl from a neighbouring Quaker family, taking his illegitimate child to live with them. For some time past Tiffinthwaite had already been a centre of worship for local

PLATE 624. *Nine-inch solid dial of a thirty-hour single-handed clock, this one also showing half-quarter notation. Original and most unusual blued steel hand. Note the serpentine flourishes to the last letter of each line.*

PLATE 625. *Detail of the dial of Plate 624. Note arrowhead indicators for half- and quarter-hours. His name and other lettering are expertly engraved.*

Quakers and John Sanderson himself followed the Quaker faith for some years, travelling here and there to spread the word, as was the Quaker custom. We know he attended Quaker meetings as far distant as Edinburgh in 1713, and probably also in 1715, which latter event is mentioned in more detail below.

John Sanderson's clocks are every bit as enigmatic as the life of their maker. Each time he made a clock it is as if he was making his very first one. His style and work practices seem so erratic that it is exceptionally difficult to set examples of his work into a sensible date sequence, though I have attempted to do this with those illustrated. He must have been a prolific maker as I have recorded over forty clocks by him, all weight-driven clocks of the 'longcase' type except for a solitary example of a bracket clock. Most of his clocks are of thirty-hour duration; eight-day examples are known by him but they are uncommon, only five having come to my attention plus a sixth made during his brief and apparently unsuccessful partnership (lasting only a year or two in the 1730s) with a Quaker named Joseph Wilkinson, who may have been his former apprentice. All (except the partnership clock and the bracket clock) have square dials.

The question arises with Sanderson's clocks as to whether they were intended to be housed in long cases. His eight-day clocks are no problem in this respect, as these are mostly still in their original cases. However, his numerous thirty-hour clocks have mostly either been housed in a case at a later date or still have no case today and sit instead on a wall shelf or bracket. I know many Sanderson thirty-hour clocks but only one in its original case. The late John Penfold, author of *The Clockmakers of Cumberland,* was well aware

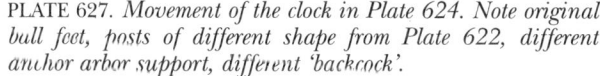

PLATE 626. *The same dial with the chapter ring removed and showing practice engraving of Remember Man, clearly by a far less skilled hand.*

PLATE 627. *Movement of the clock in Plate 624. Note original bull feet, posts of different shape from Plate 622, different anchor arbor support, different 'backcock'.*

of this problem, but ascribed it to the toll of time by virtue of the sheer antiquity of Sanderson's clocks. But as we know a much higher proportion of clocks of similar age by other makers which *are* in their original cases (as well as a very high proportion of Sanderson's admittedly limited eight-day work), this argument does not seem to hold for me. It seems likely to me that Sanderson's clocks were the Cumbrian equivalent of Walter Archer's Gloucestershire hook-and-spike clocks, which we know formed the great bulk of Archer's output, and which were sold *caseless* (see page 414). Sanderson's thirty-hour clocks probably sat on a wall bracket, maybe a quite simple home-made affair, which would be unlikely to have been preserved.

Why would Sanderson have gone to the vastly time-consuming trouble of casting decoratively-shaped pillars, and often feet too (though today a number have had their feet removed later), if they were to be housed unseen inside a longcase? For such a purpose he could have used much cheaper iron rod. For Sanderson was not a maker inclined to waste time on fancy shaping or filing of movement parts – in fact his clocks are totally devoid of the sort of hammer spring decoration found on some early posted-movement clocks. Posted-movement longcase clocks did not need feet and only rarely had them. Richard Sill, a member of the 'Wigton school' of clockmaking as I call it, used plain rod pillars for his posted-movement cased examples, as did rival Aaron Cheasbrough at Penrith, as even did John Ogden at Bowbridge. It seems to me that Sanderson intended his cast brass posts to be seen, which may also explain why they vary so very much in shape, for very seldom are the posts of any two of his clocks quite alike. This might also account for no arched dial examples being known, the arch only being of value if the clock was to be cased — Walter

COLOUR PLATE 45. *A most unusual dial for John Sanderson with engraved tulip corners. Here the chapter ring is riveted to the solid dial sheet at XII and VI, unusual practice for this maker. The ruling lines for the dial centre lettering show – in more careful work these would have been polished out. Original blued steel hands.*

Archer's hook-and-spike wall clocks all have square dials too.

Of those thirty-hour clocks I have seen only two were single-handers, those being amongst his very earliest work, so that he appears to have moved to the two-handed form for preference very much earlier than was the case with most clockmakers. Just when this was is difficult to say with certainty as his clocks are difficult to date, but it must have been within the first few years of the eighteenth century. Three examples of his work are said to be dated (1714, 1715 and 1717), though personally I have never come across a dated example.

His thirty-hour movements are all of the posted type with brass, lantern-style pillars. His dials usually attach to their movements lantern-clock fashion, i.e. by a lower lug with two top pins, but some examples have dial feet which pin through the front posts. This variation in method is typical of Sanderson's erratic methods. No plated example has yet been recorded, a highly significant difference from the work of most northern makers, who moved early to the plated-movement form. This may be a further factor indicating that these clocks were intended to show on a bracket rather than being cased. When we compare the constructional features of his thirty-hour clocks, each seems to have been made as if it was his first clock. Little apparent progression is obvious but rather his clocks each seem to have randomly varying size,

COLOUR PLATE 46. *Nine-inch hook-and-spike clock signed 'Archer, Stow' (the old w often looks like n) with rudimentary foliate cross centre design and ring-and-dot half-hour markers. Plain corners are unusual for this maker. The blued steel hand is original and of a design he often used. Date about 1700. See page 413.*

shaping or positioning of pillars, cross bars, countwheels and to some degree of the backcocks and front anchor-arbor supports and of the shape, number and positioning of the spokes in his cartwheel dial castings! Why he did not settle down to a progressive system, as most clockmakers did, is a mystery.

This apparent hotchpotch of varying component styles, shapes and sizes must indicate that Sanderson did his own brass casting. There would have been no brassfounder near or far who would have supplied this varied assortment of shapes and sizes. In fact we might draw quite the contrary conclusion, that a maker whose parts were of a consistent pattern probably did buy them in from a brassfounder.

His early dials begin at nine inches square and soon progress to twelve. His very earliest ones are solid dials, but he soon turned to the cartwheel form of casting with an occasional solid-sheet throwback, as for example with the tulip-corner dial (Colour Plate 45). He signed his clocks on the chapter ring either side of the VI numeral, initially as 'John Sanderson of Wigton Fecit'. The inclusion of the word 'of' is normally an early sign for any clockmaker. He soon began to omit the 'of', then, shortly after, he dropped the 'fecit'. Sometimes he dropped the town, signing just with his name. One or two of his later clocks are signed at Carlisle, though it is difficult to say at what period he

PLATE 628. *Eleven-and-a-half-inch dial of an early two-handed thirty-hour clock signed John Sanderson Wigton Fecit, again with his typical verse. Here he uses a very strange 'half-quarter' style marker to mark every 2½ minute unit between minute numbers. Note the engraving slip through the radial lines on IIII with another slip on the V of VIII. Original blued steel hands. Ring decoration inside and outside chapter ring. Date c.1700?*

PLATE 629. *Movement of the clock in Plate 628, showing four-spoke dial casting, practice engraved 3 behind dial right of centre, absence of decoration on hammer spring and hammer stop, front anchor support and 'backcock' of yet another different design. The rear left foot (closest to camera) is of deformed shape from an unsuccessful casting, yet was still used.*

PLATE 630. *Another thirty-hour dial of the same type. This time the engraver runs a line between the minute numbers – 4/0, 4/5, etc. About 1700?*

PLATE 631. *Apparently original oak case of the clock in Colour Plate 45 made of oak and standing 6ft.10in. Date about 1700?*

was at Carlisle. One clock is signed at Kirkbride, near Wigton, its date uncertain. John Sanderson's death date is uncertain, but probably occurred not long after 1740.

His early dial corners on his thirty-hour clocks have no spandrels. Instead his earliest have Memento Mori (Bear death in mind) engraved in the upper corners. More often his corners are completely blank, a feature preferred by some Quaker clockmakers (Sanderson was a Quaker at one period) and one which happened to be conveniently cheaper than spandrels. Eight-day dials were a different kettle of fish and usually had normal corner spandrels – perhaps because cost-saving was less important with these.

His earliest thirty-hour dial centres had polished grounds on to which was almost always engraved an admonitory verse with the same constant theme, warning man to repent his sins before it is too late – a lesson Sanderson had learned from personal experience early in life! Occasionally this verse overflowed into the corners too. The commonest verse runs: 'Remember man, That die thou must, And after that, To judgement just'. Other verses known are: 'Here I am set O man to tell ye plain, Yt [that] time once spent cannot be called again. Therefore let every stroke upon my bell, Put thee in mind of judgement, death and hell.'; 'As the hours do pass away, So doth the life of man decay, So death will come and die we must, And so be brought to judgement just.'; 'As time and clock and all things pass away, Amend your live for hear we must not stay.'. One example is even known with John Ogden's own verse: 'Behold this hand, Observe ye motion's tip, Man's pretious hours, Away like these do slip'.

Later he changed his style to use a matted centre on his thirty-hour dials, often making ringed dummy 'winding holes' to relieve the plainness. His eight-day dials normally have matted centres without any verses, though one example has a verse engraved into the matted centre. Some of his earlier thirty-hour dials, especially those with blank corners, have a series of circles scribed into the brass inside and outside the chapter ring. This is probably just for decoration to relieve the blandness.

His clocks almost always have calendars, earlier ones mostly as a square box,

399

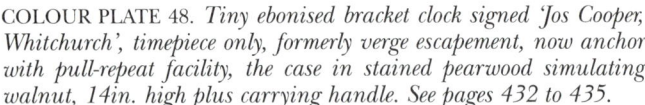

COLOUR PLATE 47. *Unsigned nine-inch dial of a thirty-hour longcase clock signed and dated on the hidden surface of the calendar wheel 'Richard Snow 1726'. This is the only clock by Richard Snow so far identified. See page 430.*

COLOUR PLATE 48. *Tiny ebonised bracket clock signed 'Jos Cooper, Whitchurch', timepiece only, formerly verge escapement, now anchor with pull-repeat facility, the case in stained pearwood simulating walnut, 14in. high plus carrying handle. See pages 432 to 435.*

occasionally a round box, but later as mouth calendars and even a full open circle 'mouth', an odd style sometimes used by other Cumbrian makers. He used the offset Westmorland calendar disc on thirty-hour clocks but was usually forced into using the normal London style inside-toothed ring on eight-day clocks, as the Westmorland system was unsuited to eight-day work.

His thirty-hour clocks have locking plate strikework, as we would expect. Some of his eight-day clocks also have this, but he was well aware of, and capable of making, rack strikework which he did sometimes use on eight-day clocks surprisingly early in the eighteenth century if the clock was required to be a repeater, as did John Ogden in the late seventeenth. It is surprising that such a primitive maker as Sanderson and such an isolated maker as Ogden were using rack strikework on longcases earlier than many London makers.

Sanderson was fined in 1715 for selling his clocks in Edinburgh, where he was prohibited by local law from doing so. It may be he had been visiting Quaker 'friends' there (as we know Quaker Isaac Hadwen did at one time – see page 418) and had taken the opportunity to earn his travel costs. Rural clockmakers often sold their clocks in market towns, even sometimes in distant ones and even though local bylaws usually existed to try to prevent this. It is probably for this reason they sometimes left such clocks unsigned or signed without any place name – a factor which would help prevent local authorities from tracking down the makers.

A handful of clocks are known of about 1720-30 (three to date) which are clearly from the same workshop and are signed 'Jer. Sanderson, Wigton', short

PLATE 632. *Dial of later type (c.1720-30) with matted centre and decorative roundels to break up the plainness. Original blued steel hands. Note the now very different and much plainer form of signature.*

PLATE 633. *Thirty-hour dial of similar period and style to Plate 632 but signed 'Jer. Sanderson Wigton'. Jeremiah seems to have been an invented name sometimes used by John Sanderson. The blued steel hands look like later replacements. The movement is signed John Ismay, No. 5.*

for Jeremiah. Extensive research has established that there was no such person. These clocks appear to be the work of John Sanderson, but the reason for this mysterious signature is puzzling. My thought was that perhaps John Sanderson used this invented name to prevent authorities from tracing him as the maker of goods sold illicitly in distant markets. Jeremiah was the prophet who lamented the decline of morals and if Sanderson were to coin a pseudonym this would certainly have been an appropriate one. It was only my opinion that John Sanderson concealed his identity behind the name of Jeremiah but very recently a John Sanderson clock dismantled for cleaning was found to have practice engraving behind the dial and, amazingly, the words included two attempts at engraving the name Jeremiah (Plate 634).

John Sanderson founded what I refer to as the Wigton school of clock-making in so far as his very idiosyncratic style, influence (or even handiwork) appears in the work of two other local makers – John Ismay and Richard Sill. Their work is indistinguishable from Sanderson's apart from the fact that their clocks bear their own names. Unlike Jeremiah Sanderson, Ismay and Sill were real people who did make clocks, though they were not nearly as prolific as Sanderson.

John Ismay is known through only five clocks, all thirty-hour posted-movement examples of the John Sanderson type and style. A movement made by him and marked as 'Number 5' has a dial signed by 'Jeremiah Sanderson', evidence of the close connections between this Wigton group. Another clock (Plate 646) has 'No.1.' engraved on the calendar wheel. Ismay was born at Thursby, a village near Wigton in Cumberland, in March 1699, was apprenticed to John Ogden at Bowbridge in 1712, but may have ended his term prematurely as he was working in Thursby by 1718. One Ismay clock is signed at Oulton (near Wigton), the others at Wigton. He is recorded as living at Tiffinthwaite at the time of his death in 1755, though he was not a Quaker.

John Ismay was related to John Sanderson in a complicated way, probably

PLATE 634. *Practice engraving behind the dial of a John Sanderson thirty-hour clock including the names John and Jeremiah.*

PLATE 635. *Solid sheet twelve-inch dial of an eight-day longcase clock by John Sanderson. This has the offset Westmorland calendar wheel not really suitable for a key-wound clock as the calendar wheel spokes obscure the left winding square for a two-day period twice a month. The dial is (made and ?) engraved by James Hendrie, which puts the clock post 1718. This clock has London style finned pillars.c.1720? Hands later.*

PLATE 636. *Rear of the dial of Plate 635 showing engraving 'Ja Hendrie fecit sculpt'.*

more by adoption than blood, but a relationship which equated to that of nephew and uncle. They regarded each other as kin, and it seems as if Ismay worked for Sanderson for most of his life, which probably accounts for the small number of surviving clocks by him. But for the signatures on the dial we would be unable to recognise the work of one from that of the other.

John Ismay's only child, Sarah, married William Lamonby, whose only daughter (another Sarah) married Joseph Hendrie, the Wigton clockmaker and son of James Hendrie, clockmaker, about whom more below.

Richard Sill arrived in Wigton some time shortly before 1704, the date he was married there to a local girl. His origins are unknown, but the name is not a local one. He died at Wigton in 1729, succeeded by his son, Joseph, who also

PLATE 637. *Twelve-inch dial from an eight-day clock signed 'John Sanderson' without the place name. c.1720-30?*

PLATE 638. *Dial of an eight-day clock by John Sanderson with 'Wigton' added as an apparent afterthought. This has his verse engraved most unusually on to the matted centre. c.1720?*

PLATE 639. *Movement of Plate 638 with dial removed to show his interesting form of rack striking, using a steel rack of unusual shape.*

PLATE 640. *Twelve-inch dial of an eight-day longcase clock by John Sanderson. Here he uses a full circle calendar with twelve-hourly knock-on, a system which saved the work of a twenty-four hour changing wheel yet avoided the problem of the Westmorland calendar system (seen in Plate 635) fouling the winding squares on certain days of the month. This clock has baluster pillars. Blued steel hands believed original. Date 1730s?*

PLATE 641. *Thirty-hour dial with original steel hands by Richard Sill of Wigton. Date c.1710? The penny moon is an unusual feature on these early Wigton clocks. This clock is clearly from the same workshop as John Sanderson's.*

PLATE 642. *The movement of Richard Sill's clock in Plate 641. Pin and lug fastening of dial to movement, lantern clock style. Very solid in construction and highly typical of the early Wigton school. The penny moon disc drives in exactly the same way as the twelve-hour calendar and from the same pin/wedge.*

PLATE 643. *Dial from another thirty-hour clock by Richard Sill of Wigton, dating about 1715. Original steel hands.*

PLATE 644. *Dial from another thirty-hour clock by Richard Sill of Wigton, dating perhaps from 1720. The dial by now is more conventional using the matted centre and some cup-and-ring decoration. Original steel hands.*

became a clockmaker. Only seven or eight clocks appear to be known by Richard Sill, all of them of thirty-hour duration. They are mostly of the John Sanderson posted-movement type, and this is a small surviving output for a man with twenty-five years in the trade, so that it seems likely Sill worked for part of the time at least for John Sanderson. Like John Ismay's clocks the only thing that distinguishes most of Sill's work from Sanderson's is the signature on the dial.

However, Sill's work includes some clocks with more conventional posted movements with rectangular posts and some with plated movements. This may suggest that for clocks which were to be *cased* Sill used the more conventional type of movement, keeping his more ornate, Sanderson-style examples to be exposed on a wall bracket. No eight-day clock is known by Richard Sill.

The dials by Sanderson, Ismay and Sill (and Jeremiah Sanderson too, though we assume this was also John Sanderson's work) have many features in common suggesting the work was that of one man who engraved for all three. It is very difficult to determine who the engraver was. It might have been John Sanderson himself. It could not have been Ismay or Sill as some of Sanderson's work pre-dates their arrival on the scene.

I know of two Sanderson clocks which have practice engraving on hidden parts of the dial. Brass was costly and could not be wasted for such practice work so that it is not uncommon to find unused and unseen areas put to this purpose. The clock in Plate 626 is pictured with its chapter ring removed showing practice work of the words 'Remember Man…'. The engraving on the dial is of a naïve and charming nature, yet executed by an expert who could do with a graver what most could not do with a pen. The practice work is much more crude, like a schoolboy's first efforts at joined up writing, carefully done but inexpertly and surely by a very different hand from the dial surface work. It may have been the work of an apprentice or journeyman just trying to acquire the engraver's skills. Or could it have been Sanderson himself attempting to copy the engraver's work?

The practice engraving of Jeremiah and other words in Plate 634 is in a very

PLATE 645. *Movement of the clock in Plate 644, showing this to be a conventional posted movement with rectangular pillars and made to be housed in a case. Much simplified from his earlier work. Lug and pin dial fastening.*

PLATE 646. *Thirty-hour clock by John Ismay of Wigton, engraved behind calendar disc No.1 (perhaps his first clock?). The verse runs: 'Shun sin least thou lament, When precious time is spent, And death to the[e] is sent, No time then to repent'. The clock was made without spandrels, the holes being where some later owner fitted them. Date about 1720? Steel hands probably original.*

PLATE 647. *Movement of the John Ismay clock in Plate 646. Typical Wigton school work but posts unusually decorative. The threads below the plates may once have had feet.*

PLATE 648. *Three Wigton school movements by (left to right) John Ismay, Jeremiah Sanderson (signed internally as John Ismay No. 5), and Richard Sill. Note the many differences in styles, sizes and shapes of each part, although each is based on the same principle.*

uncertain and unskilled hand, suggesting it is not the hand of the man who engraved the dial but of a learner. On the other hand, the dial of the eight-day Sanderson clock in Plate 635 shows highly skilled engraved work, but if we are to believe what is engraved behind the dial the clock (or just the dial?) was made and engraved by James Hendrie. Yet how unskilled the back of dial engraving appears when compared to the face side, even though this claims to be done by the engraver. The result is that it is difficult to draw any sensible conclusions from the fact that hidden engraved work is less perfect than that intended to show.

We can see from Plate 635 that this eight-day clock (dial?) was made and its engraving performed by James Hendrie, the inscription 'Ja Hendrie fecit sculpt' being an attempt at Latin for 'James Hendrie made this and engraved it'. James Hendrie arrived in Wigton about 1718, probably from Falkirk, and worked there till his death in 1768. This seems to be a higher order of engraving than on some of Sanderson's earlier clocks, so perhaps Sanderson turned some time after 1718 to Hendrie as an engraver more skilled in the formal styling.

I once saw a much-modified Sanderson dial on the back of which was engraved various practice lettering including the partly obliterated word 'Harri...', which must surely have been the name Harriman. Edward Harriman worked in Workington from about 1735 till he died in 1776. This suggests that Harriman engraved or perhaps even made clocks for John Sanderson. Such tantalising glimpses are all that is available to us of this unknown aspect of such work.

PLATE 649. *The only bracket clock so far known by John Sanderson of Wigton, cased in walnut standing 15in. high. The movement has pull quarter repeat on two bells. c.1700. Such work is quite exceptional for an early rural maker. (Photograph by courtesy of Aspreys, London.)*

I have seen only a few of the six eight-day clocks known so far by John Sanderson, but those I have seen are of two quite different types. One type has the normal finned (London-style) pillars we might expect in early eight-day work. The other type has quite different baluster shaped pillars tapering towards one end, a style I cannot recall seeing in the work of any other maker in that locality. Both types are much more finely made than the crude and erratic cast posts of his thirty-hour posted-movement clocks. We might expect the maker to have taken more trouble with his castings for his more costly eight-day clocks, just as his eight-day dial styles are more formal and conventional than his more quirky thirty-hour dials, but why there should be two quite different forms is puzzling. Does this mean that Sanderson had his eight-day work made out of house by more than one maker – such as Hendrie and Harriman for example?

A solitary bracket clock is known by Sanderson (Plate 649), an exceptionally costly item relative to most of his work and therefore one we would expect to be more conventional, along the London principles, as in fact it is to some degree. However, the dial still bears his 'Memento Mori' warning set into the herringbone border at the top, either side of his strike/silent lever. The chapter ring is relatively conventional, but his dial corners carry finely engraved floral work rather than the usual cast spandrels. The clock has pull-quarter repeating on two bells as well as hourly striking. I have not seen the movement and cannot say whether Sanderson made this himself.

PLATE 650. *Verge escapement lantern clock signed 'Archer Stow' dating from the 1690s. The lion and unicorn fret seems to have been a favourite with this maker. The dial sheet unusually has blank corners, yet the centre is engraved with the rudiments of what later became his foliate cross design. Iron hand believed original.*

WALTER ARCHER OF STOW ON THE WOLD

A variety of interesting clocks dating from the late seventeenth and early eighteenth centuries have been known for some years now by a clockmaker who signed himself variously 'Walter Archer', 'Archer, Stow' and 'Walter Archer, Stow-on-the-Wold', a town in Gloucestershire. It is uncertain whether his different forms of signing have any special significance; if they have, then we cannot make any consistent analytical sense of them, except that his few eight-day longcases all seem to have merited the cost of engraving his own name and town in full, an extra cost he no doubt recouped within the price of these, the most costly items he produced. The absence of any clear signature sequence is not altogether surprising, as we can hardly expect any clockmaker

PLATE 651. *Movement of Plate 650, showing retention of early features such as: tapered arbors; some wheels squared directly on to arbors without collets; limited decorative filing to hammer spring and hammer stop.*

PLATE 652. *Another lantern clock of the 1690s, this one signed 'Walter Archer' and here using his foliate centre, an extension of the early tulip centre. The movement has been replaced by a later spring-driven one with two hands.*

PLATE 653. *Detail of the engraved signature on Plate 652 showing some hesitancy in the lettering. Note the strange ring-and-dot fleur-de-lis half-hour markers.*

PLATE 654. *Ten-inch dial of hook-and-spike wall clock signed 'Walter Archer' and dating from about 1700. Here he used a matted centre, unusual in his thirty-hour work, but retained his earlier ring-and-dot half-hour markers. The hand looks later.*

PLATE 655. *Movement of the hook-and-spike clock in Plate 654, showing that his work is somewhat simpler than on his early lantern clocks. Here the hammer stop is missing, as are the spikes, though the original swept hook is preserved.*

to set himself in advance a pattern of signature to last throughout his lifetime.

Recent researches by Archer enthusiast Barnaby Smith have established some positive dates for this formerly ill-documented maker. We now know that Walter Archer was born in 1674 at Moreton-in-Marsh, Gloucestershire, the son of a blacksmith. By about 1695 he had moved to Stow-on-the-Wold, being first documented there in a list of inhabitants of 1699 with his brother, Charles (born 1676), who followed the same trade. I know of no clocks signed by Charles, and it seems to me he is likely to have worked for his elder brother. It is possible that those clocks signed by just the family surname 'Archer, Stow' may have been intended to signify a joint product. It seems throughout, however, that Walter was the major name behind their clockmaking activities and for our purposes at least we will regard the Archer work as being by Walter. Walter is believed to have died about 1744-45, after which his son, Richard, seems to have continued the business.

Forty-one Archer clocks have so far been recorded made within Walter

PLATE 656. *Single-handed thirty-hour longcase clock signed 'Archer, Stow' with his typical foliate cross centre (or dolphins?) with two 'eyes' and the strange blank zone he regularly left between centre and chapter ring. Original hand of his favourite pattern of this period, which is about 1710 or a little earlier.*

PLATE 657. *Another single handed thirty-hour longcase signed 'Archer, Stow', now with slightly wider chapter ring and dating about 1715-20. Here the dolphins have four 'eyes'. The same pattern of original blued steel hand. For a time he used a diamond half-hour marker.*

PLATE 658. *Another thirty-hour single-handed longcase clock signed 'Archer, Stow', this one somewhat later (1720-30?) and now without any half-hour marker, giving much plainer overall appearance. The hand could be later. The rivet mark at II is a dial foot fixing.*

Archer's lifetime, including thirteen lantern clocks (Plates 650 to 653), three converted in more modern times to spring movements, sixteen hook-and-spike clocks (Plates 654 and 655 and Colour Plate 46, page 397) and eight thirty-hour longcase clocks (Plates 656 to 668), the latter all single-handed thirty-hour examples with posted movements (I don't think I know of a two-handed thirty-hour made by him), but also four eight-day longcases. No bracket clock is yet known by him. This is to some degree what we would

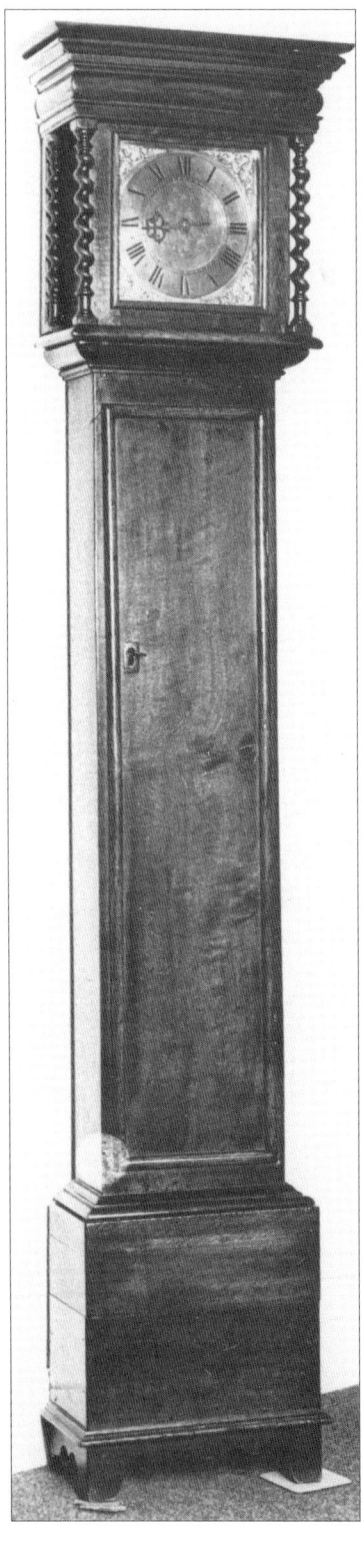

PLATE 659. *Posted movement from a single-handed thirty-hour longcase clock of about 1720-30 signed 'Archer, Stow'. Everything is now simplified in style. The dial sheets of these small dials were usually solid, being rather thin castings, probably too weak to make cartwheel fashion.*

PLATE 660. *Interesting and primitive oak case of the thirty-hour clock in Plate 656. Note the cushion mould below the hood topmould, a feature based on some London work of the day but popular with some early country casemakers. Note too the shallow and convex top-of-trunk mould, and the fact the the door has no separate 'frame' but is hinged on to and closes into the case side timbers. Date about 1710 or a little earlier.*

expect in a rural town in southern England from an early maker who originated in the clocksmithing tradition. However, his output of hanging wall clocks seems surprisingly high (twenty-nine if we add together lanterns and hook-and-spike clocks) relative to only eight thirty-hour longcases. This is probably down to price alone, because these were cheaper by the price of the case. It also helps to explain why many of the Quaker zigzag hook-and-spike clocks from the neighbouring county of Oxfordshire are caseless. This very

PLATE 662. *Twelve-inch dial from an eight-day longcase clock of about 1700-1710 signed intriguingly 'Walter Archer, Stow on ye Old', whether intentionally or an engraver's error I cannot say. Cup-and-ring decorations around the ringed winding holes. Original steel hands with most unusual pattern for the hour hand.*

PLATE 661. *Primitive case in elm from the thirty-hour clock in Plate 657. Sound holes drilled into the hood sides. No pillars. External wrought-iron H hinges. Date about 1715-20.*

strong weighting towards caseless clocks may have been even greater than the figures suggest as the destruction rate in those was probably higher than in cased versions.

It is interesting that on at least three of his hanging clocks he used the tic-tac escapement, a form of escapement not used by many clockmakers but especially suited to wall clocks, which were after all his speciality.

His thirty-hour longcase dial sheets were usually very thin and therefore were made as solid castings to retain strength, though he did use cartwheel castings for his larger, thicker and stronger eight-day dials (and on one hook-and-spike clock). Clearly both methods were known to him and his choice of which to use was determined by practical factors (and perhaps cost) rather than representing a progressive development in technique such as we saw in the work of certain other, principally northern, clockmakers such as John Ogden (page 383). His work includes two most unusual hybrid clocks part way between the lantern and the iron-framed hook-and-spike clock – see Plates 97 and 98 in Chapter Five.

All his clocks seem to display that idiosyncrasy of styling we might expect from a country maker. His lantern clocks sometimes have a scrolling centre based on flowers, an extension of the tulip theme. This soon develops into a foliate cross, a design which he continued for dial centres of his hook-and-

PLATE 663. *Twelve-inch dial from an eight-day longcase clock signed more correctly 'Walter Archer, Stow on the Wold'. Dated 1711. The style is now a little more formal.*

PLATE 664. *The clock in Plate 663 with the chapter ring removed to show the cartwheel casting. Casting faults have left two of the spokes dangerously thin, the more so as these carry two chapter ring feet. Archer only just managed to find a firm fixing for the left calendar roller.*

PLATE 665. *Detailed view of a corner of the dial of Plate 663 from behind showing the calendar roller, the casting fault in the spoke, and the date 1711 engraved on the chapter ring back.*

spike and thirty-hour longcase dials, sometimes with an 'eye' in the pattern making an overall resemblance to four curled dolphins. Occasionally his hook-and-spike and thirty-hour longcase dials offered the alternative of a matted centre (which style he seems to have always used for his more 'formal' eight-day dials). His earlier clocks have a strange half-hour marker pattern of a very simple fleur-de-lis terminating in three dotted circles, a very unusual style and almost unique to this maker (or his engraver).

PLATE 666. *View of the movement of Plate 663 dated 1711 with the dial removed to show the motionwork and rack strikework, an early use of the latter for a provincial maker. The right-hand winding square has been 'bushed' with a heavy plate.*

PLATE 667. *Oak case of the eight-day clock in Plate 662 standing a little under 7ft. About 1700-1710. The slim pillars have turned knops in their length, bamboo fashion*

PLATE 668. *Oak case of the eight-day clock in Plate 663 made in 1711. The domed top with a square dial is an unusual configuration, but occasionally found at this period in the Glou-cestershire/Oxfordshire region. Note double-D mould to door lip.*

PLATE 670. *Thirty-hour single-handed longcase signed 'Isaac Hadwen Fecit', c.1710, with the usual verse corner, housed in its original oak case in the simple cottage style of the period. Height 7ft. (Photograph by courtesy of Phillips North West.)*

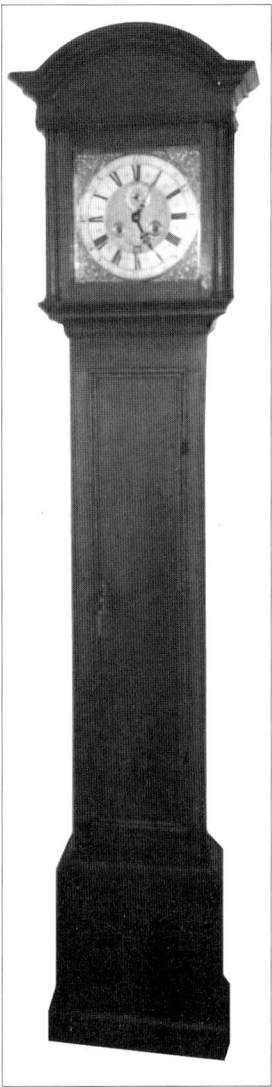

ISAAC HADWEN OF SEDBERGH AND KENDAL

Isaac Hadwen was born in 1687 into a prosperous Quaker family who lived at Side in the parish of Sedbergh in the far north-west corner of the county of Yorkshire, close to the border with Westmorland. Both his parents died when he was only ten years of age and his apprenticeship must have been arranged by his guardians. In 1699 his elder stepsister, Alice, married clockmaker Thomas Savage of Clifton, near Penrith, and it is possible that Isaac learned clockmaking under him, which would have been during the period 1701 to 1708. However, based on the evidence of his clocks, it seems to me more likely that he was trained under Quaker clockmaker John Ogden of Bainbridge, about twelve miles east of Sedbergh (see page 383). His earliest clocks very much resemble those of Ogden and are unlike those by Thomas Savage.

Like many Quakers, young Hadwen felt the urge to travel to spread the faith. Record survives of his attending meetings as far away as Edinburgh in 1711, but he was later to travel twice to America in this cause. It appears that he had enough financial resources to give him the freedom to do this, especially after he married a well-to-do bride in 1714, Sarah Moore, daughter of a prominent local Quaker.

If Hadwen was apprenticed under Ogden, that training would have ended about 1708. How soon after that he set up on his own as a clockmaker at

PLATE 669. *Thirty-hour ten-inch single-handed longcase dial signed 'Isaac Hadwen' without place name and probably one of his earliest clocks, c.1715. Diamond half-hour markers, verse corners, matted centre and style of hand all copy John Ogden's work. Excellent engraving and work – chapter ring on feet rather than riveted. (Photograph by courtesy of David Firth.)*

PLATE 671. *Twelve-inch dial of an eight-day longcase clock signed 'Isaac Hadwen Sedbergh'. He seems not to have used diamond half-hour markers for his eight-day (therefore larger) dials. Tiny holes at XII and VI are not for chapter ring rivets but for pins to hold down the ring during engraving. The corner verse now reads 'The motion's tip' instead of his earlier 'Ye'. Fine original steel hands. The winding squares are filed into four leaf clover shapes, a touch of quality. Date about 1715.*

PLATE 672. *Original oak case of the eight-day clock in Plate 671, probably reduced in the base. Note the strangely small lenticle glass, the typical heavy topmould overhang with cushion mould below it. Internal iron blacksmith hinges. The shaping of the D-mould-edged door top suggests northern styling, but the shaped bottom of the door is an extraordinary quirk of the casemaker.*

Sedbergh is not clear, as a trained clockmaker would often work for a year or two as journeyman for some other master to gain experience. However, we can assume he was in business in a serious way by about 1710 and certainly by the time he married (1714). It may not be coincidence that Ogden himself moved about this very time (between 1711 and 1715) away from Bainbridge further east to Darlington, about thirty-five miles distant of Hadwen. A young clockmaker about to settle down to married life close by was probably a serious commercial threat. It is more than likely that Ogden felt himself forced to move to avoid competition from Hadwen, who made similar clocks for a similar market in a proximity too close for both men to make a living. Moreover, it seems likely that Hadwen had the greater financial resources and could more easily ride out such an economic war.

In May 1718 Hadwen travelled with fellow Quakers to visit 'friends', as members of the faith termed themselves, in America. He probably also went on family business, as his uncle, his mother's brother Stephen Sands, had emigrated to Pennsylvania. Isaac kept a journal of his travels amongst friends in America, extracts of which still survive. He returned to Sedbergh in July 1719, but his American trip seems to have unsettled him, and not long after he moved to live

PLATE 673. *Eight-day longcase dial signed 'Isaac Hadwen Sedbergh' and bearing the initials SSM of the unidentified first owner and wife. Fine engraving and superb original blued steel hands. Note the circular grain to the matting and flower-head winding squares. Date about 1715.*

PLATE 674. *Eight-day movement of the clock in Plate 673 with pull-repeating facility for the three quarters on a single bell (inside the hour bell) driven by a self-contained spring unit, as used in pull-repeat bracket clocks and most unusual in a longcase. Note cartwheel dial and four finely finned pillars. Excellent quality.*

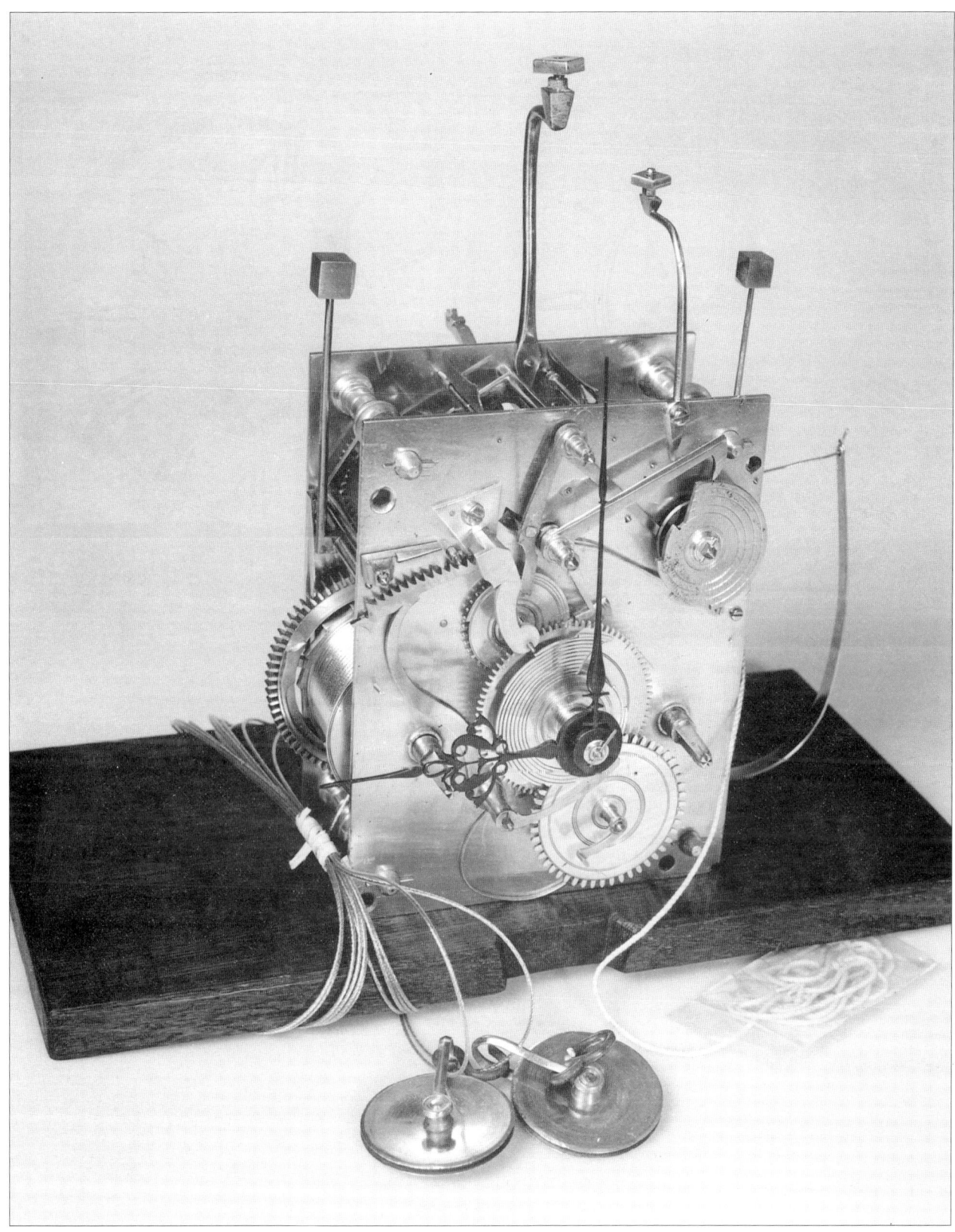

PLATE 675. *Another view of the same movement with the dial removed. The pull-repeat trigger spring is seen on the right. The workmanship is of the highest standard.*

PLATE 676. *Single-handed thirty-hour longcase clock signed 'Hadwen Kendal' and dating from about 1725-30. Note the use of spandrels for the corners. Hadwen's return to using diamond half-hour markers may have been only on thirty-hour work. Late use of a circular datebox.*

PLATE 677. *Eight-day longcase clock with centre seconds (and therefore deadbeat escapement) signed 'Isaac Hadwen Kendal'. Date about 1730. Spandrels seem to have been his normal practice after he left Sedbergh.*

at Kendal at a date I estimated between 1720 and 1722. In fact that tireless researcher Susan Stuart recently discovered a certificate of introduction, a Quaker custom, from the Sedbergh Meeting recommending him to the 'friends' of the Kendal Meeting on 24 April 1722, which pins down closely the date of his move there. He remained at Kendal till about 1735 when he seems to have moved to live at a family property at Gaile (or Gayle) in Tunstall parish, Lancashire, a few miles south of Kendal. In 1737 he again visited America and died of fever at Chester, Delaware, on 29 August of that year.

The dates of Hadwen's comings and goings are important in so far as they help us to pin down developments in his style of clockwork into certain defined periods, something far less easily done with a clockmaker who remained all his life in the one place. Regarding his clocks, I know of none by him at any period other than longcase examples – no bracket clocks, no lantern clocks. His early work is very like that of Ogden of that same period, i.e. about 1710 to 1722. Most of his early ones are ten-inch thirty-hour plated movement longcases, all of them single-handers (Plates 669 and 670) I know of only two pre-Kendal eight-day clocks and these were clearly only made for an occasional special order. Two are signed on the chapter ring simply with his name and no place name, but most are signed 'Isaac Hadwen, Sedbergh'. His earliest clocks have solid dial sheets, but before he left Sedbergh he had changed to those of the cartwheel casting type.

All his pre-Kendal clocks known to me have verses in the dial corners, Ogden-fashion. His commonest verse is Ogden's own: 'Behold this hand, Observe ye motion's tip, Man's precious hours, Away like these do slip'. He very soon changed the 'Ye' to 'the' (as Ogden did), the latter word being used on almost all his verse-corner clocks. He did occasionally coin other verses, which include: 'Winged time, Will not stay, Fleeting hours, Glide away'; 'Minutes glide o'er, Hours spend apace, Time posts away, To end our race'; 'The present hour, Like those before, Once being past, returns no more'.

Once he moved to Kendal his eight-day clocks formed a larger proportion of his output, including some of very high quality. One has a rolling moon

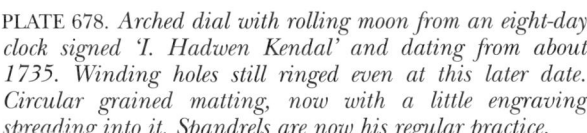

PLATE 678. *Arched dial with rolling moon from an eight-day clock signed 'I. Hadwen Kendal' and dating from about 1735. Winding holes still ringed even at this later date. Circular grained matting, now with a little engraving spreading into it. Spandrels are now his regular practice.*

PLATE 679. *Most unusual eight-day clock signed 'Isaac Hadwen Kendal' and having a ball moon in the arch and half-round brass beading around the dial edge, both of these unusual features being most often associated with Quaker clockmaker Thomas Ogden of Halifax (nephew of John of Bainbridge), who is known to have made such clocks for others in the trade as well as under his own name. It is unclear whether this is Ogden's work or Hadwen's own. Date about 1730.*

(Plate 678), one a centre seconds hand (Plate 677), one a very unusual ball moon in the arch (Plate 679). Clearly he was now making for a wealthier clientele, exactly the same thing as happened with Ogden after his move to Darlington. Another difference of style, however, is apparent in his Kendal clocks from his Sedbergh ones and that is that he now uses spandrels in his dial corners, even on his thirty-hour clocks. This development of style was ultimately followed by Ogden after his move to Darlington, though not quite so exclusively. His Kendal clocks may be signed 'Isaac Hadwen Kendal', 'I. Hadwen Kendal' or just 'Hadwen Kendal'.

Hadwen's last move was to Gaile in about 1735, within a year or two of his final fatal trip to America. I know of only one clock signed there (as 'I. Hadwen Gaill'), a single-handed thirty-hour longcase, harking back to the type he made at Sedbergh. That too has spandrels in the corners (Plate 680).

PLATE 680. *Detail from a thirty-hour single-handed clock signed 'I. Hadwen Gaill' dating from about 1735. He has now dropped the diamond half-hour marker in favour of a more conventional one of the day. The engraved work has now spread further into the matted dial centre.*

423

PLATE 681. *Single-handed dial from Will Snow's clock No. 21, showing the calendar wheel detached. The dial is signed, unusually by monogram only, but his full name, serial number and date are on the calendar disc. Note the simple scrollwork design of the centre. Chapter ring riveted at III and IX.*

PLATE 682. *Calendar disc removed from clock No. 21, showing signature, number and year. Snow often concealed the date on the calendar disc, but the number was more usually on the chapter ring.*

WILL SNOW OF PADSIDE

A clockmaker who numbered his clocks left us an excellent means of assessing the development of his style and therefore of the style of other contemporary clockmakers too. One who dated his clocks would suit the purpose even better. Unfortunately very few did both, and we could hardly expect any clockmaker to decide on a numbering or dating system from his very first days in the trade and stick with that system for what might be as much as the next forty years or more of his working life.

For some years now I have been undertaking research into a clockmaker who originated in the clocksmith tradition, William Snow of Padside, a hamlet of a few houses in North Yorkshire situated a few miles north-east of Otley in the Yorkshire Dales. He was born in 1736 and died in 1795; he numbered most of his brass dial clocks and also dated a few of them. Sometimes on his earlier clocks the number and/or date was hidden behind the dial and might escape being noticed except when the clock was dismantled for cleaning. This factor probably accounts for occasional examples noted without an apparent number, but on most of his clocks the number was normally engraved on the dial after his name. When he turned in the 1770s to making clocks with the new painted dials, he had to buy his dials from a specialist japanner such as James Wilson of Birmingham, and painted dial clocks by him are not numbered, probably because when ordering his dials he could not anticipate the sequence in which he would sell them.

I know of eighty-five numbered brass dial clocks by him, ranging from No. 9 to No. 1299, a large enough number to enable some analytical conclusions to be drawn about his working methods. There are barely half a dozen clockmakers in the land for whom such analysis is possible, so that this gives us a vital insight into the output of a rural clockmaker. It has sometimes been suggested that such high numbers on a clock were invented as serial numbers, as is known to have been the case with watch numbering. In the case of those clockmakers whose output I have studied, this is clearly not so, and the number on the clock was the actual number of the clock in true sequence.

All Snow's clocks are longcase clocks and no other type of clock is known by him. All except one are thirty-hour clocks, the one exception (No. 641) being

PLATE 683. *Mark One type of skeletonised movement from clock No. 21. The pillars and 'backcock' are of steel. Snow's dials seem to be always of the cartwheel type. Here his countwheel is also skeletonised, but later he used solid countwheels.*

a thirty-hour modified with extension wheels by Will Snow himself to run for eight days and still wound by the usual pull-chain system (Plates 686 to 688). The only other eight-day examples I know are alterations, clocks built with thirty-hour duration but converted to eight-day later by the addition of a quite different and unrelated movement. From this we can conclude that for a period of forty years in the second half of the eighteenth century Will Snow sold virtually exclusively thirty-hour longcase clocks.

From 1756 to 1764 he seems to have quite often dated his clocks. After 1764 I know of only one example which is dated, though other 'hidden' dates may yet come to light. It would seem that he tired of dating them quite early in his working life.

Will Snow's thirty-hour movements are of the plated type, but are of an eccentric construction until recently believed to be uniquely his own – but see below for details of how his clockmaking skills began. His plates are of unusual outline and are skeletonised, that is the brass plates have gaps where they do not carry arbors, pillars or other construction work. The purpose of this was for economy of this precious metal. Skeleton plates were much more time-consuming to make and finish, but contained considerably less brass than solid plates. I once weighed the brass in a Will Snow skeleton movement and compared this with a conventional plated example and the difference in cost amounted in prices of the day to one shilling, which at that time was a good day's wages. Snow was willing to spend the necessary additional working time in order to gain an extra day's wages on every clock he made – amounting to the considerable amount of about one extra month's wages per year! Of course, it may have been that he sold his clocks at a lesser price than those of competitors to gain a financial edge, which he might well have needed in

PLATE 684. *Typical later Will Snow dial, No. 750, dating from the 1780s and using the template centre pattern which he kept for many years. His chapter ring engraving of his later clocks also appears to be by the same consistent hand. Chapter ring riveted on the inner edge at III and IX and also on the outer edge at VI and XII. Original hands in blued steel.*

PLATE 685. *An example of a late 18th century arched dial clock by Will Snow with the Mark Two type of skeletonised movement, which has rectangular plates. Pillars and backcock still of steel. Solid countwheel.*

selling clocks to rural farmers who had always managed their lives quite well without them. Snow was the first clockmaker to work in this rural area away from the towns, though he may have offered his wares for sale in local markets, the closest of which was some ten miles distant.

The practice of using skeletonised plates was not unique to Will Snow. Even though this is very unusual practice, there are occasional quite unconnected clockmakers round the land who independently seem to have hit upon the same idea. Henry Mason is one such, whose work is described on page 178. Will Roberts of Otley (mentioned below) is another. Will Snow's movements, however, have very distinctive styling and can easily be recognised.

Snow's skeleton movements were of two types. Initially he made what I call his Mark One version (Plate 683)which had a very ornate and quite graceful outline with projections only where needed for such items as pillars. Later he changed to a rectangular plate form (his Mark Two model, as I call it), still skeletonised but much easier to make and finish. He made his Mark One movements only in his very first years, changing to Mark Two (Plate 685) some time in 1759 or 1760, after which time all his clocks have the later form. His rate of production varied somewhat over the years but ran not far from the national average of twenty-five clocks a year, which I have been able to calculate for half a dozen clockmakers in the land for whom such figures are calculable – roughly one every two weeks.

Production figures for any clockmaker are very difficult to assess. This is partly because his work rate would have varied considerably over the years, as would the numbers of his apprentices or journeymen, if any. In the case of Will Snow he is known to have had his three sons working in the business in his latter years. Any home-based clockmaker normally had his entire family

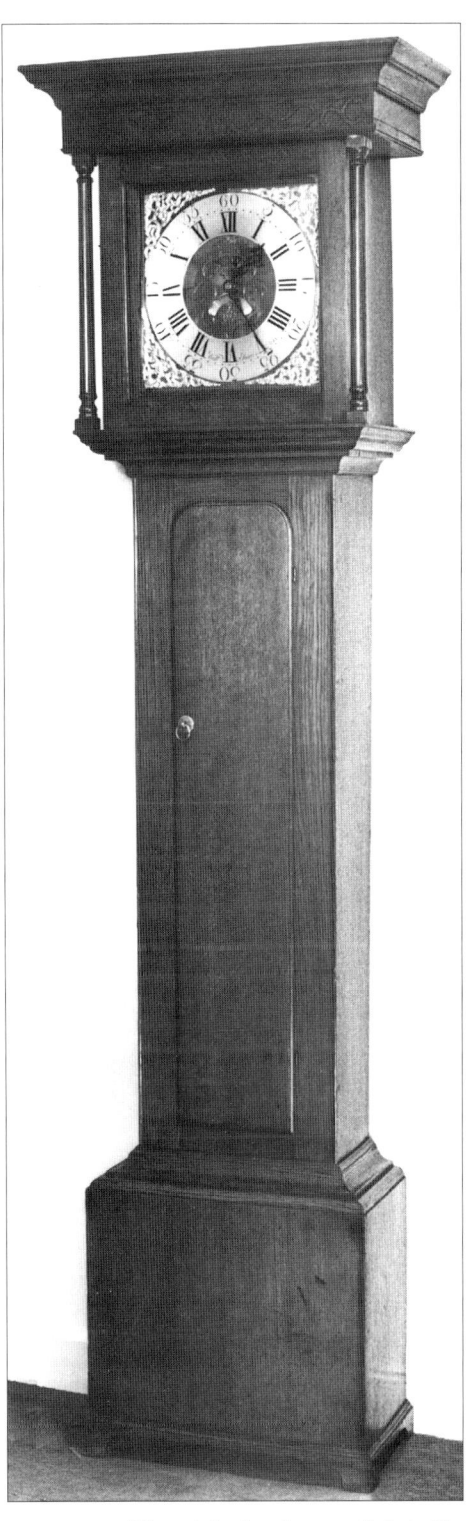

PLATE 686. *Dial of clock No. 641, c.1776, made as one of his standard eleven-inch thirty-hour dials using his template centre pattern. This clock, however, was modified by Snow into a pull-wind eight-day. Chapter ring riveted. Blued steel hands original.*

PLATE 687. *Movement of clock No. 641, here seen from the front with the dial removed, showing typical thirty-hour skeleton movement modified into eight-day duration by means of extension wheels and frames, the latter also of skeletonised form.*

PLATE 688. *The original oak case of clock No. 641 is of a style Snow used for many years, this being the simpler, square-trunked version. His dial mask and the backing to the hood fretwork are both typically painted in blue/green. Height 6ft.6in. Casemaker unidentified.*

PLATE 689. *The oak case of clock No. 750 is an example of his slightly grander style, having crossbanding in mahogany, key pattern moulding to the hood pediment and mahogany quarter-columns.*

428

PLATE 690. *Ten-inch single-handed dial signed 'Will Snow' with No. 17 positioned strangely between XII and I. He had begun to engrave his name here in error and altered the W into N! This dial is of a very different style from that of Will junior and proved to have been made by Will Snow senior in 1731. Original blued steel hand. The matted centre with cup-and-ring decoration did not require engraving skill.*

assisting, including all infants from their early teens. Of course he may well have spent much of his time on other kinds of repair jobs rather than on the actual making of clocks, as we know was the case with most rural clockmakers. His output may therefore have been limited by the number of orders he could obtain and he may have been working for much of his life at less than full production capacity.

The proportion of steelwork on his clocks was greater than was normal. For example, instead of the usual cast brass backcock, which involved quite a complicated casting, he used a solid steel rod on which to hang his pendulum, attaching his crutch direct to the anchor arbor which was pivoted between the plates. This was easier to make and cheaper, but it meant that adjustments to the pallets required the entire movement to be dismantled as opposed to simply unscrewing the backcock. His movement pillars (and occasionally his dial pillars) were also of steel. Steel was usually harder to work than brass, though for pillars and a pendulum support round steel rod was much more easily made than brass castings. The main point, however, was that steel cost only one tenth of the price of brass. For a rural clockmaker time spent on work cost nothing; materials were costly and meant laying out ready money.

His earliest clocks have a single hand. His first two-hander was made in 1756, but two-handers appear only sporadically amongst his output in gradually increasing proportion until after about 1770 the great majority have two hands. This shows that, although he was familiar with two-handed clocks and quite capable of making them from the time he first set up in business, his demand must have initially been for single-handers – exactly what we would expect in a rural population not yet used to clocks.

As we might also expect, the great majority of his clocks had square dials. I know of only seven with arched dials (the earliest made in 1763), two of which had rolling moons in the arch. Arched dials were always in the minority in thirty-hour clocks, where price was paramount and square dials cost less. Rolling moons (in the arch area, of course) were therefore unusual in thirty-hour clocks.

It seems likely that Will Snow could not engrave, at least in his earliest period. Instead he probably had his engraved work done by an unknown out-

PLATE 692. *The movement from clock No. 17, of Mark One type. Steel pillars, dial feet and backcock. Solid countwheel.*

of-house engraver. His early dial centres have scrolling leaf decoration of a relatively sparse nature on to a polished ground. Clocks numbered between No. 21 (dated 1757) and No. 322 have this sort of floral centre, varying slightly in design but all of the same general concept. By the time he made clock No. 473 he had changed to a quite different engraved centre design with much busier and more detailed scrollwork, a design he kept until the end of his brass dial clockmaking. This later design (Plate 684) was done by using a template through which the features were pricked out in dots, then joined together by 'freehand' engraving. This is obvious from a study of these dial centres where it can be seen that the engraved lines sometimes miss some of the individual dots, something we would not expect from a truly professional engraver and something not seen in his earliest dial centres. It seems possible that Will Snow himself engraved this pattern on his later dials, which would explain how he came to have the same design for about a twenty-five year span – something very unlikely if his engraving over that length of time was done by a number of different engravers.

Until recently it was not known how or where Will Snow learned his very individualistic clockmaking skills. There was a clockmaker working nearby as early as 1713 (William Roberts, a whitesmith of Fewston, near Otley, who lived

PLATE 691. *The calendar ring removed from clock No. 17 (also skeletonised for economy) and engraved with the year of making, 1731, five years before Will Snow junior was born.*

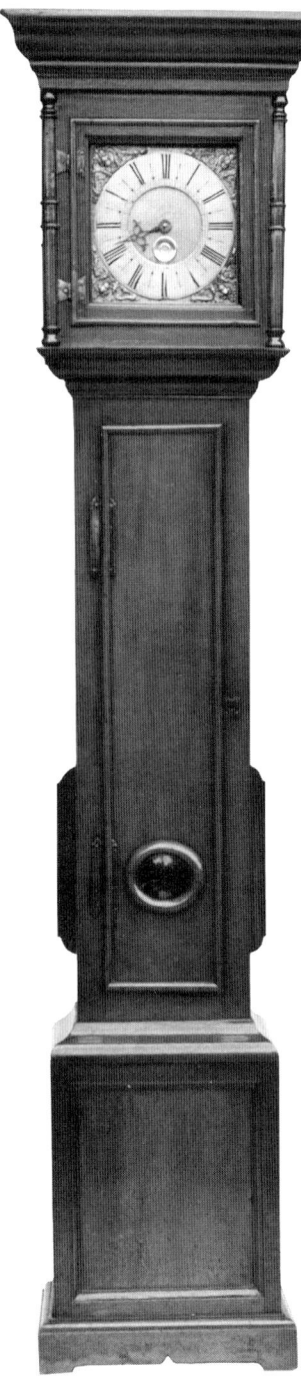

PLATE 694. *The oak case of the 1726 Richard Snow clock stands about 6ft.6in. Sidepiece additions to the trunk are to allow for a wider pendulum swing than was originally intended. Such modifications often occurred after some past restorer adjusted the pallets to give the clock a wider 'healthy' swing.*

until 1773), whose thirty-hour longcase clocks had skeletonised movements and some emphasis on a greater proportion of steelwork to brass, though his movements were plainer and heavier than Will Snow's. Although Roberts was working as a whitesmith by 1713, he was not described as a clockmaker till 1724, and no clocks seem to be known by him earlier than mid-century. However, a solitary clock has come to light recently of an even more primitive Mark One type signed by Richard Snow and dated 1726 (Colour Plate 47, page 400, and Plates 693 and 694). Richard was a yeoman farmer at Padside and the uncle of Will Snow (junior), born about 1700 and dying in 1754 when he left his books to nephew William, the clockmaker.

A further clock is known signed Will Snow No. 17 and dated 1731 (Plates 690 to 692), and another unsigned one is clearly from the same workshop and dated 1730. The identity of this first Will Snow is a puzzle. He may have been the Will Snow, born in 1655 and dying in December 1731 at Padside, who was the father of Richard, and therefore grandfather of Will junior. He was a shoemaker by trade, but may have undertaken clockmaking in his later years. Or the first clockmaking Will Snow may have been the younger brother of Richard, who was born in 1701 and who moved away to Harewood to become a weaver in 1731, presumably making clocks with brother Richard in just his first few years at Padside between about 1720 and 1731. It seems that Will Snow junior must have learned clockmaking from his uncle Richard (who was making occasional clocks in association with the first William Snow (Richard's father or brother). Whether the two earlier Snow clockmakers learned or copied from Roberts of Otley can only be speculation. In the light of the latest findings it seems more likely that Roberts copied from the Snows. The Mark One movement is unique to the Snows, as all Roberts' clocks are based on the Mark Two form.

The clocks made by both William Snow senior and Richard clearly came from the same workshop and used identical components such as spandrels and hands. They differ from those by Will junior in several stylistic respects – not surprising considering the twenty-five year gap between them. Their dials had matted centres with circular ringed calendar holes bordered by cup-and-ring decoration; Will junior's had floral-engraved polished ground centres with mouth calendar. Their spandrel patterns were an earlier type not used by Will junior. Their half-hour markers were based on the diamond mark (otherwise known as a lozenge). The oldest Snow clock yet known is that made in 1726 by Richard (Colour Plate 47, page 400, and Plates 693 and 694). Its movement has a much cruder serpentine outline than the more refined, 'improved' version as used in 1731 by William senior and as continued from about 1756 by Will junior.

The very primitive clocks made in the clocksmith tradition in the 1720s and 1730s by Will Snow senior and his son (or brother?), Richard, and continued in modified form by Will Snow junior, are poles apart from the sophisticated eight-day clocks made at this very same time in London and even in provincial cities such as Leeds. Yet they are a vitally important factor in the overall background of the very beginnings of country clocks, a tradition responsible for a high proportion of rural output though probably the least understood and least documented of any. Amazingly, it is only by the slender thread of evidence of these three clocks of the 1720s and 1730s that we have any inkling of the origins of an entire 'school' of clockmaking practised by Will Snow junior for nearly half a century and continued by his three clockmaking sons and some of their sons until the late nineteenth century (and by the Roberts family too, father and sons). I had a conversation a few years ago only a mile or two from Padside with an old lady, herself a descendant of the Snow family,

PLATE 693. *Movement of the 1726 Richard Snow clock illustrated in Colour Plate 47 (page 400) showing prototype form of skeletonised plates and other individualistic features, as improved later by Will Snow junior.*

who personally remembered the last clockmaker in the Snow dynasty, yet incredibly knew nothing of their origins, which were quite forgotten until I researched their genealogy.

The following is a list of clocks known to me by Will Snow junior with their serial numbers. A = arched dial. Dated clocks have the date in brackets.

9 (1756), 10 (1756), 21 (1757), 30 (1758), 56 (1760), 57 (1760), 59 (1760), 63 (1760), 83 (1761), 96 (1760), 116 (1762), 125, 141 (1763) A+moon, 156 (1763), 171, 172 (1764), 174, 181 (1764), 195, 277, 280, 299, 318, 319, 320, 322, 331, 343A, 349, 380, 390, 401, 410, 413, 418, 421, 432, 444, 459, 473, 495, 502, 509, 530, 533A, 544, 566, 584, 596, 599A, 600A, 618, 625, 627A, 629, 634 (1776), 639, 641, 652, 658, 660, 681, 693, 695, 699, 709, 718, 731A+moon, 745, 750, 758, 765, 766, 791, 813, 839, 846, 851, 884, 923, 937, 949, 997, 1059, 1299.

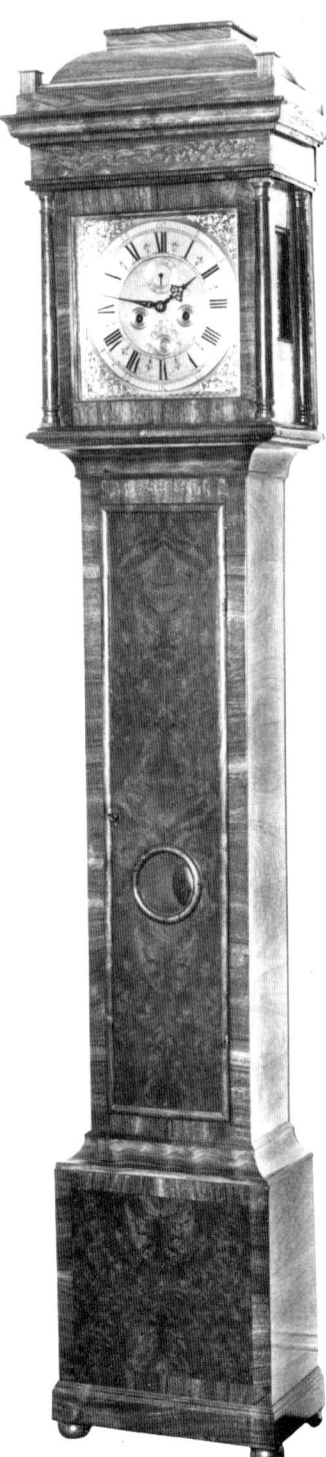

PLATE 695. *Twelve-inch dial of an eight-day longcase clock c.1705 by Joseph Cooper of Whitchurch. Fine workmanship and excellent engraving, all in the London fashion. Here he uses the first form of twin cherub spandrel. The blued steel hands look later.*

JOSEPH COOPER OF WHITCHURCH

Ten years ago the name of Joseph Cooper of Whitchurch was barely known at all, but, intrigued by the surprising quality of one or two of his clocks which came to light, the late Douglas Elliott and I undertook a little research. We now know that Joseph Cooper was working at Whitchurch on the border of Cheshire and Shropshire from at least 1707, when his first child was baptised, and that he died there in 1752. Nothing is known about his origins or training, but his clocks are of exceptional quality, clearly made for the upper market eight-day customer, rather surprising for an almost unknown early clockmaker in a small provincial town.

At least two of his clocks are signed at Malpas, adjacent to Whitchurch, including an eight-day ten-inch single-handed longcase, a most unusual product for any maker anywhere. I know of about a dozen clocks by this maker, all of them of eight-day duration (or longer). He must have made some thirty-hour longcase examples, but none has so far come to my notice, nor any lantern clocks, but a single bracket clock (Colour Plate 48, page 400) came to light recently. It is not surprising that those early provincial makers by whom bracket clocks are known seem to be those who catered principally for the eight-day market.

A selection of his eight-day clocks is illustrated in Plates 695 to 705. All are twelve-inch longcase clocks. All have a form of herringbone border, which was clearly popular with this maker, but each border has a different herringbone pattern. Only his earliest one has half-quarter markers, which he clearly dropped shortly after 1700. His half-hour markers begin with a variation of the trident device but soon evolve into a strange scrolly form of his own, each similar to the

PLATE 696. *Walnut veneered case (on softwood) of the clock in Plate 695. Long D-mould-edged door with lenticle glass, integral hood pillars, hood side windows, caddy top are all features based on the London case of the day. Height about 7ft.*

PLATE 697. *Twelve-inch eight-day dial of about 1710 by Joseph Cooper of Whitchurch, now having the second form of twin-cherub spandrel with large crown. Note his strange scrollwork half-hour markers, a regular feature of his fine quality style. No longer does he mark half-quarters. Blued steel hands probably original.*

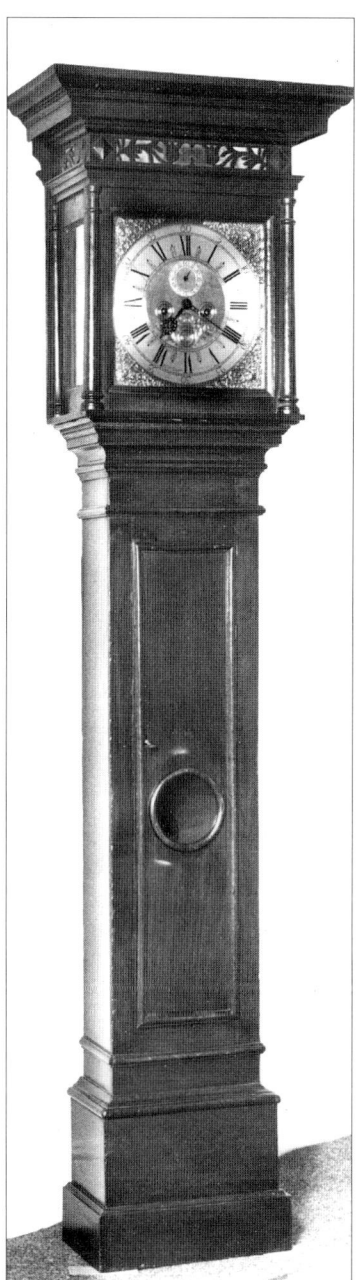

PLATE 698. *Oak case of the clock in Plate 697, standing 6ft.8in. This is clearly a provincial interpretation of the London principle, retaining many of the features of the day yet in country interpretation. Note the decorative lateral mould which breaks up what would otherwise be an unduly wide space above the door top. Top of trunk mould is ogee in shape, an early use of that form.*

PLATE 699. *Twelve-inch eight-day dial by Joseph Cooper of Whitchurch, showing further stylistic progression – a new form of female head spandrel. The date could be 1710-15 but the symbolic rose and thistle beneath the crown may symbolise the Act of Union between England and Scotland passed in 1707, so perhaps the clock dates a little earlier than might first appear. Fine quality.*

PLATE 700. *Delightful primitive oak case of the clock in Plate 699, showing some archaic features such as full-size hood side windows, very wide topmould overhang, wide space above and below trunk door cross-beaded to avoid blandness. The case is original to the clock but perhaps slightly old fashioned in its styling, or is the clock earlier than it might first seem – 1707 rather than 1715?*

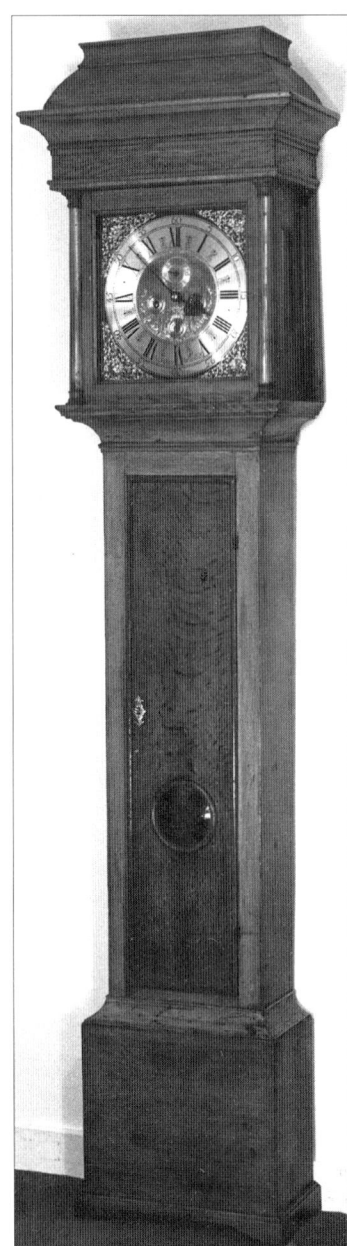

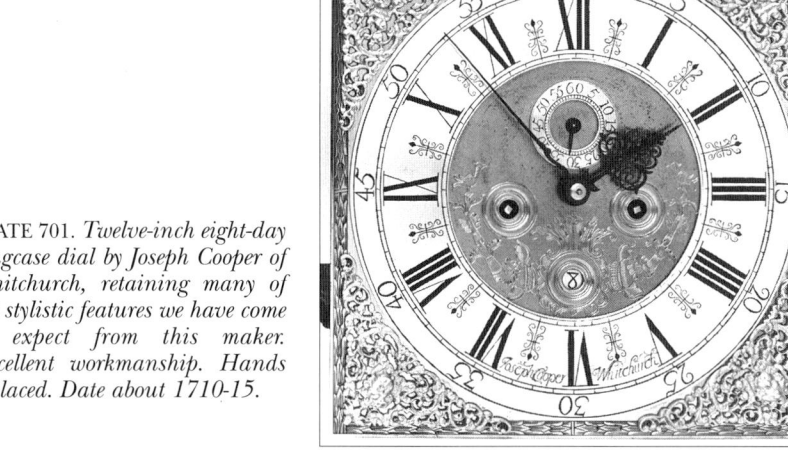

PLATE 701. *Twelve-inch eight-day longcase dial by Joseph Cooper of Whitchurch, retaining many of the stylistic features we have come to expect from this maker. Excellent workmanship. Hands replaced. Date about 1710-15.*

PLATE 703. *The case of the clock in Plate 701 in pine. The door is oak, as was often the case, to help avoid shrinkage on this, the largest, single panel. The case was originally painted or perhaps ebonised, when such a mixture of woods would not show. Caddy top, side windows, integral pillars, lenticle, D-mould-edged door are features based on London styling, though the execution is a little heavier than in London work. Height 7ft.1in.*

PLATE 702. *Movement detail of the clock in Plate 701. Note cartwheel dial sheet (completely unfinished on the back), finned pillars (four in number), fancily shaped hammer spring and decorative hammer stop spring (anti-rattle spring). Most clockmakers did not bother with the latter, hence a sign of persisting quality. Rack striking.*

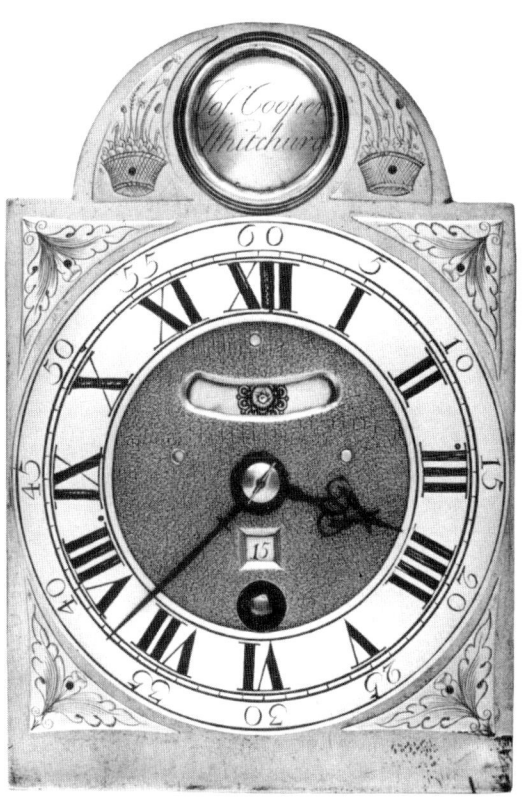

PLATE 704. *Twelve-inch dial of a later eight-day clock by Joseph Cooper of Whitchurch dating from perhaps 1720 or a little later, yet retaining many of his preferred earlier features. The minute numbers are now noticeably larger than before. Blued steel hands believed original. Cartwheel dial. Rack striking.*

PLATE 705. *The dial of the Cooper bracket clock illustrated in Colour Plate 48 (page 400) is an odd shape, being taller than wide (width 4¾in.). Spandrels purpose made in solid brass engraved on the surface – probably because no standard spandrels existed this small. Backplate signed 'Cooper'. Original blued steel hands.*

previous one yet each different. His spandrel pattern changes as time goes by along with the national trend in spandrel pattern development. Minute numerals grow gradually larger with time, again with the national trend. His winding holes continue to be ringed and his calendar box is usually circular and ringed, both these features being retained somewhat later than was the general fashion. The engraving around his calendar box is very variable and usually much more extensive than was the general fashion. His steelwork is usually ornately shaped and filed, his pillars usually finned in the London manner.

Other examples reported to me include a month-duration longcase with *six* latched pillars and an eight-day longcase with five latched pillars and tulip-engraved dial signed at Malpas, both of these having outside countwheel strikework. Clearly these must be amongst his earlier products. He did move on to the rack striking system by 1710 or so (unlike many clockmakers who were late in adopting this change) and this indicates that he was up to date with mechanical developments. He also moved on quite early, perhaps by 1720, to four-pillar construction from his earlier five-pillar method (sometimes even six). This was typical of the attitude of many northern clockmakers, who realised early that five (or more) pillars were unnecessary. The question of latching was a variable one for this maker, who sometimes latched all and sometimes only some of his pillars, a regular method of his being to latch just the centre one, as London makers often did – the purpose being to enable easy assembly. His later clocks are said to be of considerably lesser quality than his earliest ones, and this is to be expected as the degree of ornate shaping of mechanical parts reduced in the work of all clockmakers as the eighteenth century progressed.

PLATE 707. *Rear view of the eight-day dial showing cartwheel casting, hammering marks, a little practice engraving and the lining-up marks to get his calendar square.*

PLATE 706. *Eleven-and-a-half-inch dial of a longcase clock, signed 'Jonas Barber Bouilingbrigdg', probably his first ever eight-day clock. The engraved border to the winding and calendar holes is most unusual, as are the two small spandrels beside the seconds dial. Original blued steel hands. Note the winding squares sit left of centre. Date about 1720.*

JONAS BARBER THE ELDER OF WINSTER

Until recently the name of Jonas Barber was a puzzle to clock enthusiasts, the reason being that there were in fact three clockmakers by that name whose names and dates were hopelessly intermixed. In undertaking research for my book *Westmorland Clocks & Clockmakers* (published in 1974) I managed to unravel the Barber genealogical knot.

Jonas Barber of Ratcliffe Cross, London, was born in 1652 at Menston in the parish of Otley, West Yorkshire, youngest son of a blacksmith. He was the first of the Barber family to take up clockmaking, moving to London by 1672 and working there till his death in 1698. His kinsman John Williamson of Leeds may have worked with or for him – see page 371. Jonas Barber of Ratcliffe Cross made high quality (London) clocks, mostly eight-day longcases, but one lantern clock is known. Although he was born into a rural blacksmithing family, his work indicates that he was trained in the strict and highly stylised discipline of London clockwork.

Jonas' elder brother, Thomas Barber, followed his father's trade as a blacksmith. His second brother, John, born about 1645, became a clockmaker at nearby Skipton (West Yorkshire), where he was working by 1685 and was last heard of in 1700. However, no clock is known to exist bearing the name of John Barber of Skipton and it may be that he spent his life anonymously working as a journeyman to some other clockmaker.

It was at Skipton that I finally uncovered the baptism in 1688 of Jonas Barber the second, son of John and nephew of Jonas Barber of Ratcliffe Cross, London. Where he trained we cannot say, but his uncle had died in London by the time he was ten years old and his father probably died before he was much older than twelve, so any influence they had must have been before he was of the usual age of apprenticeship, which was thirteen or fourteen.

PLATE 708. *The eight-day movement with the dial removed showing the four latches holding the movement pillars. Note internal countwheel for strike control. The S-shaped lever is a trip for re-setting the countwheel when needed. The hour-pipe wedge is to drive the twelve-hour knock-on 'Westmorland' calendar wheel.*

Jonas Barber set up his first workshop at Bowland Bridge, a hamlet in Westmorland, about the year 1708-1710 when he was aged between twenty and twenty-two. Why he went there remains unknown, as do the places of his training and residence during the preceding ten years or so, though we do know he was at Ulverston in Lancashire in 1707 and it is just possible he worked there under some other clockmaker. (The only one recorded in that area then was Thomas Townson of Pennington, near Ulverston, blacksmith and clocksmith, known only through a single, very crude posted-movement thirty-hour longcase clock dated 1704. He is believed to have worked there from at least 1702 till his death in 1744.)

Barber worked at Bowland Bridge (from which place he married in 1717)

PLATE 709. *Primitive nine-and-a-half-inch dial with cup-and-ring centre of a single-handed thirty-hour longcase clock signed simply 'Jonas Barber'. Meeting arrowheads form the half-hour markers. The steel hand looks later. Probably from his Bowland Bridge period, c.1715-20.*

until 1727 when he moved to the village of Winster not far away. There he worked till his death in 1764, a self-styled 'clocksmith', succeeded there by his son of the same name. His work begins very much in the clocksmith style of simple, rustic thirty-hour clockwork, well made but modest.

Jonas Barber's earliest clocks (both those made at Bowland Bridge and at Winster) were mostly signed without any place name. The reason for this practice, common with many of the first rural clockmakers, is not known with certainty, but it is thought it was because the place name of a small hamlet would go unrecognised at a distance of more than a few miles and a clockmaker who might take his goods to sell in the markets of more distant towns might not want to pin himself down to a single location.

A few of his Bowland Bridge clocks *did* carry the place name, including the one illustrated in Plate 706 which spells the place erratically as 'Bouilingbrigdg' and is his earliest eight-day clock yet known. To date only four clocks have come to light bearing that place name. Two are thirty-hour examples, two are eight-days. One thirty-hour spells the place 'Boulanbridg' and clearly Barber had trouble with his spelling. One nine-inch by eight and a half-inch dial thirty-hour clock, known to us today only through an old record of it, has the engraved corner verses 'Behold this hand, Observe the motion's tip, Man's precious hours, Away like these do slip'. This verse was coined and first used in the 1690s by John Ogden of Bainbridge (see page 383), and it is intriguing to wonder how Jonas Barber came to know of it and to copy it. It seems unlikely that Barber worked with Ogden, as his work seems to have no other Ogden influence but this one instance of the Ogden verse.

His earliest eight-day clock (Plate 706) has a dial eleven and a half inches square of the cartwheel casting type. His calendar ring (now lost) was obviously a spoked version of the Westmorland type and would have made winding impossible on occasional days during the month as the passing spoke obscured the winding square. This was an inherent problem when using this type of calendar on eight-day work (found also on John Sanderson's clock in Plate 635) and was solved by most clockmakers by using the twenty-four hour intermediate drive wheel and a large inner-toothed ring, as in London work.

PLATE 710. *Ten-inch single-handed thirty-hour longcase dial signed 'Jonas Barber' shown with the chapter ring detached to reveal the date 1742 on the dial sheet spoke. By now his centres are matted with a circular grain. The spoked Westmorland calendar wheel shows well.*

PLATE 711. *Ten-inch single-handed thirty-hour longcase dial signed 'Jonas Barber' not unlike the previous one and probably dating from the 1740s. The blued steel hand is original. Note the circular matting 'grain'.*

His movement pillars on this eight-day clock (and on three other thirty-hour clocks of his earliest period) have latches on all four pillars. Later this maker (and his son after him) began the unusual practice of alternate latches and pins, that is two alternate pillars latched and two pinned. Alternate pins/latches were such an unusual combination that this was almost a Barber trade mark, though one or two other clockmakers elsewhere did independently hit upon this same method.

This eight-day clock was probably Barber's first. Although he marked out most carefully his dial plate and train layout, in the end he got it slightly wrong and his winding squares sit left of centre within the winding apertures – the dial photograph exaggerates this somewhat because of the camera position. His dials were virtually always of the cartwheel cast type, though two or three examples are known with solid dial sheets, principally, as we might expect, from his Bowland Bridge period.

At all periods the great bulk of his work consisted of plate-movement thirty-hour longcase clocks. No posted-movement clock is known by him, perhaps not surprisingly, since by the time he became established the posted movement was passing from fashion in the North. No bracket or lantern clocks are known by him – again not surprising in view of his location and period. The earliest of his thirty-hour clocks have nine-inch dials (sometimes a fraction under), followed by ten-inch dials. Some of his earliest ones use the cup-and-ring decoration principle (Plate 709), an easy alternative to the usual tulip-style engraved centre of the period for a clockmaker unable to engrave. Later he used a matted centre (Plates 710 and 711). In the late 1740s he set about numbering his clocks, the number, and sometimes the year too, engraved on the movement frontplates, and he is one of the earliest provincial clockmakers to do this.

APPENDIX

THE 'A' TEAM

A list by county of those clockmakers known to have been working by 1700 by whom household clocks are known to survive (whether the surviving work was signed at that place or at a previous or subsequent place of work, and whether or not that surviving work itself pre-dates 1700). Some makers who moved around may therefore appear in more than one county. The lists do not include makers whose place of work has not been identified. Not counted are those known only for turret clocks or only for watches or by record only. Surviving examples of work by some makers have been erroneously dated too early (for example in Britten, where a number of eighteenth century lantern clocks are dated to the mid-seventeenth century) and these have been omitted. Dates for some makers may be approximate but proving pre-1700 working. Clocks merely 'estimated' at c.1700 have largely been omitted as unproven. Those clockmakers known to have been London trained are so designated. All dates are working dates unless otherwise specified.

```
Abbreviations and symbols
* = longcase clock maker
Q = Quaker
b. = born
c.= circa (i.e. estimated working dates)
d. = died
m. = married
lc(s) = longcase clock(s)
br(s) = bracket clock(s)
lant(s) = lantern clock(s)
```

BEDFORDSHIRE
*William Carter, Ampthill (?ex London). 1698-d.1737. lcs & brs.
*Jonathan Chambers, Shefford. 1669-93. lc.
James Ford, Bedford (Goldington). 1684-d.1697. lant.
Thomas Palmer, Shefford. 1694-d.1732. lant.
*Thomas Russell, Wootton (ex London 1691+). 1700-d.1745. lc.
William Tarry, Woburn. m.1692-1699 lant. (Maybe later
   Saffron Walden, Essex, where one such m. 1702-d.1729.)
Ex London: 2
Total 6. 3 made lanterns, 3 longcases, 0 both, 1 brackets.

BERKSHIRE
*John Davis, Windsor (?ex London). 1678-1689. lant & lc.
*William Flagget, Newbury. 1680-1707. lc.
*John Hocker, Reading. 1682-d.1729. lant & lcs.
John Holloway, Newbury (ex Devizes?). 1680s-1690s. lant.
John Knapp, Reading (ex London). c.1690-c.1730. lants & br.
*Joseph Norris. Abingdon (ex London). 1698, d.1727.
   lants, lcs & brs.
Bryan Rumball, Newbury. m.1670, d.1685. lants.
Ex London: 3.
Total 7. 6 made lanterns, 4 longcases, 4 both, 2 brackets.

BUCKINGHAMSHIRE
*Richard Carter, High Wycombe. c.1690-c.1700. lcs.
George Chandler, Wingrave. b.1654 Drayton Parslow,
   d.1729. lant.
*John Ford, Aylesbury (ex Oxford,qv). 1712-1725. lc.
Thomas Ford, Buckingham. c.1680-c.1710. lants.
*Joseph Harley, Wingrave. ?(dubious) late 17c. lc.
*John Hill, Risborough. late 17c? lc.
*John Knibb, Hanslope. (ex Oxford, lant & brs). lc.
Thomas Seward, Newport Pagnell. 1695. lant.
*John Welsh, Chesham. c.1700. lant & lc.
*Hugh Willis, Stewkeley, ?b.1656-1680s. blacksmith & lc.
Ex London: 0
Total 10. 5 made lanterns, 7 longcases, 2 both, 1 brackets.

CAMBRIDGESHIRE
Thomas Loftus, Wisbech. late 17c. lant.
Ex London: 0
Total 1. 1 made lanterns, 0 longcases, 0 both, 0 brackets.

CHESHIRE
*John Buck, Chester. m.1662-1679+. lants & lc.
*George Clayton, Marple. c.1680-d.1716. lcs.
*John Hough, High Leigh/Knutsford. c.1695-1728. lcs & brs.
John Kent, Congleton. late 17c. lant.
Thomas Poole, Audlem. (17c?) lant.
*John Smallwood, Macclesfield/Chelford. 1685-d.1715. lc.
*Gabriel Smith, Barthomley/Nantwich. b.1656-d.1741. lant, lcs & brs.
*Thomas Talbot, Nantwich. 1695-d.1717. lant & lcs.
*Edward Wrench, Chester(Gloverstone). free & m.1694-d.1714. lcs.
Ex London: 0
Total 9. 5 made lanterns, 7 longcases, 3 both, 2 brackets.

CORNWALL
Zero.

CUMBERLAND
*Aaron Cheasbrough, Penrith. 1689+. lcs.
*Q John Sanderson, Wigton. 1691-1740+. (lants?) & lcs & brs.
Ex London: 0
Total 2. 1 made lanterns, 2 longcases, 1 both, 1 brackets.

DERBYSHIRE
*Joseph Kirk, Harstoft & Clay Cross (later Nottingham & Skegby). 1690s-1740s. lcs & br.
*Daniel Tantum. Derby. c.1700. lcs & br.
*Francis Tantum, Loscoe (?& Derby). c.1700. lcs.
*John Young, Derby. c.1700. lc.
Ex London: 0
Total 4. 0 made lanterns, 4 longcases, 0 both, 2 brackets.

DEVON
Thomas Adams, Plymouth. c.1685. lant.
John Bennett, Plymouth. 1668. lants.
*Edward Clement, Exeter (?ex London). c.1690s-d.1720. lcs & lant.
*William Clement, Totnes (?ex London). 1695-d.1736. lcs & brs.
*Q. Abel Cottey, Crediton. 1690s (to Philadelphia 1700 lc). lant.
Q. Arthur Davis, Tiverton/Westleigh/Cullompton/ Kentisbeare. c.1700-1723. lant & br.
*Ephraim Dyer, Bideford. 1689-1720s. lcs
*Ambrose Hawkins, Exeter (ex Wells 1692-5). d.1705. lants & lcs.
*William Hunt, Exeter. 1697-1730s. lc.
*William Kent, Barnstaple. ?c.1700. lc.
*Lewis Pridham, Sandford. 1687-d.1749. lant & lcs.
Joseph Reepe, Plymouth. c.1650-c.1670. lant.
John Reynolds, Barnstaple. late 17c. lant.
Thomas Savage, Exeter. b.1644 (ex London?)-1690s. lant.
Walter Trout, Exeter. 1697. lants.
Ex London: 3.
Total 15. 11 made lanterns, 8 longcases, 4 both, 2 brackets.

DORSET
*Q? Thomas Baker, Blandford. (?m.1683) c.1690-c.1715. lcs.
*Lawrence Boyce, Puddletown. 1690s-d.1738. lcs.
*Henry Bunston, Lyme Regis. m.1691-d.1732. lc & lant.
Ralph Cloud, Beaminster. b.1663-1714. lant.
James Gray, Shaston (Shaftesbury). 1688-1695. lants.
Joseph Gray, Shaftesbury. 1680s-1690s. lant.
Nicholas Hancock, Shaftesbury. 1690s-1702+. lant.
John Hoskin, Lyme Regis. 1685. lant.
*Q Richard Howe, Dorchester. b.1666-d.1713. lc.
James Michell, Chardstock. 1670s. lant.
*John Michell, Chardstock. b.1675-1690s. lant & lc.
John Mintern, Beaminster. late 17c-1712. lant.
*Q James Norman, Charminster. 1690s-1730s. lc.
*Richard Pinney, Beaminster. m.1661-d.1696. lc.
*Q William Smith, Dorchester. 1670s-1730. lc.
Richard Stephens, Bridport. late 17c. lant.
EX LONDON: 0
Total 16. 10 made lanterns, 8 longcases, 2 both, 0 brackets.

DURHAM
*Ninyan Burleigh, Durham. 1705 (from London & Pontefract?). br, lc & lant.
Ex-London: 1.
Total 1. 1 made lanterns, 1 longcases, 1 both, 1 brackets.

ESSEX
*William Bacon, Colchester. c.1680-?. lants & lcs.
Thomas Clay, Chelmsford. c.1650?. lants.
John Fordham, Dunmow (?ex London). late 17c-c.1720. lant.
*John Groome, Colchester.c.1658-?d.1691. lcs.
*Thomas Holborough, Colchester. 1698-to Ipswich 1706, d.1727. lcs.
John Lee, Ashen. c.1690. lant.
Q Stephen Levitt (ex Sudbury), Chelmsford. 1689-m.1708. lant.
*Q Jeremy Spurgin, Colchester. m.1690-d.1699. lant & lc.
*John Stevens, Colchester (ex London). 1691-c.1710. lant & lc.
James Wheeler, Colchester. free 1627-c.1635. lants.
Ex London: 2.
Total 10. 8 lanterns, 5 longcases, 3 both, 0 brackets.

GLOUCESTERSHIRE
*Walter Archer, Stow-on-the-Wold. b.1674-1742. lant & lc.
John Bicknill, Cirencester. 1679-d.1704. br.
John Holloway, Stroud. 1690s?-d.1711. lant.
*William Holloway, Stroud. (b.1638) 1658-d 1693. lant & lc.
Jasper Lugg, Gloucester. b.1656-d.1685. lant.
*S. Moore, Tewkesbury. c.1690. lc.
*Joseph Thompson, Cirencester. b.1667-d.1739. lc
*George Voyce, Dean. 1694-d.1722. lc.
*John Washbourne, Gloucester. late 17c-d.1731. lcs.
William Wood, Nailsworth (& Avening). 1698-d.1723. lants.
Ex London: 0
Total 10. 5 made lanterns, 6 longcases, 2 both, 1 brackets.

HAMPSHIRE
*Elias Bernard, Southampton. c.1700. lc.
*Francis Harding, Portsmouth (ex London post 1687). c.1700. lc.
*Peter Knibb, Farnborough (ex London). 1679. br cl.
Ex London: 2
Total 3. 0 made lanterns, 3 longcases, 0 both, 1 brackets.

HERTFORDSHIRE
*Francis Berry, Hitchin. 1692-c.1700. lant & lc.
Humphrey Clarke, Hertford (ex London 1680+). late 17c lant.
John Justin, Watford. c.1700. br.
*Zachariah Mountford, St. Albans (ex London 1704+). lant & lc.
John Read, Bishops Stortford. late 17c. lant.
*John Walthall, Hatfield (ex London post 1691). c.1695. lc.
Ex London: 3
Total 6. 4 made lanterns, 3 longcases, 2 both, 1 bracket.

HUNTINGDONSHIRE
*Phillip Carter, Huntington. c.1685-95. lcs.
Ex London: 0
Total 1. 0 made lanterns, 1 longcases, 0 brackets.

KENT
*John Bishop, Maidstone. late 17c-d.1710. lant & lcs.
*John Brookstead, Tonbridge (ex London). c.1695. lc.
*Joseph Cottam, Maidstone. free 1694-1698. (hood cl)=lc.
*Thomas Deale, Ashford. m.1680-d.1687. lc.
John Dodd, Faversham. m.1693. lant.
John Greenhill, Ashford. 1680s-d.1706. lant & lcs.
*John Greenhill, Maidstone. m.1680-d.1712. lant & lc.
Richard Greenhill, Ashford. d.1687. lants.
*Richard Greenhill, Canterbury. free 1676-d.1705. lc.
Edward/Edmund Gregsby, Sittingbourne. c.1673-c.1685. lant.
*John Knight, Faversham. c.1700. lc.
*Simon Lamb, Rochester (ex London 1670+). 1677-c.1700. lc.
Thomas Shindler, Romney. free 1676-1728. lant.
Ex London: 2
Total 13. 7 made lanterns, 9 longcases, 3 both, 0 brackets.

LANCASHIRE
*John Ball, Liverpool. m.1699-d.1716. lc.
*James (& John?) Barton (JB), Ormskirk. free (Chester) 1696-d.1718. lants & lc.
*George Battersby, Manchester. m.1694. lc.
*Richard Breckell, Holmes/Lancaster. 1690s-d.1724. lants & lcs.
*Thomas Bulman, Liverpool (ex Newcastle). 1701+. lcs & br.
*Q John Cooper, Warrington. 1698. lcs & br & lant.
*Robert Davis, Burnley, 1695-1723+. lcs.
*William Grice, Cronton (Warrington) (?ex London). c.1700. lc.
John Lyon, Warrington. 1668-1678. lant.

Joseph Pryor, Liverpool. 1676-d.1719. br.
*John Waller, Preston. c.1689-d.1724. lc.
Henry Webster, Aughton. 1680s-d.1697. lants.
*James Whittaker, Middleton (Manchester). late 17c-d.1720. lcs.
*James Worthington. Warrington. 1699-d.1749. lc.
Ex London: 1
Total 14. 5 made lanterns, 10 longcases, 3 both, 3 brackets.

LEICESTERSHIRE
William Bates, Leicester. b.1671, free 1692. lant & lc.
Christopher Carter, Galby. b.1676-c.1700. lants.
*Thomas Clarke, Husband's Bosworth. c.1700. lc.
*Roger Lee, Leicester. b.1670, free 1691-d.c.1730/34. lant & lcs.
*Joseph Pickering, Lutterworth. c.1700. lc.
Nathaniel Rogers, Melton Mowbray. c.1690. lant.
John Spence, Leicester. 1688. lant.
*John Wilkins, senr, Leicester. b.1639, free 1660-d.1721. lant & lc.
*Thomas Wilkins, Leicester. free 1698-c.1740. lc.
*John Worth, Humberstone. b.1656-1702. lc.
Ex London: 0
Total 10. 6 made lanterns, 7 longcases, 2 both, 0 brackets.

LINCOLNSHIRE
*Boniface Bywater, Stamford. c.1700-d.1752. lc & br.
*John Ingram (I), Spalding. 1674-d.1715. lcs.
*John Stockeld, Lincoln. c.1700-d.1750. lcs & br.
*John Watts, Stamford (?ex London) (?earlier at Northampton). c.1690s-d.1719. lant & lc.
Ex London: 1?
Total 4. 1 made lanterns, 4 longcases, 1 both, 2 brackets.

NORFOLK
Peter Amyot, Norwich. c.1660-1720. lants.
William Barlow, King's Lynn. late 17c. lant.
*Jeremiah Hartley, Norwich. free 1706-d.1717. lant & lcs.
Nicholas Howard, Yarmouth. c.1690. lant.
*Daniel Manley senr., Yarmouth (ex London). 1686-d.1701. lant & lcs.
*Samuel Reeve, Diss (ex Suffolk?). late 17c. lants & lcs.
*Thomas Reeve, Harleston. late 17c, m. 1700-1734. lant & lcs.
William Shaw, Diss. supposedly 1676 (dubious). lant.
Thomas Tue, Kings Lynn. 1640s-d.1710. lants.
John Smith, Kings Lynn. 1610. lant.
Ex London: 0
Total 10. 10 made lanterns, 4 longcases, 4 both, 0 brackets.

NORTHAMPTONSHIRE
*Timothy Goodman/mer, Towcester. c.1680 (or later?). lant & lcs.
*Sampson Jackson, Thorpe Malsor (and Newton and Woodford) (ex London). (b. Countesthorpe, Leics, 1639) 1678-d.1711. lcs & lants.
*Richard Martin, Northampton. late 17c. lc & lant.
*Thomas Power, Wellingborough. c.1680-d.1709. lants & lcs.

*John Pursell, Towcester. c.1700. lc.

D(aniel) Robinson. Northampton (?ex London). c.1700. lant.

*Henry Simcock, Daventry. 1693-1716. lcs.

Joseph Townsend, Hellidon. late 17c. lants.

Francis Wright, Oundle. 1674-1681. lants.

John Wright, Mansfield. late 17c (?d.1709). lant.

Ex London: 2.

Total 10. 8 made lanterns, 6 longcases, 4 both, 0 brackets.

NORTHUMBERLAND

*Thomas Bullman, Newcastle (& London?). 1690-1695, then L'pool. lcs & brs.

*Abraham Fromanteel, Newcastle (ex London). late 17c. lant & lcs.

*Robert Ilderton, Alnwick. 1680s/Newcastle. 1707+. lc.

*George Mills, Alnwick/Sunderland. 1692-c.1710. (then Ripon). lc.

*Samuel Ogden (ex Halifax, where b.1669-1712), Benwell (N'castle). d.1728. lant & lc.

*William Prevost, Newcastle (ex London). 1680s-90s. lcs.

William Shafto(e), Newcastle. m.1664-89. lant.

*Deodatus Threlkeld, Newcastle. 1680s-d.1732. lcs & brs.

Ex London: 2(or 3?).

Total 8. 3 made lanterns, 7 longcases, 2 both, 2 brackets.

NOTTINGHAMSHIRE

John Flint, Mansfield. c.1660. lant.

*Owen Gascoigne, Newark. b.1647, 1670s-d.1719. lc & lant.

*Joseph Kirk, Harstoft, Clay Cross (both Derbys), Skegby, Nottingham. 1690s-1740ish. lcs & br.

Richard Roe, Epperstone. 1680-d.1720. lants.

*D(aniel) Tantum, Nottingham. c.1700. lc. & br.

John Wright, Mansfield. late 17c-d.1709. lant.

Ex London: 0

Total 6. 4 made lanterns, 3 longcases, 1 both, 2 brackets.

OXFORDSHIRE

*Samuel Aldworth, Oxford. 1689-1690s, then London. lcs, lant & br.

*Michael Bird (snr), Oxford (ex London). 1654-d.1689. lcs, lant.& br.

*John Ford, Oxford. 1691-1708. (then Aylesbury). lcs.

*Thomas Gilk(e)s, Sibford Gower. 1680s+-1743. lcs.

*Curtice Greenaway, Oxford (ex London). 1699-d.1702. lant & lc.

George Harris, Fritwell. 1640s-d.1694. lants.

Nicholas Harris, Fritwell. 1670s-d.1738. lant.

Robert Harvey, Oxford. 1588+ lants

William Kenning, Banbury (ex London). 1674. dated lant.

*John Knibb, Oxford/Hanslope. 1660s-d.1722. lants, lcs & brs.

*Joseph Knibb, Oxford. lants,lcs & brs.

Benjamin Lamprey, Banbury. 1696 d.1721. lant

John Quelch, Oxford. 1663 -d.1699. lant.

George Walker, Oxford. 1689. lant.

Ex London: 3

Total 14. 12 made lanterns, 7 longcases, 4 both, 4 brackets.

SHROPSHIRE

Francis Rowley, Bridgnorth. c.1690-d.1702. lant.

*Richard Savage, Shrewsbury. 1692-d.1728. lant & lc.

Ex London: 0

Total 2. 2 made lanterns, 1 longcases, 1 both, 0 brackets.

SOMERSET

Thomas Bilbie, Chewstoke. 1660s. lant.

Thomas Browne, Bristol. free 1643-d.1680. lant.

Robert Champion, Wells. c. 1630. lant.

John Clarke, Bristol (1 and/or 2). free 1650-1679. lant.

William Cockey, Warminster (Wilts.) & Wincanton. 1690s-1730. lant.

*John Culliford, Bristol. free 1692-d.?1718. lc & lants.

Joseph Curtis, Chew Magna (& Axbridge?). c.1680-1720?. lants.

Lawrence Debnam, Frome. c.1690. lant.

*James Delaunce (ex London 1685+) Frome c.1690, later Downton (Wilts). lant & lcs.

Richard Gilbert, Bristol (ex London). 1673-1711. lant.

John Harford, Bath. c.1658. lant.

*Ambrose Hawkins, Wells (later Exeter). 1690-1692 (d.1705). lants & lcs.

John London, Bristol. free 1675-c.1690. lants.

*William Martin, Bristol. m.1689-1696. lcs & lants.

Robert Mills, Taunton. ?late 17c. lant.

John Mogg, Shepton Montague. c.1690. lant.

Samuel Turner, ?Bristol. late 17c. lant.

Thomas Veale, Chew Magna/Bristol. free 1652-1697. lants.

*Solomon Wasson, Bristol. free 1642. one lant & lcs.

*Edward Webb (sig'd Church Stoke) Chewstoke. 1678-d.1694. lant & lcs.

James Webb, Bristol. c.1680-1711. lant.

John Webb, Wells. 1684. lant.

Mathew Webb, Chewstoke. 1688. lant.

William Webb, Taunton. c.1695. lant.

John Westover (?Wedmore). 1675. lant.

*Benjamin Willoughby, Bristol (ex London 1691). free 1693-1700+. lcs, lants & brs.

Ex London: 3.

Total 26. 26 made lanterns, 7 longcases, 7 both, 1 brackets.

STAFFORDSHIRE

Ralph Barber, Lichfield. late 17c. lant

*Peter Stretch, Leek. b.1670 (to Philadelphia 1702). lc & lant.

Samuel Stretch, Leek. b.1657-1692 (to Philadelphia c.1711). lants.

Thomas Swinnerton, Newcastle under Lyne. free 1674-d.1708. lants.

Ex London: 0

Total 4. 4 made lanterns, 1 longcases, 1 both, 0 brackets.

SUFFOLK

Thomas Baley, Bury St. Edmunds. late 17c. lant.

*Isaac Blowers, Beccles. m.1690-d.1719. lants & lcs.

John Chamberlain, Bury St. Edmunds. (b.1644) c.1670-d.1726. lants.

Francis Coleman (senr), Ipswich. 1655-d.1709. lant.

*Richard Copping. Bury (ex London?). m.1662-d.1689. lant & lcs.

*William Goodwin, Stowmarket (ex London 1682). 1688-d.pre 1734. lants & lc.

*Mark Hawkins (I), Bury St. Edmunds. c.1694-d.1750. lcs & lants.

*Thomas Holborough, Ipswich (ex Colchester 1690s). 1706-d.1727. lcs.

Isaac Hurst, Bury St. Edmunds (ex London?). c.1680. lant.

Stephen Levitt, Sudbury. 1680s (Chelmsford post 1689-m.1708). lant.

*Richard Marsh, Ipswich. b.1636, m.c.1670-d.1707. lc.

*Henry Mayhew, Parham (& Hacheston). 1686-d.1720. lant & lcs,

John Maynard, Long Melford. late 17c.-?d.1720. lants.

Thomas Miller, Gazeley. 1690s-d.1701. lant.

*Roger Moore, Ipswich. m.1687-d.1727. lant & lcs.

*Samuel Reeve, Stonham (also Diss?). b.1648-d.1718. lant & lcs.

*Stribling, Benjamin. Stowmarket. m.1693-d.1720. lant & lcs.

Thomas Webb, Sudbury (?ex London). late 17c. lant.

Ex London: 4?.

Total 18. 16 made lanterns, 10 longcases, 8 both, 0 brackets.

SURREY

*John Aylward, Guildford (later at Brentford). late 17c-c.1720. lants, lcs. (& brs?).

Nicholas Fry, Godstone. c.1690. lant.

*Thomas Lodge, Farnham (ex London 1695). c.1713. lc.& br.

William Risbridger, Dorking. m.1680-1724. lants.

Ex London: 1

Total 4. 3 made lanterns, 2 longcases, 1 both, 2 brackets.

SUSSEX

*Thomas Barret, Lewes. 1690-d.1752. lant & lcs.

*John Gilmore (senr.), Battle. b.1660-d.1717. lant & lc.

*John Gilmore (junr), Battle. b.1677,m.1700-d.1726. lant & lcs.

*Samuel Hammond, Battle. c.1688-d.1736. lcs.

*John Harman. Horsham. 1681-d.1711. lant & lcs.

*Thomas Muddle, Rotherfield. b.1670-d.1756. lant & lcs.

John Russell, Lewes. c.1635. lant.

Joshua Smyth, Steyning. 1680. lants.

*Abraham Weston, Lewes. m.1699-d.1746. lant & lcs.

Ex London: 0

Total 9. 8 made lanterns, 7 longcases, 6 both, 0 brackets.

WARWICKSHIRE

*George Guest, Aston (Birmingham). late 17c. lc & lants.

*Edward Norton (Warwick?). late 17c. lcs.

*Nicholas Paris senr. Warwick. c.1669-d.1716. lant, br & lcs.

*Samuel Watson, Coventry (London too). b.1650-d.1740. br & lc.

Ex London: 0

Total 4. 2 made lanterns, 4 longcases, 2 both, 2 brackets.

WESTMORLAND

*Richard Washington, Kendal. b.c.1655-d.1698. lant & lc.

Ex London: 0

Total 1. 1 made lanterns, 1 longcase, 1 both, 0 brackets.

WILTSHIRE

*Edmund Card, Codford St. Peter. (ex London 1688-1698)d.1702. lc

*Edward Cockey, Warminster. c.1690-c.1710. lcs.

William Cockey, Warminster. c.1689-c.1730 & Wincanton/Frome. lant.

*James Delaunce. Downton (ex London/Frome 1690s). c.1700-1736. lant & lcs.

John Holloway, Lavington (later Gloucester). b.pre-1638-d.1676. lants.

John Holloway, Devizes (b.Lavington). 1682-1684, later Newbury. lants.

*Richard Holloway, Lavington. late 17c. lc.

*Humphrey Marsh, Highworth. c.1690-1708. lc & lants.

George Newton, Seend. c.1625-1679. lants.

John Snow, Salisbury. c.1630-d.1661. lants.

Nicholas Snow, Salisbury. free 1629-d.1644. lants.

*William Snow, Marlborough. b.1638-d.1722. lant & lcs.

*Samuel Tinham, Salisbury. 1680s-m.1704. lant, lc & br.

*Thomas Wentworth (1st), Salisbury. 1675-1690. lant, lcs & br.

*Simon Worley, Starton (=Staverton). c.1700. lcs.

Ex London:2

Total 15. 11 made lanterns, 9 longcases, 5 both, 2 brackets.

WORCESTERSHIRE

John Baxter, Conderton (Overbury). 1670. blacksmith & lant.

Samuel Cox, Worcester. c.1690. lant.

*John Frasor, Worcester (ex London 1681+). post 1688. lc & lant & br.

Richard Monkland, Worcester. c.1690. lant.

Ex London: 1

Total 4. 4 made lanterns, 1 longcases, 1 both, 1 brackets.

YORKSHIRE

*William Banks, Sheffield. c.1690. lc & lant.

*Ninyan Burleigh, Pontefract (ex London 1692+). c.1695-1700. lc, lant & br.

Q George Canby, Selby. m.1665-d.1705. lant.

*Thomas Cruttenden, York (ex London 1679). d.1698. lant & lcs & br.

Q Thomas Etherington, York. free 1684-d.1728. lant.

*Abraham Farrer. Pontefract. c.1695. lcs.& lants.

*Jonathan Farrer, Halifax. 1695-1702. lc.

*John Green, Skipton. 1696-d.1742. lc.

*George Mills, Ripon, c.1700-c.1730. (from Alnwick & Sunderland) lc.

*Q James Ogden, Halifax. late 17c-d.1715. lants & lcs.

*Q John Ogden, Bowbridge. (b.1665) c.1690-d.1741. lcs

*Samuel Ogden, Halifax (& Ripponden?). b.1669-1712 (then Benwell, Newcastle on Tyne 1712 - d.1728). lant & lcs.

*William Raynes, York (ex London). post 1680-d.1694. lant & lcs.

John Rooksby, York/Hull. free 1647-c.1690. br cl.

*Ralph Rowntree, York. b.1673, m.1696. lcs.

*Joseph Shepherd, Sheffield. 1696-d.1714. lcs

John Stringfellow, Halifax (ex London 1693). 1694-d?1718. lant & lc.

*John Terry, York. 1690s-d 1757. lcs.

*William Tipling, Leeds. 1692-d.1712. lcs.

*John Williamson, Leeds (ex London). 1683-c.1748. lant & lcs & br.

Ex London: 5.

Total 21. 11 made lanterns, 17 longcases, 8 both, 4 brackets.

WALES

*Richard Joynson, Wrexham. 1700-d.1711. lc.

Ex London: 0

Total 1. 0 made lanterns, 1 longcases, 0 both, 0 brackets

SCOTLAND

John Alexander, Edinburgh. free 1671-1694. lant.

*Andrew Brown, Edinburgh. free 1675-d.c.1711. lcs.

*Patrick Kilgour, Aberdeen 1672/Edinburgh 1690s. lcs.

*Thomas Kilgour, Inverness .1690s. lc.

Humphrey Milne/Mills, Edinburgh. free 1660-d.1693? lant.

*Paul Roumieu junr., Edinburgh. free 1682-d.c.1711. lcs.

Ex London: 0

Total 6. 2 made lanterns, 4 longcases, 0 both, 0 brackets.

IRELAND

Walter Bingham, Dublin. free 1680-d.1727. br.

*Q Ezekiel Bullock. Lurgan (Co. Armagh). b.1650, m.1684-1714.lants & lcs.

Thomas Parker, Dublin. free 1693-d.1751. br.

Ex London: 0

Total 3. 1 made lanterns, 1 longcases, 1 both, 2 brackets.

AMERICA

*Samuel Bispham, Philadelphia. 1696 - ?. lc.

*Q Abel Cottey (from Devon), Philadelphia. 1700-d.1711. lant & lc.

Ex London: 0

Total 2. 1 made lanterns, 2 longcases, 1 both, 0 brackets.

INDEX

Page numbers in bold refer to illustrations